Advances in Ceramic Armor

Advances in Ceramic Armor

*A collection of papers presented at the
29th International Conference
on Advanced Ceramics and Composites,
January 23-28, 2005,
Cocoa Beach, Florida*

Editor
Jeffrey J. Swab

General Editors
Dongming Zhu
Waltraud M. Kriven

Published by

The American Ceramic Society
735 Ceramic Place
Suite 100
Westerville, Ohio 43081
www.ceramics.org

Advances in Ceramic Armor

For information on ordering titles published by The American Ceramic Society, or to request a publications catalog, please call 614-794-5890, or visit www.ceramics.org

ISSN 0196-6219

ISBN 1-57498-237-0

Contents

Dynamic and Static Testing to Predict Performance

Damage Characterization: Observations, Mechanisms, and Implications

vii

Preface

The 29th International Conference on Advanced Ceramics and Composites marked the third year for the "Topics in Ceramic Armor" session at this conference. This session was created in 2002 to provide an annual forum for the presentation and discussion of unclassified information pertaining to the development and incorporation of ceramic materials for armor applications. It was hoped that this session would create an environment that fostered collaboration between government, industry, and academic personnel leading to the development of novel ceramic materials and concepts for armor applications to protect our soldiers.

It is not surprising that this session has grown each year in light of the current global threats facing our defense forces. This year approximately 150 people attended the session which was comprised of over 60 oral and poster presentations covering a wide range of ceramic armor-related topics. These topics included:

Impact and Penetration Modeling
Dynamic and Static Testing to Predict Performance
Damage Characterization: Observations, Mechanisms, and Implications
Non-Destructive Evaluation
Novel Material Concepts

On behalf of the entire organizing committee I would like to thank all the authors and presenters for their contributions to this effort. I extend my personal thanks to the members of the organizing committee, Lisa Prokurat Franks, Jim McCauley, Jerry LaSalvia, Scott Schoenfeld, Janet Ward, Dave Stepp and Andy Wereszczak for their assistance in developing and organizing the session. Finally a special "thank you" goes to The American Ceramic Society staff who provided guidance and administrative support to keep things running smoothly and efficiently.

See you in 2006!

Jeffrey J. Swab

Impact and Penetration Modeling

SOME OBSERVATIONS ON THE STRENGTH OF FAILED CERAMIC

Gordon R. Johnson
Network Computing Services, Inc.
P. O. Box 581459
Minneapolis, Minnesota, 55415

Timothy J. Holmquist
Network Computing Services, Inc.
P. O. Box 581459
Minneapolis, Minnesota, 55415

ABSTRACT
This article presents some observations on the strength of failed ceramic under conditions of high-velocity impact. Included are results of recent computations that provide good agreement with a variety of test data, as well as an explanation of the techniques used to represent the damage and strength of the ceramic as it transitions from intact to failed material. For most of the examples noted in this article the damage and failed strength are determined from impact and penetration computations, and are not measured directly from laboratory tests. Some direct test data for failed ceramics have been generated and reported in the literature; they tend to show a great deal of scatter, they tend to not cover the range of pressures and other variables experienced during high-velocity impact and penetration, and they are generally not in agreement with the corresponding data obtained from the computations. Some observations are presented to explain some of these apparent discrepancies and to show the relative effects of intact strength, failed strength and damage.

INTRODUCTION
Ceramics are very strong materials, especially in compression. They are well suited for armor applications when subjected to high pressures during impact and penetration. It is generally agreed that ceramic materials exhibit strength after they are failed, and that this strength is pressure dependent. There is not good general agreement, however, about the magnitude of the strength of this failed material. This is due to the lack of test techniques to directly measure the strength of the failed material under the conditions (strain, strain rate, pressure, temperature, particle size, shape, arrangement) of interest. There have been several techniques used to determine the strength of the failed material, but for direct testing of failed material the pressures are generally lower than those experienced in high-velocity impact and penetration. Another approach has been to determine the strength of the failed material by performing computations to match the results of penetration tests by using assumed strength characteristics. This approach also has its problems inasmuch as it requires an accurate description of the intact strength, the damage model and the assumed form of the failed strength. The computational algorithms (finite elements, meshless particles, sliding/contact) must also be accurate. Sometimes the strength of the failed ceramic has been taken to be the only important variable, but the strength of the intact material and the strength of the partially damaged material can also be important. This article presents computational results, test results for high-velocity impact, test results for failed ceramic, together with some possible explanations for these data.

RECENT COMPUTATIONAL RESULTS THAT INCLUDE FAILED CERAMIC

The authors have developed three similar computational models for the response of ceramics subjected to large strains, strain rates and pressures. They include an intact strength, a failed strength, a damage model for the transition from intact strength to failed strength, and a pressure model that includes bulking. This discussion will be limited to these models as they are well understood by the authors and they illustrate some of the issues of interest. The JH-1 model [1] does not soften the intact material during the damage process, but allows it to drop suddenly to the failed strength when the damage is complete ($D = 1.0$). The JH-2 model [2] softens the material gradually as the damage is accumulated ($0.0 < D < 1.0$). The JHB model [3] treats the damage and failed material in a manner similar to that used in JH-1, with the differences being that the JHB model uses an analytic form for the strengths of the intact and failed material, and it allows for a phase change.

Figure 1 shows intact strength (at two strain rates) and two failed strength levels for the JHB model for silicon carbide [4]. The lower failed strength level of $\sigma^f_{max} = 0.2$ GPa was determined from computations to match test data [4] and the higher (Walker) failed strength of 3.7 GPa is discussed later. The intact strength is well represented by the model and does not exhibit significant softening at the high pressures. This characteristic supports (but does not prove) the assumption of the JH-1 and JHB forms regarding the lack of softening for partially damaged material. Other arguments for this form are that it is well-suited to represent dwell and penetration, it has been used to accurately simulate a range of test data [4], it does not allow gradual softening that can introduce numerical inaccuracies, and it enables the constants to be obtained with a straightforward process.

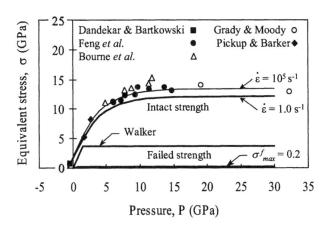

Figure 1. Strength versus pressure for silicon carbide test data and the JHB model.

The top portion of Figure 2 shows interface defeat ($V = 1410$ m/s), dwell and penetration ($V = 1645$ m/s), and penetration ($V = 2175$ m/s) for a tungsten rod impacting a confined silicon carbide target [4,5]. The damage constant ($D_1 = 0.16$) is determined (from the $V = 1645$ m/s data) to match the time at which dwell ceases and penetration begins (about 20 μs) and the maximum strength of the failed material ($\sigma^f_{max} = 0.2$ GPa) is determined from the penetration rate after penetration begins. There is also good agreement between the computed results and the test data for the molybdenum rod and the various velocities.

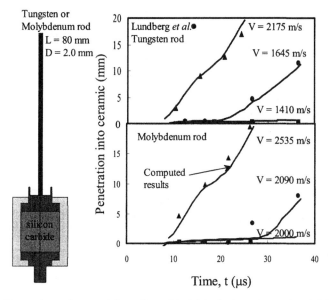

Time, t (μs)

Figure 2. Comparison of computed results and test data for tungsten and molybdenum rods impacting a confined silicon carbide target.

The damage is represented by $D = \Sigma (\Delta\varepsilon_p / \varepsilon^f_p)$ where $\Delta\varepsilon_p$ is the increment of equivalent plastic strain during the current cycle of integration and $\varepsilon^f_p = D_1(P^* + T^*)^n$ is the plastic strain at failure for a dimensionless pressure, P^*, and a dimensionless hydrostatic tensile strength, T^*. The other constant is assumed to be $n = 1.0$ such that the failure strain is a simple linear function of the pressure. It is well established that the ductility increases as the pressure increases, but the form of the relationship is unknown. In a similar manner, the failed strength uses an assumed slope (σ^f vs P^*) for the low-pressure region, and the failure strength is essentially determined by a single constant (σ^f_{max}).

The same constants are used for the computational results in Figure 3, where P is the total penetration and L is the initial length of the tungsten rod. Here, for a different set of tests involving total penetration for a range of higher velocities [6], there is again very good agreement [4]. Computational results are also shown for an assumed case of no failure of the ceramic (such that the intact strength is used for the entire computation), and it can be seen that the computed penetration results are much too low [1]. It is clear from these results that the ceramic does fail, and that the failed strength is much lower than the intact strength. A final set of computed results is for the assumed case of no failure strength and no ductility (the material fails as soon as it experiences plastic strain). Here the penetration is much too high and it has not stopped. This indicates that the ceramic must have some ductility and/or some strength after failure. An intermediate case with ductility ($D_1 = 0.16$ in the damage model), but no failure

5

strength, gives slightly too much penetration for $V = 1500$ *m/s,* but essentially the same penetration (as for the $\sigma^f_{max} = 0.2$ *GPa* computations) at the higher velocities.

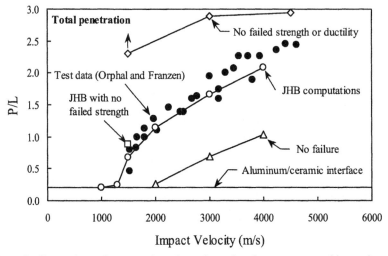

Impact Velocity (m/s)

Figure 3. Comparison of computed results and test data for a tungsten rod impacting a confined silicon carbide target at various impact velocities.

A final computation is shown in Figure 4, where a steel projectile impacts and perforates a thin, layered target of silicon carbide over aluminum. Even at the lower impact velocity (670 *m/s*) the agreement (exit velocity) is in good agreement with experimental data [4].

ISSUES AND UNCERTAINITIES REGARDING FAILED CERAMIC

Although the ability to accurately simulate a range of high-velocity conditions is encouraging, there are some issues and uncertainties. Two major assumptions are that the damage model is represented by a single constant (D_1) and the failed strength is essentially represented by two constants (a slope at low pressures and σ^f_{max}). Even simple strength and failure models for metals contain on the order of 10 to 20 constants. It would appear that failed ceramic, with the particle size, shape and arrangement changing under high-pressure deformation, would be at least as difficult to model. It is unlikely that the failed ceramic behaves in as simple a manner as the models assume.

Figure 5 shows some additional computations compared to the same test data as shown previously in Figure 3. The Walker strength model [7] consists of a Drucker-Prager model with a strength cutoff. It is included in Figure 1 and it has a maximum strength of 3.7 *GPa.* This model does not explicitly account for intact material, and therefore, one possible interpretation is that it represents an average of both failed and intact material. It falls between the intact strength and the failed strength for the JHB model. The Walker results in Figure 5 are slightly below the test data, and this is because the constants were determined from the rates of penetration rather than the total penetration. The CTH Eulerian computations presented by Walker were essentially duplicated with Lagrangian EPIC computations performed by the authors.

6

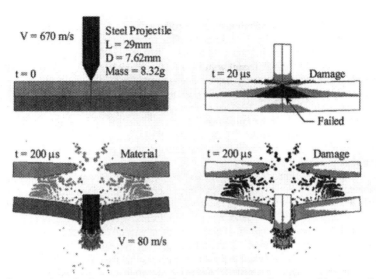

Figure 4. Computed results for a steel projectile impacting and perforating a thin, layered target of silicon carbide over aluminum

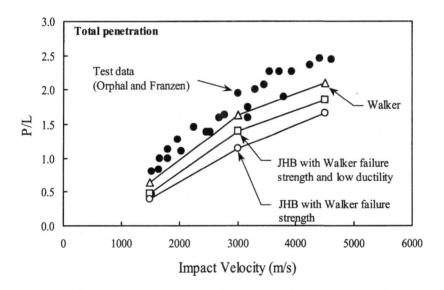

Figure 5. Penetration versus impact velocity for a tungsten rod impacting a confined silicon carbide target, for various computational models.

Even though the penetrating tungsten rod is primarily in contact with failed ceramic material, the damage model and the strength of the intact material have a significant effect. The second set of computations in Figure 5 assumes a JHB type of model, with the intact strength shown in Figure 1, a damage model with $D_1 = 0.16$, and the Walker model as the failed strength (adjusted to zero strength at zero pressure, and subjected to the strain rate effect). It can be seen that the computed penetration results are much too low, and that the intact strength and ductility have an important effect. The final set of computations uses the same intact strength and the adjusted Walker model for the failed strength, but the ductility is essentially eliminated by using $D_1 = 0.001$ in the damage model. These results fall between the other two sets of computations, and it appears that the intact strength and the damage model (ductility) have significant and approximately equal effects.

There is a real need for strength data for failed ceramics, especially under the conditions experienced during penetration. Although there have been numerous efforts to generate such data, there are no generally accepted models and constants for failed ceramic under the conditions experienced during penetration. As an example, recent work at Southwest Research Institute [8] examined the strength of two states of damaged/failed silicon carbide, as shown by the two thicker lines in Figure 6. The comminuted material was pre-damaged using a thermal shock procedure and this resulted in a pattern of cracks that weakened the material (although it remained intact). The powder material has much less strength. This tends to illustrate the wide range of behavior that can be attributed to failed ceramic.

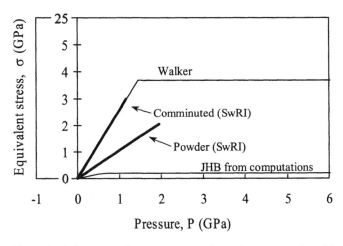

Figure 6. Failure strength versus pressure for various tests and models

The slope of the comminuted material is very similar to that used in Walker's model ($b = 2.5$), but it was not possible to test to high enough pressures to achieve strengths above about 3.0 *GPa,* or to show a decrease in the slope as the pressure increased. It would appear that this (partially damaged) comminuted material (for pressures up to about 1.0 *GPa*) is again an average of the intact material and the (fully damaged) powder material. Based on the

8

computations in Figure 4, if the intact strength and ductility are included, then the failed strength must be much lower than that provided by the comminuted data.

The most disturbing aspect of the data in Figure 6 is that the strength of the powder material goes to about 2.0 GPa, at a slope of about 1.0. This strength of 2.0 GPa is significantly larger than the value (σ^f_{max} = 0.2 GPa) determined from the computations. One possibility is that there are errors in the computations and/or the models for the intact strength, damage model, pressure model, etc. (which must also be correct for the determined failed strength to be correct). Also, the experimental procedures and analyses were recently developed and there could be issues with the experimental technique and/or the associated interpretation.

Another possibility is that the failed strength is somehow lowered under the conditions that exist during penetration. The simple Bernoulli pressure for tungsten penetration into silicon carbide is 0.78, 3.12 and 12.5 GPa for impact velocities of 1000, 2000, 4000 m/s, respectively. These pressures are much higher than those generated during the aforementioned tests. The strains and strain rates are also significantly higher during penetration. Even though it intuitively appears that the ceramic could not become much weaker than the powder used in the tests, perhaps the combination of large strains, high stain rates and high pressures could alter the size, shape and arrangement of the particles, thus producing a lower strength. Although it does not appear that the ceramic will melt under these conditions, the possibility has not been eliminated.

Figure 7 shows some responses for the strength as a function of plastic strain and damage, under a constant pressure and strain rate. The JH-1 and JHB models provide instantaneous failure (when D = 1.0), and the JH-2 model provides a gradual failure (for $0.0 < D < 1.0$). The authors have chosen to use the JH-1 and JHB forms for the reasons stated previously, although it is possible that the JH-2 form is as good or better. Note that the intact strength and damage for the JH-2 model would be defined differently than for the JH-1 or JHB models. The response for the combined model provides a possible explanation that is consistent with the penetration data, the comminuted and powder data, and the computational results. Unfortunately, there do not appear to be any existing experimental techniques that can be used to produce such data (intact to damaged to failed, under high pressures and deformations).

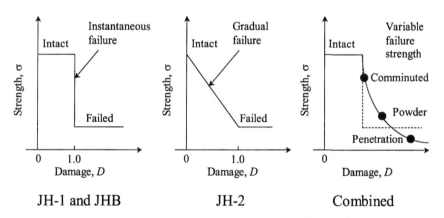

Figure 7. Damage and failure responses for various models

9

SUMMARY AND CONCLUSIONS

It has been demonstrated that it is possible to accurately simulate a wide range of impact and penetration conditions by using computational ceramic models that include intact strength, damage, and strength of failed material. These computations use a failure strength that is derived from the penetration experiments, however, and this failure strength is not always consistent with data obtained explicitly from failed material. It would appear that this discrepancy is due to errors in the computations, errors in the material models, and/or lack of appropriate data for the damage and failure strength for the conditions experienced under high-velocity impact and penetration.

ACKNOWLEDGEMENT

This work was performed under prime contract F42620-00-D-0037-BR02. The authors would like to thank D. W. Templeton (U. S. Army Tank-Automotive Research, Development and Engineering Center) and C. E. Anderson Jr. (Southwest Research Institute) for their contributions to this work.

REFERENCES

[1]T. J. Holmquist and G. R. Johnson, "Response of silicon carbide to high velocity impact," *J. Appl. Phys.*, **91**, 5858-66 (2002).

[2]G. R. Johnson and T. J. Holmquist, "Response of boron carbide subjected to large strains, high strain rates, and high pressures," *J. Appl. Phys.*, **85**, 8060-73 (1999).

[3]G. R. Johnson, T. J. Holmquist and S. R. Beissel, "Response of aluminum nitride (including a phase change) subjected to large strains, high strain rates, and high pressures," *J. Appl. Phys.*, **94**, 1639-46 (2003).

[4]T. J. Holmquist and G. R. Johnson, "Characterization and evaluation of silicon carbide for high velocity impact," *J. Appl. Phys.*, accepted for publication (2005).

[5]P. Lundberg, R. Renstrom and B. Lundberg, "Impact of metallic projectiles on ceramic targets: transition between interface defeat and penetration," *Int. J. Impact Engng.*, **24**, 259-75 (2000).

[6]D. L. Orphal and R. R. Franzen, "Penetration of confined silicon carbide targets by tungsten long rods at impact velocities from 1.5 to 4.6 km/s," *Int. J. Impact Engng.*, **19**, 1-13 (2000).

[7]J. D. Walker, "Analytically modeling hypervelocity penetration of thick ceramic targets," *Int. J. Impact Engng.*, **29**, 747-55 (2003).

[8]K. A. Dannemann, S. Chocron, A. E. Nicholls, J. D. Walker and C. E. Anderson Jr., "Ceramic phenomenological experiments–compressive strength of SiC," Final Report, Southwest Research Institute, San Antonio, Texas, June 2004 (Revised November 2004).

MODELING DYNAMICALLY IMPACTED CERAMIC MATERIAL EXPERIMENTS

B. Leavy, B. Rickter, Dr. M.J. Normandia
U. S. Army Research Laboratory
ATTN: AMSRD-ARL-WM-TA
Aberdeen Proving Ground, MD 21005

ABSTRACT

A number of new experimental techniques were developed to simplify the process of calibrating ceramic constitutive models. Current ceramic model calibration techniques for the Johnson-Holmquist One (JH1) model require the use of complicated penetration experiments to tune the damage evolution to a specific experiment. Application of these ceramic models to different experiments often illustrates discrepancies in the results.

This paper will illustrate the use of data from these new experimental methods to simplify the calibration process for hot-pressed silicon carbide (SiC), and thus more accurately capture the behavior of different ceramic materials. Dynamic sphere impacts as well as kinetic energy rod penetration rate study programs have been established. Comparison of the experimental data with the corresponding simulation results will highlight areas for improvement in ceramic modeling. Specifically, the macroscopic behavior the constitutive model attempts to encompass and its relation to relevant static and dynamic properties.

INTRODUCTION

Recent work has been underway to characterize the performance of ceramic materials in simplified dynamic impact events. The goal is to determine the potential armor performance of ceramics in a relatively easy and reproducible set of experiments. These screening experiments will allow time for more analysis of specific materials of interest, and eventually connect the performance to basic material properties.

Two different experimental techniques are currently being refined at the Army Research Laboratory to characterize ceramic performance. The first method is known as the Dynamic Sphere Impact experiment[1], and is the one that will be focused on in this paper. The second method is the Projectile Target Interaction (PTI) study, which looks at direct-ballistic time-dependent performance of long-rod penetrators against confined ceramics. The time-resolved reverse-ballistic experiments by Hauver[2], Lundberg[3], Orphal[4] and others are the inspiration behind the current PTI efforts. These previous reverse ballistic tests were the only data available to calibrate ceramic models in a dynamic regime. As the ability to simulate ballistic events improves, more well designed and instrumented experiments are required to gain insight into the subtle effects different component properties have in a complex armor design.

Figure 1. (A) Experimental Gas Gun Facility. (B) Target Geometry and Fixture.

EXPERIMENTAL PROCEDURES

Two separate test facilities were used to obtain the data over a wide range of velocities. The first facility uses either a 0.300 caliber or one inch powder gun with Lexan sabots for the higher velocities. The second facility includes a gas gun launch tube with foam obturator, laser velocimeters, and dual x-rays, and covers the lower velocity regime. Figure 1 shows the gas gun facility as well as a cutaway of the titanium alloy (Ti6Al4V) fixture confining the target. All metal components are Ti6Al4V circular stock. The 3.125 mm thick front plate has an entry hole to accommodate the tungsten carbide (WC) spherical projectiles. The 9.525 mm thick back plate confines a 19.05 mm thick, 76.2 mm diameter ceramic, held to tight tolerances and enclosed in a 101.6 mm diameter ring. The hot-pressed silicon carbide (SiC-N) ceramic target material was purchased from Cercom, Inc. The same material is being tested in different thicknesses and configurations in a variety of armor applications.

Two different spheres were used as projectiles. Earlier tests utilized a 2g, 6.34mm diameter, WC spherical penetrator purchased from Superior Graphite (SG) through Machining Technologies Inc, which were Grade 25 (.000025" roundness tolerance), Class C-2 material (Rockwell A 92) and contained 6% Co binder with a 14.93g/cc measured density. More recent tests have utilized a 2g, 6.35mm diameter WC spherical penetrator purchased from New Lenox (NL) Machining Co, Inc., with the same grade and cobalt content as above, and a density of 14.81g/cc.

The test procedure consists of firing the spheres embedded in a flared sabot from the launch tube, discarding the sabot prior to impact, and observing (when desired) the penetrator in flight, prior to impact, with two 150keV x-radiographs, which also served for velocity measurements. A significant benefit of this technique is material recovery for microscopic analyses.

NUMERICAL PROCEDURES

The CTH[5] hydrocode utilizing the typical Eulerian meshing scheme as well as Adaptive Mesh Refinement (AMR) was used to simulate the test results. AMR allows refinement in areas of interest. Currently, a density difference or shockwave refinement indicator is used along with a kinetic energy error indicator to refine material interfaces. The constitutive model used for the target ceramic was based on a modification of the Johnson and Holmquist One[6] ceramic model for SiC-B[7]. The model has a separate constitutive behavior for intact and fully-damaged material, and results are very sensitive to both a strain-based transition criterion and the pressure-shear dependence of the damaged ceramic. Ballistic data are traditionally required to calibrate the damage behavior of the ceramic model by modifying the damage strain (EFMAX) and the failed material strength (S3) parameters[8]. The model constants were then modified for the Cercom silicon carbide known as SiC-N (Table I). Measured quantities of density, moduli, acoustic and shear wave speeds, Poisson's ratio, spall strength, etc. from the actual ceramics used in the experiments were taken[9,10].

A similar technique was used to develop the WC sphere ceramic models used, based on the Johnson-Holmquist One and Johnson-Cook models developed by Holmquist[11]. Once again, the material properties for the actual materials used for the spheres were measured by Wereszczak[12] and Weerasooriya[13].

The titanium alloy surrounds for the experiments were modeled using the default Johnson-Cook plasticity and fracture parameters[14] in CTH. The equation of state model used was the Sesame Ti6Al4V model[15] developed by Kerley.

12

Table I. Johnson-Holmquist One ceramic model for SiC-N.

	Symbol	Units	SiC-N
Mass and Elastic Constants			
Density	ρ	g/cm^3	3.227
Shear Modulus	G	GPa	195
Young's Modulus	E	GPa	454
Specific Heat	Cv	J/Kg K	840
Poisson's Ratio	υ		0.164
Strength Constants			
Hydrostatic Tensile Strength	T	GPa	-0.498
Intact Strength 1	S1	GPa	7.10
Intact Pressure 1	P1	GPa	2.50
Intact Strength 2	S2	GPa	12.20
Intact Pressure 2	P2	GPa	10.00
Max Failure Strength	S3	GPa	0.90
Failure Strength Slope	α		0.40
Strain Rate Constant	C		0.009
Pressure Constants			
Bulk Modulus	K	GPa	225
Pressure Constant 2	K2	GPa	361
Pressure Constant 3	K3	GPa	0
Bulking Factor	ΔP		1
Damage Constants			
Damage Slope	φ		0.0120
Damage Strain	Efmax	GPa	1.20
Damage Pressure	DP1	GPa	99.502

The addition of this sphere impact data to the suite of ballistic data permitted better model refinement, and allowed for the timely development of SiC-N and WC specific Johnson-Holmquist One ceramic models to be used in a variety of armor configurations.

RESULTS & DISCUSSION

The dynamic sphere experimental procedure generates final penetration depths, crater diameters and surface damage extents which are used for model validation in numerical or analytic design tools.

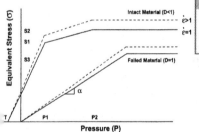

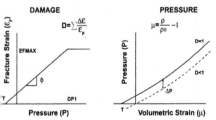

It does not provide time resolved data or load histories like the direct ballistic PTI or other reverse ballistic experiments. The test firings conducted below 300 m/s produced no visible damage to the ceramics impacted, other than radial cracks. For SiC-N, impact at 393 m/s produced the first visible signs of damage, depicted in Figure 2. At the point of impact, the ejected material produces a small ring of damaged material, which exits through the cover. For the hot-pressed SiC-N impacted at various impact velocities and shown in Figure 2, the damage extent can be identified as a function of impact velocity. This is carefully measured with a robotic reverse engineering tool capable of measuring distances and diameters with an accuracy of approximately 0.02mm. The craters are well defined,

13

as is the radial extent of the damage regions on the front surface. The rear surface of the front plate captures the damage features in the ceramic showing radial and circumferential cracks. There was no indentation into the rear backing plates, which depicted radial cracks, until the highest velocities, when circumferential cracks clearly appeared. The rear surface of the ceramic remained smooth.

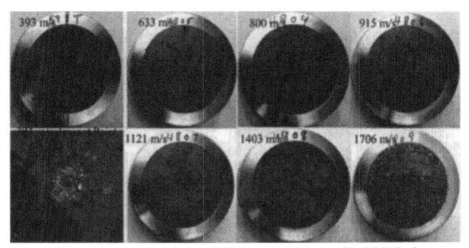

Figure 2. Photographs of confined Cercom SiC-N ceramic disk impacted by SG WC sphere at impact velocities shown. Close up of 393 m/s impact area at lower left.

After careful removal from the fixture, the front plate was removed, photographs taken, and careful excavation of the surface revealed the deepest point of penetration and the penetration directly under the impactor, which were not always identical at the varied velocities. Well defined craters were observed and indentation/penetration depths and crater diameters were measured, as were the radial extent of the surface circumferential damage. We have preserved all targets, and plan to analyze them microscopically prior to, and after, sectioning.

Measurements of penetration depth for varying velocities against SiC-N are shown in Figure 3. These data were then used to calibrate the damage parameters in the JH1 model to match the max depth of penetration of the experiments. Several different meshing schemes and cell sizes were utilized, and the subsequent differences in performance were noted. As with all materials, the varying of the cell size upon which the ceramic constitutive model is applied greatly influences the performance once the material is damaged or failed. Current efforts by Brannon[16] and others are underway to create mesh-independency material models through a variety of methods.

These experiments and simulations allowed for a ceramic model for SiC-N to be developed, utilizing previously developed models and measured properties. Subsequent efforts by Lundberg[17] to determine an interface defeat or dwell transition velocity have completed the scope of information required to model a ceramic over a wide range of ballistic configurations.

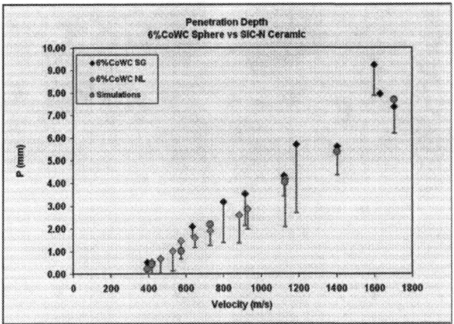

Figure 3. Penetration under impact point (Lower Error Bars) and max depth (Diamonds) for SiC-N along with corresponding simulation results (Circles).

Since during the target measurement we are excavating loose material, the question has arisen whether or not the completely damaged region below the crater should be included in the simulation depth. When damage equals one in the JH1 model (shown as red in Figure 4), the ceramic uses the failed material strength corresponding to the lower curve in the model. By decreasing the size of the cell used in the SiC-N model, the higher velocity simulations under-predicted the depth of penetration. This depth plus an additional depth where the material damage is complete may make a more realistic prediction of the ceramic performance (Figure 5).

Figure 4. Ceramic Simulations of SG WC Spheres of varying velocities using CTH with AMR and JH1 ceramic models.

15

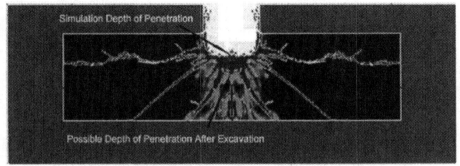

Figure 5. Closeup of 1403m/s simulation using CTH with AMR and JH1 ceramic models. Shown are current penetrations used to calibrate model, versus the possible excavation depth

Analytical code analysis by Normandia[18] clearly shows the velocity regime where the spheres no longer perform as a rigid penetrator, which is noted as a leveling off of the penetration depth over a range of velocities. This behavior should be very helpful in the calibration of our WC sphere models, in both ceramic experiments as well as our baselines into titanium and others materials like steel[19]. Figure 6 shows the results of our titanium alloy baselines, along with some earlier ARAP data[20]. The areal density is normalized by multiplying the penetration depth by the density of the target and dividing this by the diameter of the sphere multiplied by its density. More data will help resolve the changes in penetration performance at the higher velocities. Also shown are the CTH simulations with AMR using the before mentioned models for WC and Ti6Al4V.

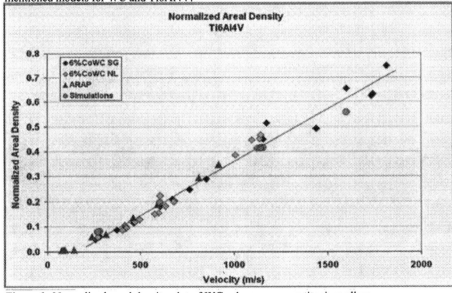

Figure 6. Normalized areal density plot of WC spheres versus a titanium alloy.

16

CONCLUSIONS

Ballistic response curves generated for a silicon carbide ceramic variant known as SiC-N illustrate the capability of simple dynamic sphere experiments to characterize the performance of ceramic materials for use in armor designs. Numerical models were used to capture the response curve features and these models have subsequently been utilized for more complex and varied ceramic armor simulations. These experiments, along with time-resolved ballistic experiments will allow us to better understand the behavior of candidate ceramic armor elements, and eventually relate the dynamic performance back to static material properties.

ACKNOWLEDGEMENTS

Gratitude is extended to: Dave MacKenzie for development of the low velocity test range and testing, and Dave Schall and technical crew for conduct of the high velocity experiments at ARL ballistic test facilities. Finally, the authors wish to thank Tim Holmquist for his ceramic and material model development work, upon which all the models shown were based.

This research was supported by mission funding at the U.S. Army Research Laboratory.

REFERENCES

[1]M. Normandia, B. Leavy, "Ballistic Impact of Silicon Carbide with Tungsten Carbide Spheres," *Proceedings of the 28th International Conference on Advanced Ceramics and Composties*, (2004).

[2]G. Hauver, et al., "Enhanced Ballistic Performance of Ceramic Targets," *Proceedings of the 19th Army Science Conference*, (1994).

[3]P. Lundberg, R. Renstrom, B. Lundberg, "Impact of Metallic Projectiles on Ceramic Targets: Transition between Interface Defeat and Penetration," *International Journal of Impact Engineering*, 24, 259-275, (2000).

[4]D. Orphal, R. Franzen, "Penetration of Confined Silicon Carbide Targets by Tungsten Long Rods at Impact Velocities from 1.5 to 4.6 km/s," *International Journal of Impact Engineering*, 19, 1-13, (1997).

[5]J. McGlaun, S. Thompson and M. Elrick, "CTH: A Three-Dimensional Shock Wave Physics Code," *International Journal of Impact Engineering*, 10, 351-360, (1990).

[6]G. .Johnson, T. Holmquist, "A Computational Constitutive Model for Brittle Materials Subjected to Large Strains, High Strain Rates and High Pressures," *Proceedings of EXPLOMET Conference*, (1990).

[7]T. Holmquist, G. Johnson, "Response of Silicon Carbide to High Velocity Impact", *Journal of Applied Phyisics*, 91, 5858-5866, (2002).

[8]D. Templeton, T. Holmquist, H. Meyer Jr., D. Grove, and B. Leavy, "A Comparison Of Ceramic Material Models," *Ceramic Armor Materials by Design, Ceramic Transactions*, 134, (2002).

[9]D. Dandekar, "A Survey of Compression Studies of Silicon Carbide," *ARL Tech Report ARL-TR-2695*, (2004).

[10]J. LaSalvia, "Measured properties of ceramics," *private correspondance*, (2003).

[11]T. Holmquist, "Model constants for WC-6%Co," *presentation to ARL*, (2004).

[12]A. Wereszczak, "Elastic Property Determination of WC Spheres and Estimation of Compressive Loads and Impact Velocities that Initiate their Yielding and Cracking,"

Proceedings of the 29th International Conference on Advanced Ceramics and Composties, (2005).

[13]T. Weerasooriya, P. Moy and W. Chen "Effect of Strain-rate on the Deformation and Failure of WC-Co Cermets under Compression," *Proceedings of the 28th International Conference on Advanced Ceramics and Composties,* (2004).

[14]G. Johnson, T. Holmquist, "Test Data and computational Strength and Fracture Model Constants for 23 Materials Subjected to Large Strains, High Strain Rates, and High Temperatures," *Los Alamos Report LA-11463-MS,* (1989).

[15]G. Kerley, "Equation of State for Titanium and Ti6Al4V Alloy," *Sandia Report SAND 2003-3785,* (2003).

[16]R. Brannon, "Silicon Carbide Project Overview," *presentation to ARL,* (2004).

[17]P. Lundberg, B. Lundberg, "Transition between Interface Defeat and Penetration for Tungsten Projectiles and Four Silicon Carbide Materials," *International Journal of Impact Engineering, in press,* (2004).

[18]M. Normandia, B. Leavy, "A Comparison of Ceramic Materials Dynamically Impacted by Tungsten Carbide Spheres," *Proceedings of the 29th International Conference on Advanced Ceramics and Composties,* (2005).

[19]R. Martineau, M. Prime, and T. Duffey, "Penetration into HSLA-100 steel with Tungsten Carbide Spheres at Striking Velocities Between 0.8 and 2.5 km/s," *International Journal of Impact Engineering, 30,* 505-520 (2004).

[20]R. Contiliano, McDonough, T. "Application of the Integral Theory of Impact to the Qualification of Materials and the Development of a Simplified Rod Penetrator Model," *Aeronautical Research Associates of Princeton, Inc. Report 368,* (1978).

MODELING SPHERICAL INDENTATION EXPERIMENTS ONTO SILICON CARBIDE

Timothy Holmquist
Network Computing Services
P. O. Box 581459
Minneapolis, MN 55485

ABSTRACT

This article presents computational modeling and analysis of the Hertzian spherical indentation experiment. A spherical diamond indenter is used to indent silicon carbide to various levels of indentation. Indentation depths of $\delta = 7$ μm (experimental level) to depths greater than $\delta = 50$ μm are presented. The EPIC code is used to model the indentation process, where the indenter is assumed to be elastic and the JH-1 ceramic model is used to represent the silicon carbide. The computed result for $\delta = 7$ μm shows good correlation to the experiment for both loading and unloading. The damage process is presented and identifies when initial yielding occurs, the extent of damage at peak load, and additional damage that occurs from unloading. A noteworthy result is the determination of the pressure levels that occur. Pressures of over 10 GPa are produced, which are levels generally only attained through ballistic impact. Computed results are also presented for very deep indentations ($\delta = 52$ μm). The results show a damage process that includes the development of ring cracks and material failure. The effect of strength and damage is also presented by varying JH-1 model parameters. The indentation response is sensitive to the intact strength when large plastic strains are produced. It appears that much larger indentations ($\delta >> 7$ μm) are needed to understand the damage and failure processes in silicon carbide.

INTRODUCTION

Over the past 20 years there has been much interest in understanding the response of ceramic materials for large strains, high strain rates, and high pressures. Because ceramics are very strong in compression it is difficult to perform laboratory tests that produce direct measurements of stresses, strains, etc. such that the strength and damage can be determined. Instead it has been necessary to infer the strength of the failed ceramic (and damage process) using ballistic impact experiments. Although this approach has been used with some success [1], it would be desirable to have other experiments to infer/validate the material response for similar, high-pressure, high strain conditions. This work investigates the feasibility of using Hertzian indentation experiments to help understand the strength and damage behavior of ceramics.

Recently, there has been significant work reported on Hertzian indentation applied to ceramic materials [2, 3]. The Hertzian indentation test uses a high strength indenter (typically tungsten carbide or diamond) to permanently indent the material. Recent developments allow for the indentation force, as a function of indentation depth, to be measured with a single experiment [3]. Two desirable features of these tests are the high pressures (10 GPa – 20 GPa) and permanent strain (damage) produced in the material. The experiments are also relatively inexpensive, repeatable, and performed under quasi-static conditions. The disadvantage is that the stresses, strains, and pressures are not explicit outputs of the experiment (similar to ballistic experiments). The output is generally presented in the form of a force vs. indenter-depth-of-penetration (IDOP) relationship that represents the applied force on the indenter as a function of the ceramic displacement.

This work presents an initial study in modeling the Hertzian indentation experiment with the primary objective being to assess the usefulness of the data to provide strength, damage, and failure behavior in ceramics. The remainder of this article will summarize the JH-1 model, present computed results for small indentations ($\delta = 7$ μm) and large indentations ($\delta = 52$ μm), investigate ring crack evolution, and evaluate the effect of strength and damage on indentation response.

DESCRIPTION OF THE JH-1 CERAMIC MODEL FOR SILICON CARBIDE

The JH-1 constitutive model, and the associated constants for silicon carbide, are presented in Figure 1. The constants are for the specific silicon carbide known as SiC-B. It is produced by Cercom and is hot pressed (pressure assisted densified). The model consists of an intact strength and a failed strength that are functions of the pressure, the strain rate, and the damage. Pressure, bulking and damage are other aspects of the model. This is the first of three closely related models, JH-1 [4], JH-2 [5], and JHB [6] presented by Johnson and Holmquist. One of the primary differences between the three models is that the JH-2 model allows the strength to degrade gradually as the damage is accumulated, rather than soften/fail instantaneously after it is fully damaged, as is done in the JH-1 and JHB models. For SiC-B the JH-1 and JHB form appears to be better suited to represent the constant strength characteristics of SiC-B [1]. The JH-1 model is used for all the computations herein; as it was previously used to successfully model a large range of impact problems for silicon carbide [1].

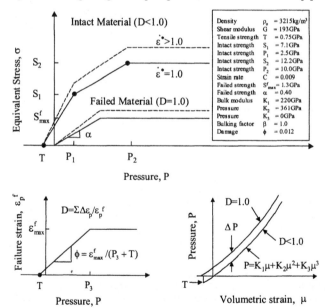

Figure 1. Description of the JH-1 constitutive model including the constants for SiC-B.

20

COMPUTED RESULTS

The following computations are performed with the EPIC code, using 2D Lagrangian finite elements [7] and the JH-1 model for silicon carbide as previously defined. The initial geometry used for all the indentation computations is presented in Figure 2a. The diamond indenter is 2.25 *mm* in diameter and is assumed to remain elastic. The elastic properties for diamond are E = 850 *GPa* (Young's modulus) and $v = 0.20$ (Poisson's ratio) as provided by Wereszczak and Kraft [3]. The initial computational grid uses 25 sets of crossed triangular elements (four triangles in a quad) across the radius of the indenter, and similar sizes in the target. To minimize the size of the target, nonreflective shell elements are applied to the sides and rear of the target [7]. The rear of the target is also fixed in the axial (loading) direction and friction between the indenter and target is assumed zero.

Computed results for experimental indentation depths

Figure 2b presents computed and experimental results for an indentation of $\delta = 7$ μm. Both loading and unloading are presented and show good agreement. Damage contours are presented that show initial yielding at approximately $\delta = 1.5$ μm, increased damage at an intermediate load, and the damage at peak load. Note that the maximum damage occurs below the contact surface of the indenter, which is similar to what is observed experimentally [2]. The unloading process also produces damage as is shown in the lower right portion of Figure 2b. After unloading a permanent displacement of $\delta = 1$ μm occurs, which produce small maximum plastic stains of $\varepsilon_p \sim 0.02$. A comment should be made regarding the change in slope (cusp) that occurs in the computed result at approximately $\delta = 0.5$ and 3 μm. The cusp is a result of using a series of finite elements to represent a spherical surface. Each cusp is the result of an additional finite element coming into contact with the ceramic surface. This produces a step increase in the surface area resulting in a step increase in resistance. In reality, the indenter is spherical and the surface area increases in a continuous manner as the load increases.

Figure 3 presents the JH-1 strength model for silicon carbide, plate impact test data from various researchers, and the computed equivalent stress-pressure range that occurs in the indentation experiment. The peak pressures under the indenter are very high, exceeding 10 GPa, which is significantly above the pressure at the Hugoniot Elastic Limit (HEL). This is an important result because ceramic behavior, for high pressure conditions, is very difficult to produce quasi-statically and generally has only been produced through ballistic impact experiments. It appears that indentation tests (for $\delta = 7$ μm) produce high pressures, but it also appears that the amount of plastic stain and damage produced is small with no material failure.

Computed results for deep indentation

As shown in Figure 2b, typical indentation experiments produce IDOP on the order of $\delta = 10$ μm. Although these tests produce damage in the ceramic, the plastic strains are small and none of the material fails ($D = 1.0$). It is of interest to determine the level of indentation, and maximum load required to produce material failure within the ceramic. Figure 4 presents computed results for an indentation of $\delta = 52$ μm and a peak load of $F = 6500$ N. The loading and unloading response is presented, including material damage. At $\delta = 30$ μm a ring crack is initiated. As the load is increased, damage continues to accumulate, until approximately $\delta = 47$ μm when the first material failure occurs. As the load continues to increase, the zone of material failure continues to grow, and the force-IDOP response becomes softer and less stable. The peak load and maximum IDOP for this example is $F = 6500$ N and $\delta = 52$ μm, respectively. The load

21

is gradually removed, producing a permanent indentation of $\delta = 27$ μm. Material damage and failure also increase during the unloading process. Note the damage and failure contours produced at peak loading. The failed zone ($D = 1.0$) occurs well beneath the indenter in a conoid shape, less damage occurs around the failed zone ($0.5 < D < 1.0$), with less damage yet ($0 < D < 0.5$) occurring in a spherical shape that extends to the surface. The damage and failure patterns are very similar to those produced from projectile impact where ceramic dwell occurs [1]. The pressure levels are also similar, exceeding 19 GPa. The results indicate that indentations approaching $\delta = 50$ μm are required to produce material failure, and if experimentally feasible, could provide valuable insight into the damage process and its effect on strength.

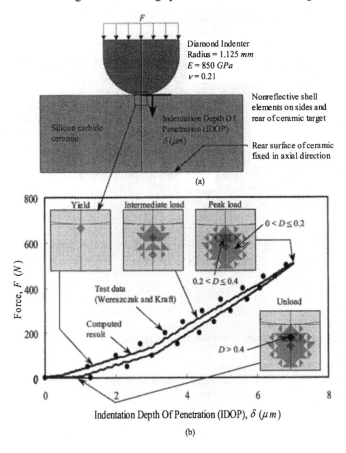

Figure 2. Computed and experimental results for a Hertzian indentation onto silicon carbide. (a) Initial 2D geometry. b) Loading and unloading response for both the computed and experimental results including damage contours at initial yield, intermediate load, peak load and after unloading.

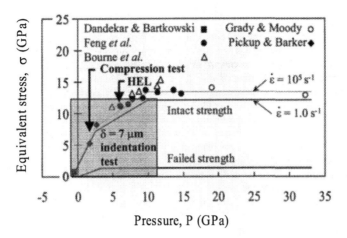

Figure 3. Strength verses pressure for test data and the JH-1 model for silicon carbide. Also shown is the pressure-stress region which results from a $\delta = 7$ μm indentation computation.

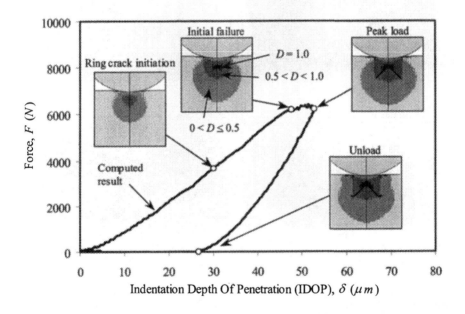

Figure 4. Computed results for a Hertzian indentation onto silicon carbide to a high peak load. Damage contours are shown at ring crack formation, initial sub-surface failure, peak load and after unloading.

Ring (cone) crack evolution

Hertzian fracture (ring and cone cracks) has been studied for many years, particularly in glasses. Hertzian fracture begins as a surface ring crack outside the indentation contact zone, and at a critical load, propagates downward in a cone configuration [2]. Figure 5a shows an example of surface ring cracks in alumina ceramic from Lawn [2] where two ring cracks are clearly evident. Figures 5b-d present the computed formation of surface ring cracks that occur from the indentation computation presented in Figure 4. In the computation, ring cracks are defined by material failure (D = 1.0) initiated on the surface. Figure 5b shows damaged material on the surface, but no failed material. Figure 5c shows the initiation of the first ring crack at a load and indent of F = 3680 N and δ = 30 μm respectively. The ring crack occurs due to large radial tensile stresses. A second ring crack occurs at F = 6350 N and δ = 52 μm, as presented in Figure 5d, and again occurs due to the development of large tensile radial stresses. Figure 5d also shows the first ring crack beginning to propagate downward (initial cone crack development). It should be noted that the development of ring cracks is sensitive to meshing and friction. Although not specifically address herein, earlier work has indicated that rings cracks occur at lower loads when a finer mesh and friction are used.

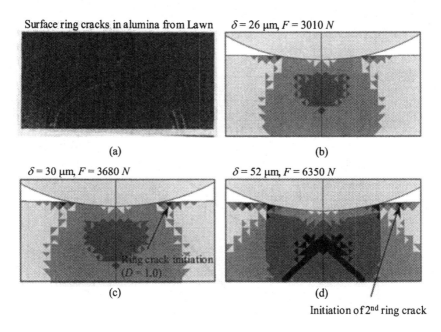

Figure 5. Computed and experimental results for a Hertzian indentation onto ceramic showing damage and the development of ring cracks. a) An example of surface ring cracks in alumina (looking down in the direction of the indent). b-d) Computed results showing damage accumulation and the development of two ring cracks in silicon carbide.

Effect of strength and damage

Here, the effect of ceramic strength and damage is investigated. The purpose of this study is to identify the sensitivity that various model parameters have on the indentation response. If large variations in model parameters produce small variations in the indentation response, then it becomes more difficult to use indentation experiments to determine or validate model constants. Figure 6 presents load-IDOP responses for four variations in JH-1 strength and damage model parameters including the elastic response. The effect of intact strength, tensile strength, failed strength and damage is investigated independently. Each parameter is reduced by approximately 50 % (the exact values are shown in Figure 6). The baseline response is presented for comparison using the JH-1 constants as presented in Figure 1. Figure 6 shows that there is no difference between the baseline response, the reduced damage slope ($\Phi = 0.006$) response, and reduced failed strength ($s^f_{max} = 0.7$ GPa) response. This is because no material fails under these low indentation conditions, so the response is only dependent on the intact strength, which is not changed for these conditions. There is a small difference in the indentation response when the tensile strength is reduced to $T = 0.3$ GPa. This occurs because the tensile strength affects the intact strength, although large changes in the tensile strength produce small changes in the intact strength. Lastly, a large reduction in the intact strength produces a large reduction in the indentation response. This is because much more of the material goes plastic (the material is much weaker) as is evident from the large deviation from the elastic response. These results suggest that for very strong materials (like silicon carbide) much deeper indentations are required (to produce more plastic deformation) for the experiments to provide helpful information regarding strength, damage and failure.

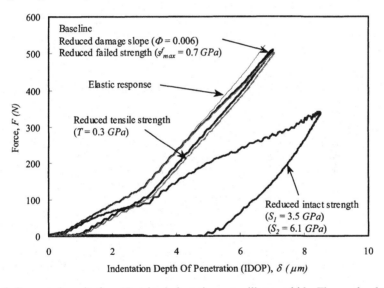

Figure 6. Computed results for a Hertzian indentation onto silicon carbide. The results show the effect of four different JH-1 model parameters on the indentation response including the elastic response.

SUMMARY AND CONCLUSIONS

This article has presented computational modeling, and analysis, of the Hertzian spherical indentation experiment. A spherical diamond indenter was used to indent silicon carbide to $\delta = 7\ \mu m$. The results were compared to experiment and showed good correlation for both loading and unloading. At peak loading, pressures over 10 GPa were produced, but the plastic stains and damage that occurred were small. Computed results were also presented for very deep indents ($\delta = 52\ \mu m$). The results showed a damage process that included the development of ring cracks, material failure, and large permanent deformation. Lastly, The effect of strength and damage was investigated by varying JH-1 model parameters. The indentation response was sensitive to a reduction in the intact strength because much more plastic deformation occurred. It appears from this study, that much greater indentation depths ($\delta >> 7\ \mu m$ when using the indenter geometry described herein) are needed to understand the damage and failure processes in silicon carbide.

ACKNOWLEDGEMENTS

The majority of this work was performed under Prime Contract No. F42620-00-D-0037-BR02, although some work was performed under contract DAAD19-03-D-0001 with the U. S. Army Research Laboratory. The author would like to thank D. W. Templeton (U. S. Army TARDEC), C. E. Anderson Jr. (Southwest Research Institute), A. A. Wereszczak (Oak Ridge National Laboratory), and G. R. Johnson (NCS) for their contributions to this work. The views and conclusions contained in this article are those of the authors and should not be interpreted as presenting the official policies or positions, either expressed or implied, of the U. S. Army Research Laboratory or the U. S. Government unless so designated by other authorized documents. Citation of manufacturer's or trade names does not constitute an official endorsement or approval of the use thereof. The U. S. Government is authorized to reproduce and distribute preprints for Government purposes notwithstanding any copyright notation hereon.

REFERENCES

[1] T. J. Holmquist and G. R. Johnson, "Response of Silicon Carbide to High Velocity Impact," *J. Appl. Phys.* **91** (9) 5858-5866 (2002).

[2] B. R. Lawn, "Indentation of Ceramics with Spheres: A Century after Hertz," *J. Am. Ceramic Soc.* **81** (8) 1977-94 (1998).

[3] A. A. Wereszczak and R. H. Kraft, "Instrumented Hertzian Indentation of Armor Ceramics," Ceramic Engineering and Science Proceedings, **23** 53-56 (2002).

[4] G. R. Johnson and T. J. Holmquist, "A Computational Constitutive Model for Brittle Materials Subjected to Large Strains, High Strain Rates, and High Pressures," *Proceedings of EXPLOMET Conference,* San Diego, (August 1990).

[5] G. R. Johnson and T. J. Holmquist, "An Improved Computational Constitutive Model for Brittle Materials," *High Pressure Science and Technology – 1993,* edited by S. C. Schmidt, J. W. Schaner, G. A. Samara, and M. Ross, (AIP, 1994).

[6] G. R. Johnson, T. J. Holmquist, and S. R. Beissel, "Response of Aluminum Nitride (Including a Phase Change) to Large Strains, High Strain Rates, and High Pressures," *Journal of Appl. Phys.* **94** (3), p. 1639-1646, (2003).

[7] G. R. Johnson, R. A. Stryk, T. J. Holmquist, and S. R. Beissel, "Numerical Algorithms in a Lagrangian Hydrocode," Report No. WL-TR-1997-7039, Wright Laboratory (June 1997).

ANALYSIS OF TIME-RESOLVED PENETRATION OF LONG RODS INTO GLASS TARGETS

Charles E. Anderson, Jr., I. Sidney Chocron, and James D. Walker
Southwest Research Institute
P. O. Drawer 28510
San Antonio, TX 78228-0510

ABSTRACT

The penetration response of tungsten-alloy long-rod penetrators into soda-lime glass targets is investigated at two impact velocities. Penetration depths were measured by flash radiography at different times during penetration. A modified experimental arrangement provided independent measurements of the residual projectile length and velocity for finite-thick glass targets. Numerical simulations using a Drucker-Prager constitutive model are compared to the experimental data and conclusions are drawn concerning the constitutive behavior of the glass.

INTRODUCTION

In 1993, Anderson, et al. [1], reported the results of experiments and simulations of long-rod projectiles into steel and glass targets at two impact velocities, 1.25 km/s and 1.70 km/s. The numerical simulations, using the wavecode CTH [2], reproduced reasonably well the experiments for the steel targets, but not the glass targets. The Johnson-Holmquist model [3] was used to model the constitutive behavior of the glass; parametric studies were conducted to assess the influence of constitutive parameters. Five computational zones were used to resolve the radius of the projectile. It was found that the treatment of mixed cells—how to homogenize the flow stress when there is more than one material within a computational cell—could have as much, if not more, influence on total penetration than the selection of the constitutive constants. Additionally, the simulations could not capture the transition to rigid-body penetration, which was observed in the experiments. The simulations considerably overpredicted projectile erosion.

The number of zones used to resolve the projectile radius was increased to 10 to minimize the influence of mixed cells in subsequent work [4]. It was shown that an elastic-perfectly plastic constitutive model, with a flow stress of 1.0 GPa, seemed to capture the early penetration-time response reasonably well; however, the simulations overpredicted projectile erosion, and underpredicted total penetration. It was then assumed that the projectile penetrated solely failed material. A Drucker-Prager constitutive model was used to describe the response of failed glass:

$$Y = \min(\beta P, \overline{Y})$$

where Y is the flow stress, β and $\overline{Y}$ are constants, and P is the hydrostatic pressure. The pressure-dependent region corresponds to comminuted pieces sliding over each other; and the cutoff, or cap, corresponds to material-deforming flow. One set of constants for both impact velocities ($\beta = 2.0$ and $\overline{Y} = 1.5$ GPa) provided approximate agreement between the simulations and experimental data.

Interest in glass has increased because of the observation of the phenomenon referred to as failure waves (for example, see Ref. [5]), and because of the importance of transparent "armor" for vehicles to provide adequate ballistic protection against a variety of threats. In

27

addition, computational power has increased tremendously over the last 12 or so years. Therefore, it was decided to re-examine the work conducted earlier. The results presented here represent "work in progress;" we summarize new insights in simulating penetration into glass, and provide comments on what we plan to do in the near future.

THE EXPERIMENTS

The experiments were conducted by Hohler and Stilp at the Fraunhofer Institut für Kurtzzeitdynamik (Ernst-Mach-Institut) [1,4]. The blunt-nose, tungsten-alloy (density 17.6 g/cm^3) projectiles had a length-to-diameter ratio (L/D) of 12.5, and a diameter of 5.8 mm. The targets were fabricated from sheets of soda-lime (float) glass, density 2.5 g/cm^3, with a compressive strength of 0.90 GPa. Ten sheets of 1.9-cm-thick glass and one sheet of 1.0-cm-thick glass were bonded together using double-sided adhesive (0.014-cm thick), for a total thickness of 20.0 cm. The glass was backed by a block of mild steel, as shown in Fig. 1. The lateral dimensions for most of the glass targets were 15 cm by 15 cm. Two of the experiments were performed with 30-cm x 30-cm wide glass plates to investigate if there was any influence of the lateral dimensions. The experimental results were independent of this doubling of the lateral dimensions.

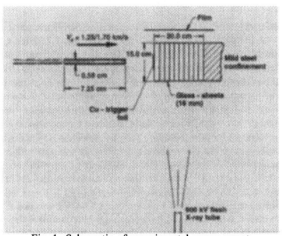

Fig. 1. Schematic of experimental arrangement.

Flash radiography was used to record the positions of the front and tail of the projectile, as well as the length of the projectile, as a function of time after impact. The nominal impact velocities for two sets of experiments were 1.25 km/s and 1.70 km/s.

The position-time histories of the nose and tail of the rods are shown in Fig. 2 for the two sets of experiments. Each pair of points (nose and tail positions) represents one experiment. The experimental spread in the impact velocities was ±20 m/s. The experimental data were adjusted by a simple proportionality to bring all data to a common impact velocity of 1.25 or 1.70 km/s. In none of the experiments did the projectile penetrate to the mild steel block.

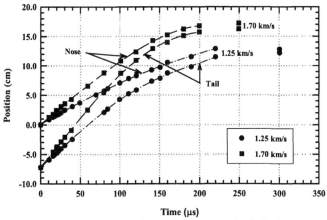

Fig. 2. Position-time data for experiments into glass. The dashed lines represent
regression fits through the data.

A modified target was constructed for some of the tests to provide an estimate of the
projectile velocity at different depths of penetration. This experimental configuration is shown
in Fig. 3. The glass portion of the target had different thicknesses, and the projectile was allowed
to exit the target through a 3.0-cm wide cavity drilled into the mild steel backup plate. A 4.5-cm
wide, 1.5-cm thick, honeycomb sheet was placed at the rear of the glass plates to provide
confinement but minimize penetration/perforation resistance. Flash radiography was then used
to obtain the residual velocity and projectile length behind the target.

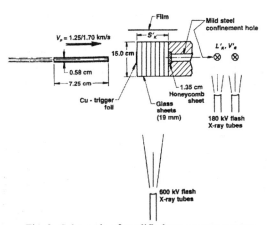

Fig. 3. Schematic of modified target arrangement.

29

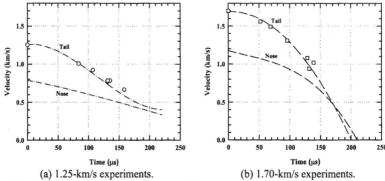

(a) 1.25-km/s experiments.　　　　　　(b) 1.70-km/s experiments.

Fig. 4. Time differentiation of the position-time regression fits in Fig. 2 gives the tail and nose (penetration) velocities vs. time.

Least-squares polynomial regression fits to the position-time data are shown as the dashed lines in Fig. 2. Differentiation of these regression fits with respect to time gives the tail and nose (penetration) velocities vs. time for the two sets of data. These are plotted in Fig. 4. Also plotted in the figures are the measured rod velocities using the experimental arrangement of Fig. 3. The residual rods nominally have nearly the same velocity as the tail since most of the projectile is moving with the tail velocity [6].

The normalized lengths of the rod vs. time are shown in Fig. 5. The closed symbols denote data using the experimental arrangement shown in Fig. 1, while the open symbols denote data using the arrangement shown in Fig. 3 (after the rods have left the target). Additionally, some of the residual rods were recovered in the impact tank; these are plotted as open symbols with a center dot beyond 300 μs. The transition from eroding to rigid-body penetration occurs around 130 μs - 140 μs for both impact velocities. Thus, penetration into glass can be divided into two distinct phases: eroding penetration and rigid-body penetration.

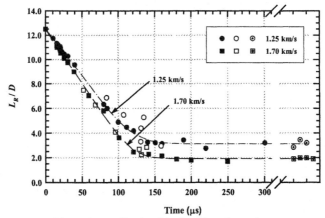

Fig. 5. Normalized residual rod length vs. time.

30

NUMERICAL SIMULATIONS

CTH [2] was again used for the numerical simulations, except now 25 zones were used to resolve the radius of the projectile. With this resolution, the treatment of mixed cells should have less than a 4% effect on the penetration results. The simulations conducted in Ref. [4] were then repeated. The comparisons between simulations and the experimental results are focused on times less than 150 μs, as this is the timeframe for eroding penetration. (Once the eroding response of the penetrator is modeled correctly, then the issue of how well the simulations reproduce rigid-body penetration can be evaluated.) The simulated penetration-time histories for $Y = 1.0P$ ($\beta = 1.0$, with no cap) are shown in Fig. 6(a), along with the experimental data.

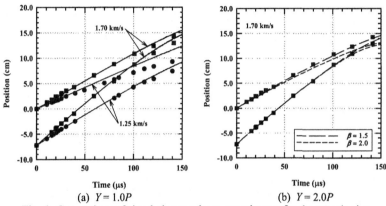

(a) $Y = 1.0P$ (b) $Y = 2.0P$

Fig. 6. Comparison of simulation results to experiments for the constitutive response $Y = \beta P$ (no cap)

Although the penetration response for the 1.70-km/s set of experiments is matched reasonably well in Fig. 6(a), the glass does not offer sufficient penetration resistance for the 1.25-km/s impact case, which implies that β needs to be increased. A parametric study was conducted on the dependence of the pressure slope, β. Two of the results, $\beta = 1.5$ and 2.0, are shown for the 1.70-km/s impact case in Fig. 6(b). As the pressure-dependent slope increases, the penetration resistance increases.

A $\beta = 1.5$ appears to do quite well for the 1.70-km/s impact case, but it still underpredicts the penetration resistance for the 1.25-km/s experiments. On the other hand, $\beta = 2.0$ provides too much resistance to penetration at 1.70 km/s. This suggests that a cap, that is, a $\overline{Y}$, is required to limit the flow stress for the higher impact velocity. Numerical simulations were conducted at both impact velocities for the constitutive case of $Y = \min(2.0P, 1.5 \text{ GPa})$. The results are shown in Fig. 7. Agreement is reasonable, but could be better. The simulations appear to reproduce the early-time penetration results, but as time progresses, the simulations diverge from the experimental data. A small error in the penetration velocity will accumulate over time, which will then be reflected in a noticeable error in the position-time response. (The fact that the simulations reproduce the tail position-time response fairly accurately implies that the constitutive model for the tungsten-alloy projectile is accurate. However, as the error in projectile length increases—because the penetration velocity is not correct—the calculated tail position will deviate form the experimental results.) Agreement is somewhat better for the 1.70-

31

km/s experiments than for the 1.25-km/s experiments. It is noted that there was little difference between the results for the previous 10-zone resolution simulations and the current 25-zone resolution simulations.

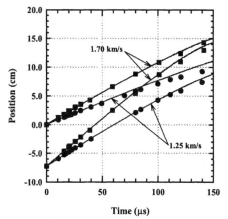

Time (μs)

Fig. 7. Comparison of numerical simulations to experimental data for the constitutive response $Y = \min[2.0P, 1.5 \text{ GPa}]$.

Further analyses were conducted on the results of the simulations. The calculated penetration and tail velocities vs. time for $Y = 2.0P$ and $Y = \min[2.0P, 1.5 \text{ GPa}]$ are compared to the experimental velocity-time data for the 1.25-km/s case in Fig. 8. It is seen, except for the first ~25 μs, that the penetration response for the two constitutive cases is essentially identical. This implies that the cap has minimal influence on the penetration response for the 1.25-km/s impact case. A larger slope β is required to decrease the calculated penetration velocity and bring it into better agreement with the experimental results.

The calculated penetration and tail velocities vs. time for $Y = 1.0P$ and $Y = \min[2.0P, 1.5 \text{ GPa}]$ are compared to the experimental velocity-time data for the 1.70-km/s case in Fig. 9. (Note that the constitutive model $Y = \min[2.0P, 1.5 \text{ GPa}]$ is common to both Figs. 8 and 9, but that the slopes are different for the $Y = \beta P$ models for the two figures.) Increasing the slope β from 1.0 to 2.0 decreased the calculated penetration velocity in Fig. 9. However, according to Fig. 6(b), where there is no cap, the change in slope would have had more influence if the flow stress had not been limited by the cap of 1.5 GPa. It can then be inferred that a slightly higher value of $\overline{Y}$ is required to decrease the calculated penetration velocity at 1.70 km/s.

Thus, it is concluded, to a first approximation, that the penetration pressure is sufficiently large that the cap controls the penetration response at 1.70 km/s. The slope β controls the penetration resistance at 1.25 km/s. The slope and cap have secondary roles in controlling the penetration resistance at 1.70 and 1.25 km/s, respectively.

SUMMARY AND CONCLUSIONS

Time-resolved experiments were conducted for long-rod penetration into glass targets at two impact velocities. It was concluded from examination of the experimental data that

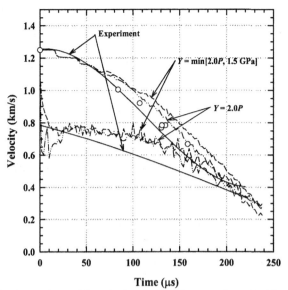

Fig. 8. Comparison of the calculated tail and penetration velocities to the experimental results shown in Fig. 4(a).

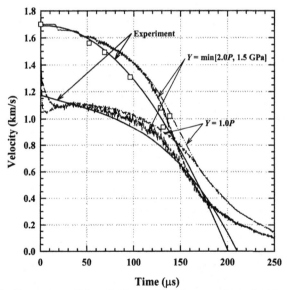

Fig. 9. Comparison of the calculated tail and penetration velocities to the experimental results shown in Fig. 4(b).

penetration could be divided into two phases, an erosion phase and a rigid-body penetration phase. Numerical simulations of the experiments were conducted. It was assumed that the projectile penetrated failed glass, which could be described by a Drucker-Prager constitutive model. The model has two adjustable constants: the slope of a pressure-dependent region and a cap that limits the flow stress. It was found that the constitutive model $Y = \min[2.0P, 1.5\text{ GPa}]$ provided reasonable agreement between calculated and experimental results, but agreement could be better.

It was determined that the calculated response at 1.70 km/s is controlled by the cap, and the response at 1.25 km/s is controlled by the slope. In the near future, we will use an analytical model [7] to iterate on the slope and the cap to optimize the estimates for β and $\overline{Y}$. These values will then be used in numerical simulations and the results compared to the experiments. It is expected that one set of parameters will provide fairly reasonable agreement between calculated response and experiment.

In principle, it is preferred to obtain constitutive constants from independent laboratory experiments, and then assess how well numerical simulations can reproduce the results of ballistic experiments. It is planned to conduct fundamental laboratory characterization experiments to obtain β and $\overline{Y}$ directly, using the techniques developed in Refs. [8-9]. It will be interesting to see how the laboratory-determined constitutive constants compare to the values inferred from the simulations.

REFERENCES

[1]C. E. Anderson, Jr., V. Hohler, J. D. Walker, and A. J. Stilp, "Penetration of long rods into steel and glass targets: experiments and computations," *Proc. 14th Int. Symp. on Ballistics*, Québec City, Canada, 24-29 September (1993).

[2]J. M. McGlaun, S. L. Thompson, and M. G. Elrick, "CTH: A three-dimensional shock wave physics code," *Int. J. Impact Engng.*, **10**, 351-360 (1990).

[3]G. R. Johnson and T. J. Holmquist, "A computational constitutive model for brittle materials subjected to large strains, high strain rates, and high pressures," *Shock Waves and High-Strain Rate Phenomena in Materials*, pp. 1075-1081 (M. A. Myers, L. E. Murr, and K. P. Staudhammer, Eds.), Marcel Dekker, NY (1992).

[4]C. E. Anderson, Jr., V. Hohler, J. D. Walker, and A. J. Stilp, "Modeling long-rod penetration into glass targets," *14th U. S. Army Symp. on Solid Mech.* (K. R Iyer and S-C Chou, Eds.), pp. 129-136, Battelle Press, Columbus, OH (1996).

[5]Y. Partom, "Modeling failure waves in glass," *Int. J. Impact Engng.*, **21**(9), 791-799 (1998).

[6]C. E. Anderson, Jr. and J. D. Walker, "An examination of long-rod penetration," *Int. J. Impact Engng.*, **11**(4), 481-501 (1991).

[7]J. D. Walker and C. E. Anderson, Jr., "An analytic penetration model for a Drucker-Prager yield surface with cutoff," *Shock Compression of Condensed Matter—1997* (S. C. Schmidt, D. P. Dandekar, and J. W. Forbes, Eds.), 897-900, AIP Press, Woodbury, NY (1998).

[8]K. A. Dannemann, A. E. Nicholls, S. Chocron, J. D. Walker, and C. E. Anderson, Jr., "Compression testing and response of SiC-N ceramics: intact, damaged and powder," *29th Int. Conf. Advanced Ceramics & Composites*, Cocoa Beach, FL, 23-28 January (2005).

[9]S. Chocron, K. A. Dannemann, A. E. Nicholls, J. D. Walker, and C. E. Anderson, Jr., "A constitutive model for damaged and powder silicon carbide," *29th Int. Conf. Advanced Ceramics & Composites*, Cocoa Beach, FL, 23-28 January (2005).

A CONSTITUTIVE MODEL FOR DAMAGED AND POWDER SILICON CARBIDE

Sidney Chocron, Kathryn A. Dannemann, Arthur E. Nicholls, James D. Walker, Charles E. Anderson
Southwest Research Institute
PO Drawer 28510
San Antonio, Texas 78228-0510

ABSTRACT

To perform numerical simulations of ballistic impact into ceramic targets, suitable constitutive models are needed for intact, damaged and powder ceramic. For ballistic applications some constitutive models are already available in the literature but they are rarely based entirely on mechanical characterization. In fact, the parameters published are usually a blend of the material properties available and some tuning to ballistic experiments.

This paper explains how a constitutive equation can be obtained from the compression testing of ceramics under confinement. Since the test procedure and results are described in another paper in this conference, this paper will focus on the analytical and numerical models performed to support the testing. The tests provide stress vs. strain and stress vs. hoop strain curves that, with a simplified mathematical model developed and described in this paper, can be plotted as stress vs. pressure curves.

Numerical models of the confined tests were also performed to both support and supplement the analytical model. First they were used to check that the analytical model was correctly implemented and then, due to the inherent limitations of the analytical model (only valid during elastic deformation), they help to explore constitutive models like Drucker-Prager with a cap and Johnson-Holmquist.

The final outcome of this paper is a constitutive model for damaged and powder SiC-N that can be used for ballistic simulations and has its origin entirely in a mechanical characterization procedure.

INTRODUCTION

The constitutive equation of SiC has already been the subject of many papers, see for example [1,2,3]. Indeed it is of major importance to have the right constitutive model when simulating ballistic perforation of ceramics since total penetration and penetration velocity are very sensitive to the model used. To the authors knowledge the constitutive models for SiC developed up to this date rely partially on ballistics data, i.e., some of the parameters of the model are set to match ballistic data.

The objective of this paper is to explain how, with a very simple experimental technique supported by a mathematical model and proper numerical simulations it is possible to construct a constitutive model for low and medium strain rates. The experimental technique is explained in detail in [4] and summarized briefly here. The mathematical model conclusions are promptly presented and used to interpret the test results and build the Equivalent Stress vs. Pressure curve for fractured (also called comminuted) SiC and SiC powder. The details of the mathematical model are delayed until the next section to better emphasize the constitutive model. Finally numerical simulations that serve as a double check of the analytical model are compared to the model and the tests.

EXPERIMENTAL TECHNIQUE

The experimental technique has been extensively explained in another paper in this same conference, see Dannemann [4], and will only be briefly summarized in this paper. The Silicon Carbide cylindrical specimen is inside a steel sleeve as shown in Figure 1. With an MTS machine or the Hopkinson Bar a known load (σ_{zz}) is applied to the specimen, while measuring the hoop strain $\varepsilon_{\theta\theta}$ with the strain gage.

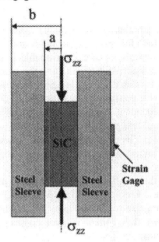

Figure 1: Experimental Set-up

The stress state in the sample is triaxial making it non-trivial to calculate the equivalent stress as well as the pressure the specimen is seeing. As it will be shown in the Mathematical Model section and double checked in the Numerical section the expressions to determine the pressure and the equivalent stress in the specimen are simple and come from the axisymmetric solution of elasticity problems:

$$P = \frac{1}{3}\left(-\sigma_{zz} + E_{steel}\, \varepsilon_{\theta\theta}(r = b)\, \frac{b^2 - a^2}{a^2}\right) \qquad (1)$$

$$\sigma_{eq} = |\sigma_{zz}| - \frac{E_{steel}}{2}\, \varepsilon_{\theta\theta}(r = b)\, \frac{b^2 - a^2}{a^2} \qquad (2)$$

Where a is the inner radius of the sleeve and b the outer radius. With the help of equations (1) and (2) the MTS tests can be "translated" into Equivalent Stress vs. Pressure curves that can later be used in the numerical models.

RESULTS AND DISCUSSION

Figure 2 puts in perspective the work performed during this research project. It illustrates the different pressure-strength curves for models currently in use for ballistics modeling (namely JH-1 and the SwRI analytic penetration model).

36

Boron carbide and silicon carbide models are already available in the literature [2,5], but, as stated by the authors, the constants that define the fractured silicon carbide behavior were not always determined by characterization of the material, but rather by trying to directly match ballistic test results with numerical simulations. Specifically, for fractured silicon carbide, Johnson and Holmquist used an unexpectedly low slope (=0.4) for the strength vs. pressure plot found in Klopp and Shockey [1], and a cap of 1.3 GPa for the strength to match the experimental results for penetration of tungsten rods into confined silicon carbide targets. The constitutive model used by Johnson is reproduced in Figure 2 with lines tagged with "JH-1" for different strain rates and for failed material.

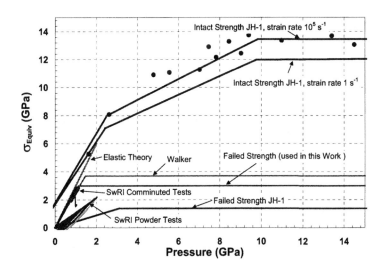

Figure 2. Constitutive model proposed in this work compared to JH model for SiC.

Using penetration data for SiC, Walker [3] produced an analytical penetration model where the slope was 2.5 and the cap of 3.7 GPa for the SiC, which is also showed in Figure 2 and tagged "Walker". Subsequently, Orphal, et al. [6], using the same slope of 2.5 and a cap of 3.7 GPa, showed that it is also possible to reproduce the experimental results using CTH simulations for penetration of tungsten and gold long rods into silicon carbide. The straight line (slope 3) with the tag "Elastic" would be σ_{eq} vs. P for an elastic material; the other curves are the experimental results for quasistatic tests performed during this work and labeled "SwRI".

A close-up view of that part of the plot is presented in Figure 3, where the slope is seen to fall between 2.5 and 2.8 for the fractured ceramic. The cap was not reached in the tests performed during this research program because the platens failed routinely around 3 GPa.

A last view at Figure 2 for the SiC powder test results, which is considered the maximum comminution possible, gives a slope of around one, still higher than the slope of the JH model for fractured SiC.

37

MATHEMATICAL MODEL

As explained previously the tests performed in this research load the specimen in both longitudinal and radial direction, making the problem triaxial and non-trivial to interpret. The following mathematical model completely solves the elastic problem and allows the calculation of quantities like pressure, confinement pressure and equivalent stress in the specimen.

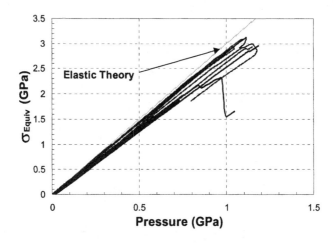

Figure 3. Quasistatic tests performed during this research project.

The problem, as presented in Figure 1, is axisymmetric and should be solved with Lamé's radial displacement field:

$$u = Ar + \frac{B}{r} \qquad (3)$$

where u is the displacement in the radial direction of a steel or SiC-N point, r the distance from the axis of symmetry to the point, and A and B are constants that depend on the boundary conditions.

There are four main assumptions in this model: 1) The strain in the vertical direction in the sleeve is zero. (i.e. the specimen slips in the sleeve without any friction). This assumption is not readily evident, but in [7] it is shown to be a good approximation to reality. 2) Radial stresses are identical on both sides of the specimen/steel interface. 3) Radial displacements are identical on both sides of the specimen/steel interface. 4) The outer boundary in the sleeve is stress free:
$\sigma_{rr} = 0$

Solution inside the specimen

Inside the specimen the displacement field constant, B, must be zero to be non-singular:

$$u_r = A_I \, r \qquad (4)$$

The axial load applied by the MTS machine is σ_{zz}, which is known. If the elastic constants of the specimen are λ_I, μ_I, then it is possible to write for the elastic part of the load using Hooke's law in polar coordinates:

$$\sigma_{rr} = (\lambda_I + 2\mu)A_I + \lambda_I A_I + \lambda_I \varepsilon_{zz} \tag{5}$$

$$\sigma_{\theta\theta} = 2(\lambda_I + \mu)A_I + \lambda_I \varepsilon_{zz} \tag{6}$$

$$\sigma_{zz} = 2\lambda_I A_I + (\lambda_I + 2\mu_I)\varepsilon_{zz} \tag{7}$$

where the strains have been replaced by their corresponding derivative from equation (4). Note that in the specimen $\varepsilon_{zz} = \text{const}\,(\neq 0)$. In these derivations the stress is assumed negative in compression. Observe also that $\sigma_{rr} = \sigma_{\theta\theta}$.

The equivalent stress is given by:

$$\sigma_{eq} = |\sigma_{rr} - \sigma_{zz}| = |2\mu_I A_I - 2\mu_I \varepsilon_{zz}| = 2\mu_I |A_I - \varepsilon_{zz}| \tag{8}$$

And the pressure in the specimen:

$$P = -\frac{1}{3}(2\sigma_{rr} - \sigma_{zz}) = = -(3\lambda_I + 2\mu_I)(2A_I + \varepsilon_{zz})/3. \tag{9}$$

It is important to note that the pressure in the specimen and the confinement pressure are different, since the confinement pressure is given by:

$$P_c = -\sigma_{rr} = -(2(\lambda_I + \mu_I)A_I + \lambda_I \varepsilon_{zz}) \tag{10}$$

Solution inside the sleeve

In the confining sleeve, Lamé's solution is still good but different from the previous one:

$$u_r = A_{II} r + \frac{B_{II}}{r} \tag{11}$$

Assuming λ_{II} and μ_{II} are the elastic constants of the sleeve material, Hooke's law gives:

$$\sigma_{rr} = (\lambda_{II} + 2\mu_{II})\left(A_{II} - \frac{B_{II}}{r^2}\right) + \lambda_{II}\left(A_{II} + \frac{B_{II}}{r^2}\right) + 0 = 2(\lambda_{II} + \mu_{II})A_{II} - 2\mu_{II}\frac{B_{II}}{r^2} \tag{12}$$

$$\sigma_{\theta\theta} = \lambda_{II}\left(A_{II} - \frac{B_{II}}{r^2}\right) + (\lambda_{II} + 2\mu_{II})\left(A_{II} + \frac{B_{II}}{r^2}\right) + 0 = 2(\lambda_{II} + \mu_{II})A_{II} + 2\mu_{II}\frac{B_{II}}{r^2} \tag{13}$$

$$\sigma_{zz} = \lambda_{II}\left(A_{II} - \frac{B_{II}}{r^2}\right) + \lambda_{II}\left(A_{II} + \frac{B_{II}}{r^2}\right) + 0 = 2\lambda_{II}A_{II} \tag{14}$$

The equivalent stress in the sleeve would be:

$$\sigma_{eq} = \left\{\frac{1}{2}\left(4\mu\frac{B_{II}}{r^2}\right)^2 + \frac{1}{2}\left(2\mu A_{II} - 2\mu\frac{B_{II}}{r^2}\right)^2 + \frac{1}{2}\left(2\mu A_{II} + 2\mu\frac{B_{II}}{r^2}\right)^2\right\}^{1/2} = 2\mu\{A_{II}^2 + 3B_{II}^2/r^4\}^{1/2} \tag{15}$$

This last equation shows that the equivalent stress in the sleeve is maximum at the inner diameter.

Boundary conditions

1) At $r = a$, radial stresses match, then from (5) and (12):

$$2(\lambda_I + \mu_I) A_I + \lambda_I \, \varepsilon_{zz} = 2(\lambda_{II} + \mu_{II}) A_{II} - 2\mu_{II} \frac{B_{II}}{a^2} \tag{16}$$

2) At r = a, radial displacements match

$$A_I a = A_{II} a + \frac{B_{II}}{a} \tag{17}$$

3) The outer stress is zero:

$$(\lambda_{II} + \mu_{II}) A_{II} - \mu_{II} \frac{B_{II}}{b^2} = 0 \tag{18}$$

The application of axial stress to the specimen upon loading by the MTS machine gives:

$$(\sigma_{zz}) = 2\lambda_I A_I + (\lambda_I + 2\mu_I) \varepsilon_{zz} \tag{19}$$

We have four linear equations ((16),(17),(18),(19)) with four unknowns: $A_I, A_{II}, B_{II}, \varepsilon_{zz}$, that can be easily solved with a few lines of FORTRAN code to give the displacement field for any applied load σ_{zz}.

It is possible to explicitly find pressure and equivalent stress in the specimen as a function of the applied load σ_{zz} and the hoop strain measured during the test:

$$P = -\frac{1}{3}(\sigma_{rr} + \sigma_{\theta\theta} + \sigma_{zz}) = -\frac{1}{3}(2\sigma_{rr} + \sigma_{zz}) \tag{20}$$

But from Equations (12), applied at r=a, and (18) it is possible to write $\varepsilon_{\theta\theta} = \frac{u_r}{r} = A_{II} + \frac{B_{II}}{r^2}$ as a function of the confinement pressure $P_{conf} = -\sigma_{rr}(r=a)$ which, together with Equation (20) leads to the desired Equation (1) (note that in the specimen σ_{rr} and $\sigma_{\theta\theta}$ are constants). Equation (2) follows simply from rewriting (8) as:

$$\sigma_{eq} = |\sigma_{rr} - \sigma_{zz}| = |-P_{conf} - \sigma_{zz}| = -P_{conf} + |\sigma_{zz}| \tag{21}$$

Onset of plasticity

It would be very tedious to solve analytically the plastic deformation of the specimen, but from the equations derived above it is straightforward to calculate the onset of plasticity. Assuming the strength of the specimen is proportional to the pressure (Drucker-Prager constitutive equation), it is possible to write for the strength of SiC:

$$Y = Y_o + \beta \, P \tag{22}$$

The specimen will begin to yield when:

$$\sigma_{eq} = Y$$

or, from   (8),

$$\sigma_{eq} = 2\mu_I |A_I - \varepsilon_{zz}| = Y_o + \beta \, P \tag{23}$$

Since the specimen expands radially, $A_I > 0$ and $\varepsilon_{zz} < 0$ because it is being loaded in compression, so

$$2\mu_I |A_I - \varepsilon_{zz}| = 2\mu_I (A_I - \varepsilon_{zz}) \tag{24}$$

Combining (23) with (24) and   (9) gives:

$$2\mu_I\left(A_I - \varepsilon_{zz}\right) = Y_o + -\frac{\beta}{3}(3\lambda_I + 2\mu_I)(2A_I + \varepsilon_{zz}) \qquad (25)$$

This last equation should replace (19) in the linear system if the plastic onset needs to be calculated.

NUMERICAL MODEL

The numerical modeling of both static and dynamic tests was performed using the finite difference hydrocode, AUTODYN 2-D. Lagrangian meshes were considered the optimum because of the slip condition between the test specimen and the confining sleeve.

Table 1. Elastic Constants used in the Analytical and Numerical Modeling.

	Intact SiC-N	Comminuted SiC-N	Steel
Elastic Modulus	475 GPa	115 GPa	210 GPa
Poisson's Ratio	0.17	0.115	0.33
Y_0	-	28 MPa	-
β	-	2.5	-
$\overline{Y}$	-	3.0 GPa	-

Two Lagrangian meshes with 0.5 mm cell size were used for these simulations. Since these are static test simulations the hourglass constant as well as the static damping constant were increased from the AUTODYN defaults to 1 and 0.02 respectively. This allowed reasonable computation time for a static simulation with an otherwise intrinsically dynamic code. Table 1 shows the material constants used in the simulations.

Simulation of the Elastic and Elasto-plastic Properties for the Fractured SiC-N test

The analytical model was used to do multiple runs and find in a short time the elastic constants that could best fit the fractured SiC-N behavior. It is important to point out that if only numerical simulations were available, this would have been a major time consuming task. With a good analytical model each run takes a few seconds, so a sweep through different Poisson's ratios is done very fast. The stress vs. hoop strain curves were found to be very sensitive to the Poisson's ratio value. The values that best matched the experimental data (both stress vs axial strain and stress vs. hoop strain) were a Young's modulus of 115 GPa and a Poisson's ratio of 0.115.

The analytical model was also extensively used for determining the constants of the Drucker-Prager constitutive model that was used with a cap ($\overline{Y}$). Since the yield depends on three constants (Y_0, β and $\overline{Y}$), various simulations were performed to determine which constants could best fit the tests. Figure 4 shows the results obtained when $Y_0 = 28$ MPa and $\beta = 2.5$, with a cap of 3 GPa. Both the numerical and analytical simulations used the same constants and provide excellent agreement with the tests.

41

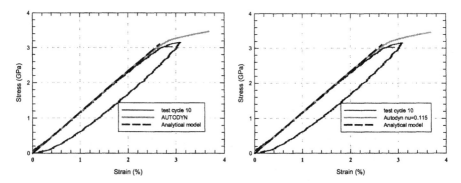

Figure 4. Elasto-plastic simulations compared with experimental results

CONCLUSIONS

Constitutive models for fractured SiC and SiC powder were presented. These models are based solely on mechanical testing and the elasticity theory.

For fractured SiC the Drucker-Prager constitutive equation obtained is:

$$Y = 0.028 + 2.5 \ P \ \text{(units in GPa)}$$

While for SiC powder the equation is:

$$Y = 1.1 \ P$$

The caps of these models were not clearly reached during the testing but for the fractured SiC it was assumed to be 3 GPa which agreed well with the limited observations at these high stresses.

ACKNOWLEDGMENTS

The authors gratefully acknowledge the financial support of the US Army Research Laboratory. Technical assistance and insights from Dr. Michael Normandia at ARL is also gratefully acknowledged

REFERENCES

[1] R. W. Klopp and D. A. Shockey, "The Strength Behavior of Granulated Silicon Carbide at High Strain Rates and Confining Pressure, *J. Appl. Phys.*, **70** (12), 1991.

[2] T. Holmquist, G. Johnson, "Response of silicon carbide to high velocity impact", *J. Appl. Phys.*, **91** (9), 2002, 5858-5866.

[3] J. Walker, "Analytically Modeling Hypervelocity Penetration of Thick Ceramic Targets", Int. J. Impact Engng., **29**, nos. 1-10, December 2003, Hypervelocity Impact Proc. of the 2003 Symposium, 747-755

[4] Kathryn Dannemann, S. Chocron, A. Nicholls, J. Walker, C. Anderson, Compression testing and response of SiC-N ceramics: intact, damaged and powder, Proceedings of Advanced Ceramics and Composites Conference, January 23-28, 2005, Cocoa Beach, Florida

[5] Gordon Johnson, Tim Holmquist, "Response of boron carbide subjected to large strains, high strain rates, and high pressures", Journal of Applied Physics, **85**, (12), 1999, 8060-8073

[6] D. Orphal, C.E. Anderson, D. Templeton, T. Behner, V. Hohler, S. Chocron, "Using Long Rod Penetration to Detect the Effect of Failure Kinetics in Ceramics", Proceedings of the 21st International Symposium on Ballistics, Adelaide, Australia, April 2004.

[7] Kathryn Dannemann, S. Chocron, A. Nicholls, J. Walker, C. Anderson, "Ceramic Phenomenological Experiments-Compressive Strength of SiC", Final Report, SwRI Project 18.06188.

DESIGNS AND SIMULATIONS OF BALLISTIC-RESISTANT METAL/CERAMIC SANDWICH STRUCTURES

Yueming Liang, R. M. Mcmeeking, A. G. Evans
Materials department, University of California
Santa Barbara, CA 93106

ABSTRACT

In this paper, we report our numerical work on the ballistic penetration of various armor designs by a 20 mm fragment-simulating projectile at a velocity of 1000 m/s. For the design with monolithic metal plate, numerical simulations show that final failure of the plate is tearing of the metallic material. Incorporation of ceramic material (i.e., metal/ceramic armor) changes the final failure mode of the armors to plastic bending and stretching of the backing plate. Two types of metal/ceramic designs are studied. One consists of ceramic layer backed by a metal plate, with or without a front metallic faceplate. The other metal/ceramic design utilizes the honeycomb structure and is made by filling the free space in a honeycomb-core sandwich panel with solid ceramics. The ballistic limits (the minimum mass required to stop the projectile) associated with all the designs are obtained and compared. Finite element calculations show that incorporation of ceramic gives a benefit of mass reduction of 20-30%. While utilizing honeycomb structure slightly improves the ballistic limits (compared with the case of ceramic layer backed by metal plate), it could remarkably reduce the deformation of the backing plate at ballistic limits. It is also found that for all metal/ceramic designs, the front faceplate does not contribute to improve the ballistic performances.

INTRODUCTION

The studies on ballistic penetration of armors have always been very active over the past three decades; and the search of efficient and lightweight armors has never stopped. The idea of modern metal/ceramic composite armors is due to Wilkins [1, 2], who proposed and developed aluminum/ceramic armors and studied the penetration mechanisms associated with these armors. The original composite armor design consists of a piece of ceramic tile mounted on a metallic plate on the outer side (side facing the projectile). As the projectile hits the armor, ceramic deforms and erodes the projectile, thus reducing the kinetic energy of the projectile. The rest of kinetic energy in the system is then consumed by the deformation of the backing metallic plate. During the entire process, ceramic fails and breaks; but it was commonly assumed that fracture of ceramic does not consume much of the energy [3, 4]. Since the pioneering work of Wilkins, various designs have been proposed as candidates of future lightweight armors. A few examples of them include ceramic armor backed by composite laminates [5, 6], ceramic armors backed by sandwich panel [7], and multi-layer multi-component (primarily ceramic) armors [4], etc. All of these armors use the same defeating mechanism, i.e., ceramic deforms and erodes the projectile and backing adsorbs the rest energy. It has been shown that each of these new designs has the potential as lightweight armor.

Present work focuses on numerical investigation of the possibility of developing lightweight armors using ceramics and all-metal honeycomb core sandwich panels, which have been found to have superior capability to mitigate damage due to blast waves [8-10]. Instead of developing multi-layer armors, we propose a new design that is made by filling the free space in a honeycomb-core sandwich panel with solid ceramic. It is expected that the cell structure associated with honeycomb can put some extra constraint on ceramic, therefore enhancing its ballistic resistance. In addition, the stiffness of the sandwich structure is much higher than that of a monolithic plate. As a result, sandwich structure filled with solid ceramic might also stop the projectile with less structural deformation. 4340 steel was chosen as the metallic material, as its material data are readily available [11]. The ceramic material used in this work is solid Boron Carbide. The ceramic model in the FE simulation is the Johnson-Holmquist model [12, 13], which has been implemented into ABAQUS/Explicit through its user subroutine feature. The chosen projectile is the 20mm fragment simulating projectile (FSP) [4], which is much bigger and heavier than most armor-piercing projectiles.

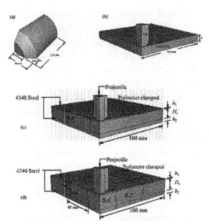

Figure 1. Schematics of the projectile and various armor designs. (a) A quarter of the 20 mm fragment simulating projectile; (b) A quarter of the monolithic plate; (c) A quarter of the metal-ceramic-metal triple-layer design, the axisymmetric model of this design is just a circular plate with radius 100 mm; (d) A quarter of the design incorporating honeycomb.

METAL/CERAMIC ARMOR DESIGNS

A military standard 20mm Fragment Simulating Projectile (FSP) is chosen as the penetrating projectile. The dimensions of the 20mm FSP is given in Figure 1(a). For comparison purpose, the initial velocity of the projectile is kept constant at 1000 m/s for all simulations.

The first armor design we studied is a monolithic metal plate, which was the very original design for ballistic protection. A schematic of a quarter of the metal plate is shown in Figure 1(b). For simplicity, the plate was assumed to be square with lateral dimension L =10cm. The same lateral dimensions were also assumed for all metal/ceramic composite designs. Two metal/ceramic composite designs are studied. The first one was a layer of ceramic material sandwiched between two metal plates, as shown in Figure 1(c). The second composite design also has a three-layer structure, but a more complicated central layer, which consists of a square honeycomb metal structure with solid ceramic filling the free space between honeycomb walls (see Figure 1(d)). Furthermore, designs similar to Fig. 1(c) and (d) but without a front face were also studied, in an effort to study the contributions of front faces to the ballistic performances of

44

proposed armor designs. For designs shown in Fig. 1(b) and Fig. 1(c), axisymmetric calculations were also performed in view of the isotropy in the plane of the panel, and the results were compared.

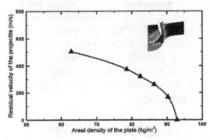

Figure 2. Residual velocity of the projectile versus areal density of the plate. Inset shows plate deformation shortly before penetration.

PENETRATION OF MONOLITHIC METAL PLATES

Penetration of the monolithic metal plates by the chosen projectile were carried out using the general purpose finite element package ABAQUS/Explicit. Isothermal conditions were assumed for the entire deformation history. Element deletion was performed to metal elements in the target; elements are removed from the calculations once they are considered "failed" according to the Johnson-Cook model. Final penetration is resulted when elements in a connected region surrounding the projectile are deleted. Numerical simulations showed the final failure mode of the plate is tearing, as shown in the inset of Figure 2. Once the projectile penetrates through the target plate, its kinetic energy becomes constant and its average terminal velocity can be calculated from the residual kinetic energy. A series of calculations with various plate thicknesses were carried out. The terminal velocity of the projectile is plotted against the plate areal density (plate density times plate thickness) in Fig. 2. It's shown that projectile terminal velocity decreases with increasing plate thickness and there exists a critical thickness of 11.7 mm beyond which the projectile cannot penetrate the plate. The areal density of the plate associated with this critical thickness is identified as the ballistic limit of 4340 steel against the given projectile at the impact velocity of 1000 m/s. Since we consider the ballistic performances of all designs on the basis of mass, this ballistic limit provides a baseline for future armor designs. Any design that gives a smaller ballistic limit is considered a better design.

PENETRATION OF METAL/CERAMIC COMPOSITE ARMORS AND COMPARISONS OF BALLISTIC PERFORMANCES

In this section, we discuss the results of ballistic penetration into metal-ceramic-metal triple-layer designs. The results from axisymmetric calculations are presented first and later three-dimensional calculations are discussed.

As the projectile hits the ceramic (or metal/ceramic composites) target, a compressive wave is initiated at the impact interface and it becomes tensile after it's reflected from the free surface at the back of the target. Conventionally it was thought that this tensile wave is responsible for the fracture of ceramic material. For metal-ceramic-metal triple-layer sandwich

45

armors, contours of ceramic damage and Mises stress during a short time period after impact are plotted in Figure 3. It is shown that as time increases, the profiles of damage in ceramics closely follow the profiles of stresses, and clearly the damage is due to high-pressure induced ceramic pulverization. In addition, a few microseconds after the initial contact, deformation of the projectile is still very small while the ceramic ahead of the projectile is almost completely damaged. This implies that the strength of fractured ceramic controls the ballistic performances of metal/ceramic armor systems. Figure 3 also shows that ceramic damage is greater in regions closer to the perimeter of impact. Numerical results indicate that hydrostatic pressure in regions near the impact perimeter is much lower than that in regions near the center of impact area, which implies lower material strengths near the impact perimeter. As a result, ceramic damage initiates from the perimeter of impact area.

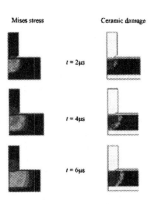

Figure 3. Evolutions of Mises stress (left column) and ceramic damage (right column) at early times of penetration. Red colour indicates ceramic pulverization.

For a metal-ceramic-metal triple-layer armor with h_1=1mm, h_2=4mm, and H_c=13mm, Figure 4(a) shows the evolutions of projectile kinetic energy and plastic dissipations in each material. One sees that the kinetic energy of the projectile drops at a very high rate at early times (up to time = 24μs) but later at a much smaller rate. Accordingly, plastic dissipations in the projectile and in ceramic increase quickly before t = 24 μs but quite slowly afterwards. In contrast, plastic dissipation in 4340 steel was quite low at t = 24 μs and it increases at a fairly low rate during the entire penetration history. It was quite surprising that the plastic dissipation in ceramic is comparable to that in metals, which is contrary to the conventional thought that fracture of ceramic during a ballistic event doesn't consume much energy. The discrepancy comes from the difference in ceramic failure mechanisms. Conventionally it was thought that fracture of ceramic during a ballistic event is brittle fracture caused by tensile waves. However, in the present study, failure of ceramic is caused by high-pressure induced plastic deformation, which is more capable of adsorbing energy than brittle fracture.

46

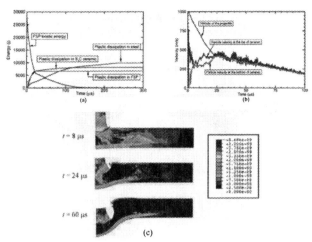

Figure 4. (a) Evolution of projectile kinetic energy and plastic dissipations in metals and ceramics as a function of time; (b) Evolution of velocities during penetration process; (c) Material deformation at different times. Colour indicates Mises stress.

Figure 4(b) shows, as a function of time, the velocities of the projectile and particle velocities at the top and bottom of the ceramic at the symmetry axis. Right after impact, the top surface of the ceramic layer immediately gained some velocity. However, the bottom face of the ceramic layer gained velocity only after the arrival of the stress wave. During the short time period after first contact, the velocity of the projectile drops quickly and the target materials right in front of the projectile gain velocities through momentum transfer. At time t = 24 μs, the velocity of the projectile, and the particle velocities in the ceramic layer are almost identical. They remain the same afterwards during the deformation history. Therefore, the entire deformation history can be divided into two stages. In the first stage, the projectile interacts with front face and with ceramic, causing projectile deformation and ceramic damage and transmitting momentum from the projectile to target materials. The deformation of the back faceplate is small during this stage, as shown in Fig. 4(c). At the end of Stage I, which is about t = 24 μs, the projectile and the target material ahead of the projectile have gained the same velocity, and the deformation goes to stage II, in which the projectile pushes the target and the backing plate bends and stretches to consume the rest kinetic energy (see Fig. 4(c)). If the total kinetic energy of the entire system at the end of stage I is smaller than the energy consumption capability of the backing plate, the projectile will be stopped. Otherwise, penetration of the armor is resulted.

It has been experimentally shown that, during penetration of solid ceramic, radial and circumferential cracks are formed. However, this cracking phenomenon cannot be predicted with axisymmetric models. In this regard, three-dimensional calculations have also been carried out to investigate ballistic penetration of metal-ceramic-metal triple-layer armors. In this case, a rectangular numerical specimen was chosen and meshed with 3-dimensional elements (see Fig. 1(c)). All materials are the same as those used in the axisymmetric calculations. Figure 5 shows the formation of ceramic damage (front and back faces are not shown for clarity) at the end of stage I. Clearly, one can see the pulverized zone and formation of radial and circumferential

47

cracks. Additional results show that the final deformations of the armors are also very similar to those from axisymmetric calculations, i.e., dramatic bending and stretching of the backing plate.

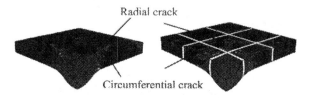

Figure 5. Crack formation in ceramics during penetration (3D calculations)

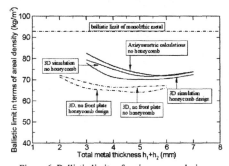

Figure 6. Ballistic limits of various armor designs

In order to obtain the ballistic limits of the proposed armor designs, series of calculations were conducted. For all calculations, the thickness of front faceplate is kept constant at $h_1=1$mm; however the thickness of the back faceplate is varied. For each combination, calculations with various thicknesses of ceramic are performed until a critical thickness (at which the projectile is stopped) is found. The areal density (which is metal density times metal thickness plus ceramic density times ceramic thickness) at this critical point is deemed as the ballistic limit of the proposed design. The ballistic limits of metal-ceramic-metal triple-layer designs are shown in Figure 6 as a function of total amount of metal in terms of total metal thickness. In general, the ballistic limits are about 25% lower than that of monolithic metal. There appear to be an optimal metal thickness at which the ballistic limit is the lowest. The results from axisymmetric calculations and those from 3-D calculations are slightly different, which might come from the difference of specimen shapes or in part is due to mesh sensitivity (failure with Johnson-Cook model is mesh sensitive). In general, the proposed designs outperform monolithic metals and they are good candidates for lightweight armors.

The ballistic performance of metal/ceramic armor with honeycomb core filled with solid ceramic has also been investigated. The numerical sample is shown in Fig. 1(d). For the present study, the thickness of the honeycomb wall was $t = 2$ mm, and the distance between adjacent parallel walls was $s = 40$ mm. The front faceplate still has a thickness of $h_1= 1$mm. For a given thickness of back faceplate, the thickness of the core (combination of honeycomb and ceramics) was varied until the ballistic limit of stopping the FSP was obtained. Damage of ceramic is also

48

shown in Figure 5. One can see that incorporating honeycomb changes the pattern of ceramic damage. With honeycomb, the radial cracks propagated along the symmetry planes; more circumferential cracks appeared near the impact region. Since the honeycomb walls strengthen the back faceplate, solid ceramic in the impact region experiences more constraint from surrounding metal. Therefore, more energy is consumed by ceramic deformation. Due to honeycomb, quite big deformations are also induced in the front faceplate, whereas for the triple-layer design, front faceplate is just slightly deformed. Figure 6 shows that the ballistic limits of the designs incorporating honeycomb are lower than those of the triple-layer designs. This benefit came from the additional constraint provided by the honeycomb. On the other hand, addition of honeycomb walls also increases the mass of the armor. As a result, for the present designs, the total benefit of adding honeycomb is not significant. In the future, more work will be done to study the ballistic performances of armors with various honeycomb parameters (e.g., wall thickness, wall spacing, etc.), and to find the optimal performances honeycomb can provide.

For armor systems, especially personal body armors, in addition to the capability of stopping the projectile, the deformation of the armor is also an important measure of armor performance. The less the deformation is, the more protection the armor provides. Figure 7 shows the deflections of the back faceplate at the center of the panel (All are 3D calculations). One sees that incorporation of honeycomb structure reduces the deflection of the back faceplate by an amount of about 20%. With honeycomb structure, the entire metal system behaves as an integral part and the panel has much stronger bending stiffness than the back faceplate itself. As a result, less deformation is induced on the back faceplate before the projectile is stopped.

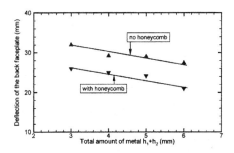

Figure 7. Deflections of the back faceplate at ballistic limits. Results are shown for 3-dimensional calculations. The thickness of the front faceplate is 1mm.

In all previously shown results, the thickness of the front faceplate was kept constant at 1mm. In these cases, it was found that the projectile easily penetrated the front faceplates. Therefore, we speculated that the contribution of the front faceplate to ballistic resistance is not significant. In order to investigate the effect of front faceplate, previous 3D calculations were repeated with the front faceplate removed. For all cases, the projectile directly hits the ceramics, causes pulverization, and finally large plastic deformation of the back faceplate. The deformations at all stages are quite similar to what have been shown. Similar procedure yields the ballistic limits of the designs without a front faceplate. As shown in Fig. 6, the designs without a front faceplate have the best ballistic performances, i.e., a 30% mass reduction. Incorporation of honeycomb core brings a bit more ballistic resistance, but still not very significant. These results imply that the front faceplate is not necessary for armor designs involving ceramics; the essentials of metal/ceramic armors are the fact that ceramic can deform the projectile and distribute the load over a larger area onto the back faceplate, then the back faceplate consumes the rest of the kinetic energy through plastic bending and stretching. Again, deflection of the back faceplate is reduced when honeycomb is introduced into the system.

CONCLUSIONS

Finite element simulations on the ballistic penetration of various armor designs (monolithic metal layer, metal/ceramic triple-layer, and metal/ceramic armor with honeycomb) by a 20 mm fragment-simulating projectile have been carried out. Numerical results showed that metal/ceramic armors generally have better ballistic performances than monolithic metals. For high speed penetration, it was shown that energy consumed by plastic deformation and fracture of ceramic is comparable to that adsorbed by metals, which is contrary to the conventional thought that fracture of ceramic doesn't consume much energy during ballistic penetration. Ceramic materials deform the projectile, distribute the load over a larger area onto the backing plate; and the final failure mode of the backing plate is plastic instability instead of tearing in the case of penetration into monolithic metal plates. Incorporation of metallic honeycomb structure in the ceramic layer brings appreciable benefit to the ballistic limit and reduces the deformation of the back faceplate as well. It has also been shown that a front faceplate is not necessary for metal/ceramic armors.

REFERENCES

[1] Wilkins, M.L., Cline, C.F., and Honodel, C.A., 1969. Fourth progress report of light armor program. Report UCRL 50694, Lawrence Radiation Laboratory, University of California.
[2] Wilkins, M.L., 1978. Mechanics of penetration and perforation. *Int. J. Eng. Sci.* 16, pp. 793.
[3] Mayseless, M. et al, 1987. Impact on ceramic targets. *J. Appl. Mech.*, 54, pp. 373.
[4] Gama, B.A. et al, 2001. Aluminum foam integral armor: a new dimension in armor design. *Comp. Struct.*, 52, pp. 381.
[5] Navarro, C., Martinez, M.A., Cortes, R., and Sanchez-Galvez, V., 1993. Some observations on the normal impact on ceramic faced armors backed by composite plates. *Int. H. Impact Engng.*, 13, pp. 145.
[6] Benloulo, I.S.C. and Sanchez-Galvez, V., 1998. A new analytical model to simulate impact onto ceramic/composite armors. *Int. J. Impact. Engng.*, 21, pp. 461.
[7] Senf, H., Strassburger, E., and Rothenhausler, H., 1997. Investigation of bulging during impact in composite armor. *J. Phys. IV France*, 7, pp. C3-301.
[8] Evans, A.G, 2001. Lightweight materials and structures. *MRS Bulletin*, 26, pp. 790.
[9] Xue, Z. and Hutchinson, J.W., 2003. A comparative study of impulse-resistant metal sandwich plates. *Int. J. Impact Engng.*, 30, pp. 1283.
[10] Fleck, N.A. and Deshpande, V.S., 2004. The resistance of clamped sandwich beams to shock loading. *J. Appl. Mech.*, 71, pp. 386.
[11] Johnson, G.R. and Cook, W.H., 1985. Fracture characteristics of three metals subjected to various strains, strain rates, temperatures and pressures. *Engng. Fract. Mech.*, 21, pp. 31.
[12] Johnson, G.R. and Holmquist, T.J., 1994. An improved computational constitutive model for brittle materials. In Schmidt, S.C., Shaner, J.W., Samara, G.A. and Ross, M. (Eds): High-pressure Science and Technology, pp. 981.
[13] Johnson, G.R. and Holmquist, T.J., 1999. Response of boron carbide subjected to large strains, high strain rates, and high pressures. *J. Appl. Phys.*, 85, pp. 8060.

CONSIDERATIONS ON INCORPORATING XCT INTO PREDICTIVE MODELING OF IMPACT DAMAGE IN ARMOR CERAMICS

Joseph M. Wells
JMW Associates
102 Pine Hill Blvd
Mashpee, 02649-2869

ABSTRACT

Traditional predictive ballistic damage models are limited in their usefulness for several reasons including their continuing inability to accurately describe the 3D morphological distribution and characteristics of the impact cracking damage in the target material(s) beyond and adjacent to the penetration cavity. Results from both destructive sectioning and advanced high resolution nondestructive x-ray computed tomography, XCT, examination reveal a distinctly non-uniform 3D damage distribution which is frequently asymmetrical along the projected penetration path. In addition to the traditionally reported conical, radial, circular and laminar cracks, surface topological step features and spiral cracking have been revealed and visualized in ceramic targets via XCT damage analysis. Techniques also have been developed to quantify the internal cracking damage fraction throughout the target sample damage volume with the use of the cracking damage distribution as a function of radius and penetration depth. Thus, XCT has the demonstrated capability for the identification, characterization and visualization of many interesting internal damage features and their volumetric distributions in armor ceramics which should be of considerable interest and utility to the predictive damage modeling community. The author advocates the collaborative incorporation of XCT as both a developmental and a verification tool for improved predictive impact damage modeling in armor ceramics.

INTRODUCTION

Prediction of the ballistic performance of armor ceramic targets is highly desirable for a variety of reasons. The traditional approach to the prediction of ballistic performance has been focused on the macro-penetration phenomenon, which is relatively straight forward to observe and measure as a function of ballistic test conditions. However, the determination of the various material factors and their effects that control the penetration process has proven more elusive. Perhaps the more rational way to understand ballistic performance is to first understand the effect that these material factors have on the material damage created by ballistic impact and then relate the damage phenomena to the penetration process. Characterization of ballistic impact *damage,* as opposed to *penetration,* in the target material surrounding the penetration cavity has been grossly under-emphasized (often ignored) in many ballistic studies of various target materials and design configurations including armor ceramic encapsulation. Damage in monolithic metal targets is perhaps somewhat easier to characterize and visualize than in the more brittle and extensively cracked armor ceramics. Yet even with monolithic metallic targets, detailed characterization of the impact damage has been limited.

Experimentally, damage characterization can be acquired through either destructive or non-destructive techniques. Destructive techniques have several liabilities including: the potential for introducing extraneous damage in sectioning, the irreversible nature of the sectioning process, and the observations of damage on a sectioned plane being physically limited to the 2D perspective. Traditional nondestructive modalities such as immersion ultrasonics and infrared thermography have been successfully utilized primarily for pre-impact target material inspection and have quite limited resolution and discrimination of the complex post-impact damage. Consequently, little detailed volumetric damage characterization and visualization at the meso-scale, in either monolithic metal or armor ceramic target materials, has been conducted and published other than the XCT work of the author and his collaborators[1-8].

Furthermore, while various predictive penetration models have frequently been described in the literature[9-11], there is an apparent absence of detailed predictive damage modeling publications. This situation has developed primarily because of the significantly greater difficulty in the understanding and modeling of the complex 3D morphology of actual impact cracking damage and the general lack of detailed physical damage characterizations and visualizations available for guidance and verification. Granted that this is a most difficult challenge, it also remains a missing link in the toolbox needed to design, screen and evaluate improved armor material and/or architectural concepts prior to their actual construction and testing on the firing range. Were such modeling tools available, both substantial savings in range expenses as well as decreased development time could be realized.

Therefore, the suggestion is advanced here that, with the recent and ongoing improvements in computational modeling tools as well as in the demonstrated capabilities of XCT 3D damage analysis, the time may be right for a serious review and assessment of the feasibility of the synergistic integration of these technologies. The potential benefit of such a technological collaboration would include the development of significant advancements in the capability for the engineering design and the predictive modeling of ballistic impact damage resistant armor ceramics.

BACKROUND

To date, neither theoretical nor computational models of detailed ballistic impact damage have emerged to accurately predict and describe the complexities and characteristics of actual ceramic ballistic target damage. Certainly, admirable progress has been made in the understanding of the micro-cracking formation in the comminuted (or Mescal) zone in ceramics in advance of an impacting projectile[12]. However, such studies have not satisfied the above mentioned modeling capability to adequately describe, much less predict, the volumetric damage details which may control the prevention, delay and/or inception and progress of penetration. One reason for this is because actual ballistic impact damage consists of both micro- and meso-cracking features. Meso-cracking is defined here as cracking features greater than 250 microns (0.25 mm). While the comminuted zone, CZ, contains essentially severely micro-cracked ceramic material directly in the path of the impacting projectile, extensive meso-scale cracking is also present in the bulk ceramic material surrounding the CZ. The detailed relationship or interdependency between these two cracking modes has not yet been adequately established. Some meso-cracks may simply be propagated extensions of CZ micro-cracks, while most appear physically unconnected to the CZ. Such meso-scale cracking has frequently been observed visually in high energy impacted targets of SiC, Al_2O_3, B_4C, TiB_2 and TiC ceramics.

The use of x-ray computed tomography, XCT, to study the meso-scale cracking damage in metallic and ceramic ballistic target materials at ARL was initiated in 1997 by the author and his collaborators. Early attempts at using immersion ultrasonics were unsuccessful in revealing the post impact damage details to the extent desired. The resolution of the XCT nondestructive technique depends on the sample density and size as well as the specific XCT facilities operational capabilities available. In general, the minimum practical feature resolution on conventional sized laboratory ceramic targets is ≥ 250 μm, ideal for examination of the meso-scale cracking. Higher resolution of features to ≤ 20 μm is possible with x-ray micro-tomography, XMCT, but, unfortunately, this modality is restricted to smaller sample sizes impractical for many, but not all, ballistic tests. In the XCT technique the target sample is rotated over 360 degrees in a series of incremental steps within an x-ray beam. The x-rays penetrate through the sample to varying extents depending upon the x-ray energy absorption or attenuation at each triangulated location within the volumetric space of the sample. The reconstruction software used in the scanning process creates a virtual density map of the scanned object in which damage features such as cracks and voids (with zero density) are easily distinguishable from the bulk ceramic or any residual imbedded heavy density projectile fragments. Results with this XCT technique were obtained and successfully utilized to characterize and visualize the volumetric impact damage in laboratory ballistic targets of TiB_2 [6,7] and TiC[8] ceramics.

Imaging versus Visualization

The process of XCT damage analysis and assessment is considerably more involved than simply obtaining x-ray images of the damaged ceramic targets. It is important to recognize that a simple image, x-ray or otherwise, is not necessarily a cognitive visualization. Most often, the acquisition of the 2D x-ray scans is simply the starting point for a process involving 2D to 3D solid object image reconstructions, virtual sectioning, point clouds and variable transparency, damage visualizations, feature isolation and segmentation, virtual metrology for damage quantification, and a healthy connection between insight, logical reasoning, creative rationalization and practical imagination. Figure 1 shows a simplified view of these interconnected aspects of the XCT damage analysis and visualization process from the author's perspective. In most all cases, these additional value added image processing and analysis steps are required to achieve an acceptable cognitive visualization of the damage details and morphology.

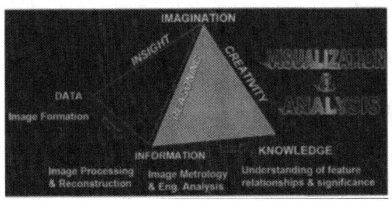

Figure 1. Outline showing the basic processes of cognitive visualization and analysis for impact damage characterization.

Summarized XCT Damage Results

Many examples of the ceramic damage characterization and visualizations obtained to date with XCT are contained in the companion NDE paper[1], also presented at this conference. A brief summary of several XCT meso-scale impact damage observations for these two ceramic target materials is as follows:

- The actual damage is complex involving multiple and superimposed or intertwined cracking modes including, but not limited to, conical, radial, and laminar cracks. The meso-cracking is relatively inhomogeneous and asymmetric through the sample volume.

- Many meso-scale cracking observations do not suggest a direct physical linkage to the micro-cracking in the ceramic comminuted zone. One example is the initiation of radial cracking from the OD inwards toward the impact point in the TiB_2 ceramic target.

- Additional damage modes not known to have been previously reported were observed including circular raised steps on the impact surface of the TiB_2 ceramic[1] and a spiral cracking morphology orbiting about the penetration path in the TiB_2 or its extended projection in the 100% dwell case for the TiC sample (see figure 2).

- Residual w-alloy penetrator fragments were observed in a fairly consolidated columnar volume extending through the thickness of the single shot TiB_2 target without pre-stress. With pre-stress and following a second impaction, the residual w-alloy fragments were more widely spread into two separate consolidated 3D wedge shapes.

- It has proven feasible to isolate and visualize the meso-cracking damage as well as the residual projectile fragments independent of the opacity of the host bulk ceramic.

- Virtual metrology of the interior target damage features has been demonstrated in-situ. Also quantitative measurements and 3D mapping of the cracking damage fraction through out the sample volume are possible and have been demonstrated.

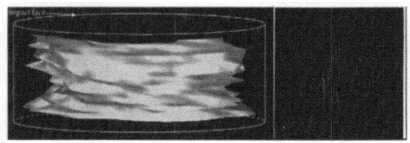

Figure 2. Transparent point cloud images of "spiral" cracking in TiB2 (left) and TiC ceramic (right)

The above mentioned results obtained with XCT are presented in the companion to this paper[1] and are summarized here to emphasize the significant progress made to date in meso-scale impact damage characterization and visualization. Such results are not necessarily the limits of potential future capabilities with this NDE technique. Undoubtedly, further advancements in the XCT equipment, software and application methodologies associated with this technology will continue to occur. The pressing issue is whether these and future advancements in XCT damage assessment and visualization will be appreciated and systematically accommodated by the predictive damage modeling community.

RELATIONSHIP TO PREDICTIVE DAMAGE MODELING

In the absence of an acceptable predictive modeling capability for ballistic impact damage and its relationship to ballistic performance, how can actual XCT NDE results provide synergistic input for future model development? Some suggestions are made here including:

- Providing realistic 3D descriptions of actual ballistic impact damage in armor ceramics and quantification of damage fraction and the volumetric mapping thereof.
- Accentuating the role of meso-scale damage to complement the evolving understanding of the micro-scale damage features.
- Obviating nonrealistic modeling assumptions such as homogeneous or axi-symmetric damage morphologies in non-applicable ballistic/target conditions. and identification of previously unrecognized/unreported damage features and morphologies.
- Creating 3D visualizations to assist in the cognitive understanding of the damage complexity and assist in the deconvolution/separation of different damage modes.
- Merging the XCT damage analyses with improved dynamic response analyses of the damaged material into the engineering toolbox to be used to design, screen and evaluate improved armor material and/or architectural concepts.
- Providing a quantitative, as well as qualitative, verification tool to test the validity of future evolving damage modeling predictions versus actually incurred damage.

Determination of the impact damage internal to the target ceramic reflects both physical manifestations of the operational damage mechanism(s) and the resultant residual structural integrity of the sample with a given damage presence. As shown by Hauver et al.[13], complete prevention of penetration can be accomplished by containing and constraining the ceramic target in such a way as to cause the projectile to be destroyed on the impact surface of the target ceramic. This phenomenon has been called "interface defeat" and involves 100% projectile dwell up to a threshold projectile velocity, V_{th}. Despite the avoidance of penetration with interface defeat, considerable cracking damage still occurs, nevertheless, within the ceramic target. This damage includes both the micro-cracked CZ as well as extensive meso-cracking in the bulk ceramic surrounding the CZ. A further increase in the projectile velocity above Vth is accompanied by the onset of penetration as well a corresponding increase in the interior cracking damage. For example, consider the schematic of a projectile impacting on the surface of a ceramic target as shown in figure 3.

The author proposes the following hypothesis relating to the influence of the meso-scale cracking damage to the prevention and onset of penetration. The fact that, at projectile velocities below V_{th}, the projectile does not penetrate through the pulverized ceramic CZ indicates that there is considerable dynamic containment pressure applied on the CZ to overcome the outward expansion pressure from the impacting projectile. That essential containment pressure or constraint on the CZ must originate from the surrounding bulk ceramic containing the meso-scale damage. If the CZ were unconstrained and free to move laterally, penetration would occur. As the extent and degree of severity of the bulk meso-scale damage increases with increasing projectile velocity, the structural integrity of the damaged bulk ceramic and the related CZ constraint are correspondingly envisioned to decrease. Consequently, the occurrence of penetration may well reflect the decrease of the dynamic constraining pressure originating with the increase in damage of the bulk ceramic.

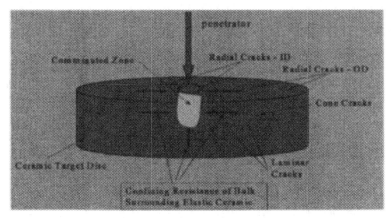

Figure 3. Schematic indicates penetration resistance created by constraining pressure on the CZ by the impact damaged surrounding bulk ceramic material.

The key to establishing the degree of accuracy of this hypothesis is the determination of the maximum level of dynamic constraint sustainable by a defined ceramic damage structure. This non-trivial computational modeling task should prove easier at this stage than predicting the damage structure itself. A direct correlation between this maximum constraint and the pressure exerted by the penetrator on the CZ should be established. To obtain a desired higher level of sustainable constraint, one would need to modify the target material/architecture to limit the resulting damage morphology and thus mitigate the overall structural stability loss of the damaged ceramic. Any proposed modification would be evaluated a priori for the level of sustainable structural constraint by the same methodology used to establish the limits of the initial damage configuration.

The essence of the above hypothesis is that the successful mitigation of the meso-scale cracking damage would increase the severity of ballistic impact required to either initiate or increase penetration. The approach to attaining the desired damage mitigation is either by extrinsic target architecture changes or by the more challenging route of intrinsic engineered material modifications. Certainly, the effect of an extrinsic target architectural approach such as ceramic encapsulation is to provide for overall dynamic and static constraint of the target ceramic. Under appropriate encapsulation constraint, the target ceramic is made extrinsically more rigid, thus preventing both dynamic bending and radial expansion. Not inconsequential, the static constraint offers both a compressive pre-stress and a post impact containment of the damaged ceramic. Various encapsulation schemes have been proposed and experimentally evaluated. Laboratory ballistic testing has been the accepted screening approach, despite its cost. The predominant acceptance criterion in use appears to be the avoidance of complete penetration and does not routinely include detailed ceramic impact damage characterization.

Especially challenging is the intrinsic engineered materials modification approach. Traditional intrinsic modification actions have considered variations of composition, microstructure, increasing interfacial area by variations in grain size or adding meso-scaled layered material geometric interfaces, and/or other approaches. These various ceramic modifications are conceptualized, formulated and processed mainly to optimize hardness,

56

strength and toughness – all properties which appear necessary but, to date, not sufficient for significantly improved ballistic performance. The suggestion being advanced here is that the mitigation of the 3D meso-scale damage morphology be explicitly included as an additional consideration in future intrinsic ceramic modification approaches as indicated in figure 4.

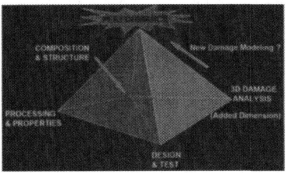

Figure 4. Schematic indicating the added dimensionality of 3D Damage Analysis and its potential effect on improving ballistic performance.

One still needs a valid screening test to initially evaluate any such proposed intrinsic ceramic modifications. Perhaps, the quasi-static ceramic indentation testing approach[14] will suffice for initial screening, but the relationship of quasi-static indentation damage to dynamic ballistic impact damage still needs to be established. Unquestionably, the use of XMCT (microtomography) on the smaller indentation ceramic samples would prove beneficial in improving both the characterization of the 3D quasi-static damage morphology and in making detailed comparative analyses with the dynamic ballistic damage of smaller size ceramic targets.

SUMMARY COMMENTS

XCT has provided the demonstrated capability for the advanced characterization, analysis and visualization of many interesting internal damage features and their volumetric distributions in armor ceramics. The dynamic behavior of this damaged material also needs to be further understood, which should be of considerable interest and utility to the predictive damage modeling community. The author advocates the collaborative incorporation of XCT as both a developmental and a verification tool for improved predictive impact damage modeling in armor ceramics. Granted the XCT analyses and dynamic response of the damaged material remain as missing links in the engineering toolbox used to design, screen and evaluate improved armor material and/or architectural concepts prior to their actual construction and test on the firing range. Were such modeling tools available, substantial savings in both range expenses as well as decreased armor ceramic development time could be realized.

ACKNOWLEDGEMENT

Acknowledgements are gratefully extended to N.L. Rupert, W.H. Green, J.R. Wheeler, H.A. Miller, J.W. Winter, Jr., W. A. Bruchey, and J.C. LaSalvia at ARL for their many and varied technical contributions and helpful discussions with the author during the course of the original XCT examinations and damage analysis of the ceramic ballistic damage. Our continuing interactions and discussions are most stimulating and appreciated.

REFERENCES

[1] J. M. Wells, "On Non-Destructive Evaluation Techniques for Ballistic Impact Damage in Armor Materials", this ACERS Cocoa Beach Conference, (2005), *In Press.*

[2] J.M. Winter, W.J. Bruchey, J.M. Wells, and N.L. Rupert, N.L. "Review of Available Models for 3-D Impact Damage", Proceedings 21st International Symposium on Ballistics, Adelaide, ADPA, v.1,111-117 (2004).

[3] H.T. Miller, W.H. Green, N.L. Rupert, and J.M. Wells, "Quantitative Evaluation of Damage and Residual Penetrator Material in Impacted TiB2 Targets Using X-ray Computed Tomography", Proceedings 21st International Symposium on Ballistics, Adelaide, ADPA, v.1, 153-159 (2004).

[4] J. M. Wells, N. L. Rupert, and W. H. Green, "Progress in the 3-D Visualization of Interior Ballistic Damage in Armor Ceramics", Ceramic Armor Materials By Design, Ed. J.W. McCauley et. al., Ceramic Transactions, v. 134, ACERS, 441-448 (2002).

[5] J. M. Wells, N. L. Rupert, and W. H. Green, "Visualization of Ballistic Damage In Encapsulated Ceramic Armor Targets" , 20th International Symposium on Ballistics, Orlando, FL, 23-27 September, 2002, ADPA, v. 1, 729-738 (2002).

[6] W. H Green, K.J. Doherty, N. L. Rupert, and J.M. Wells, "Damage Assessment in TiB2 Ceramic Armor Targets; Part I - X-ray CT and SEM Analyses", Proceedings MSMS2001,2nd International Conference on Mechanics of Structures, Materials and Systems, University of Wollongong, NSW, Australia, 130-136 (2001).

[7] N.L. Rupert, W.H. Green, K.J. Doherty, and J.M. Wells, "Damage Assessment in TiB2 Ceramic Armor Targets; Part II - Radial Cracking", Proceedings MSMS2001, 2nd International Conference on Mechanics of Structures, Materials and Systems, University of Wollongong, NSW, Australia, 137-143 (2001).

[8] J.M. Wells, W.H. Green, and N.L. Rupert, "Nondestructive 3-D Visualization of Ballistic Impact Damage in a TiC Ceramic Target Material", Proceedings MSMS2001, 2nd International Conference on Mechanics of Structures, Materials and Systems, University of Wollongong, NSW, Australia, 159-165 (2001).

[9] D.M. Stepp, "Damage Mitigation in Ceramics: Historical Developments and Future Directions in Army Research", Ceramic Armor Materials By Design, Ed. J.W. McCauley et. al., Ceramic Transactions, v. 134, ACERS, 421-428 (2002)

[10] A.M. Rajendran, "Historical Perspective on Ceramic Materials Damage Models", Ceramic Armor Materials By Design, Ed. J.W. McCauley et. al., Ceramic Transactions, v. 134, ACERS, 281-297 (2002)

[11] D.W. Templeton, T.J. Holmquist, H.W. Meyer, Jr., D.J. Grove and B. Leavy, "A Comparison of Ceramic Material Models", Ceramic Armor Materials By Design, Ed. J.W. McCauley et. al., Ceramic Transactions, v. 134, ACERS, 299-308 (2002)

[12] J.C. LaSalvia, "Recent Progress on the Influence of Microstructure and Mechanical Properties on Ballistic Performance", Ceramic Armor Materials By Design, Ed. J.W. McCauley et. al., Ceramic Transactions, v. 134, ACERS, 557-570 (2002)

[13] G.E. Hauver, P.H. Netherwood, R.F. Benck, and L.J. Kecskes, "Ballistic Performance of Ceramic Targets", Proc. 13th Army Symposium on Solid Mechanics, USA, 23-34 (1993)

[14] B.R. Lawn, "Indentation of Ceramics with Spheres: A Century after Hertz", J. Am. Ceramic Soc., 81, 1977-1994 (1998).

FAILURE WAVE PROPAGATION IN BRITTLE SUBSTANCES

M.A. Grinfeld, S.E. Schoenfeld, and T.W. Wright
US Army Research Laboratory
Aberdeen Proving Ground, MD, 21005-5069

ABSTRACT

Extensive experiments with glasses and brittle ceramic materials have been made by different research groups in Britain, Russia, and the USA in the 1990s [1-4]. The experiments showed that in addition to standard shock-wave fronts, which propagate with high, trans-sonic velocities, other, much slower, wavefronts can propagate within a substance undergoing processes of intensive damage. These moving fronts propagate into intact substance leaving behind them intensively damaged substance. The fronts have been called failure waves. The problem of failure waves demands significant progress in the relevant experiment, theory and numerical modeling, and these three pillars grow simultaneously and in close interaction with each other. Failure waves can be modeled in different ways[5-18] - in this paper we suggest modeling them as sharp interfaces separating two states: the intact and comminuted states. Here we mostly concentrate on the simplest models. However, they allow us to address several important problems. Our principal goal here is to derive a formula for the velocity of propagation of a failure wave, which is generated by oblique impact on the surface of a half-space.

INTRODUCTION

Under the action of a sufficiently intensive impact the integrity of a solid substance can be destroyed. Experimenters are inclined to see in their experiments sharp interfaces - failure waves or fronts - separating the intact and damaged (comminuted) states. They emphasize that in many respects failure waves are similar to slowly propagating combustion waves. Our model is based on this analogy in its naive version. According to this naive theoretical version of slow combustion, the internal wave "structure" is ignored and the failure front is treated as a mathematical surface (relevant discussions can be found, for instance, in the classical monographs[19, 20]). In the theory of slow waves with phase transformations such an approach has been successfully applied in the problems of low temperature physics[21] and in geophysics and celestial physics[22] about two decades ago. The discussion of possible applications and of existing difficulties when using such modeling for waves of faster phase transformation can be found in the paper[23] and monograph[24]. This discussion of the existing difficulties is relevant for modeling of failure waves also.

The concept of a failure wave is not a totally new theoretical topic. In regard to some problems of geophysics, geomechanics, and mechanical engineering, the first quantitative models have been suggested in various papers including [5,6]. More recently, several theoretical models and approaches have been suggested in [7-18]. Each of these theories correctly describes one or another feature of existing experiments, each of them faces serious obstacles in explaining others. It is too early to talk about an ultimate theory at a level comparable to the theories of combustion and detonation.

In this paper, whenever possible, we stick to the simplest models. We ignore thermal effects, viscosity, 2D-effects, even nonlinearity. Needless to say, many eventually important effects will be lost in such a limited framework. For instance, a clearly consistent understanding of standard shock-fronts is impossible without the explicit introduction of nonlinearity. We still believe that such a theory can be useful for a preliminary understanding of the physics of failure waves: this topic is still in its infancy, and it seems premature to consider possible complexities before understanding the most elementary and robust features of the relevant phenomena.

FORMULATION OF THE PROBLEM

Let us consider an initially resting uniform half-space $x \geq 0$ experiencing an impact at $x = 0$ by an oblique force P.

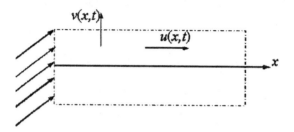

Figure 1. Oblique impact of brittle half-space undergoing intensive damage.

We assume that under the action of a sufficiently intense impact the integrity of the solid substance can be destroyed. We will use a simplified model according to which the substance can exist in two states only, the intact state and the damaged state. An oblique force P is suddenly applied at the left end of a half-space which is situated at rest initially. The impact generates a system of a longitudinal, shear and failure fronts. Their speeds and intensity is to be determined based on the magnitude of the impacting load and material parameters of the two states. This is a deeply nonlinear boundary value problem even if the two states under study are linearly elastic. The main source of nonlinearity is in the arrangement of the fronts. All potentially possible arrangements should be analyzed from the standpoint of their mathematical consistency and stability. Below, we concentrate on a single arrangement that seems most relevant to the problem under study. The corresponding arrangement of fronts is shown in the Figure 2. below.

The simple model of a two-state brittle substance adopted in this paper is discussed in the brief Appendix below (a much more detailed discussion can be found in[26]). According to this simplified model, the substance can exist in two states only, the intact state and the damaged state. Within each of the domains the energy densities e_{int} and e_{dam} per unit mass of the intact and damaged states, respectively, are given by the formulas: $\rho^\circ e_{int}(u_{m,n}) = (1/2)c_{int}^{ijkl}u_{i,j}u_{k,l}$ and $\rho^\circ e_{dam}(u_{m,n}) = (1/2)c_{dam}^{ijkl}u_{i,j}u_{k,l} + q_b$, where $u_{m,n}$ are the displacements gradients, ρ° is the mass density (assumed constant), and c_{int}^{ijkl} and c_{dam}^{ijkl} are the elasticity tensors of the two states. The positive constant q_b takes into account the energy required to produce various defects (interfaces, vacancies, shear bands, holes, etc) distributed within the bulk of damaged material. This term is analogous to the constant used in the theory of slow combustion which takes into account energy release/consumption due to chemical reactions. We assume that the constant q_b is positive whereas the authors of [17] assume that it is negative. This major discrepancy is discussed briefly in [25] and in detail in [26].

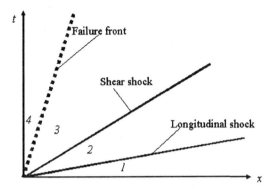

Figure 2. An arrangement of wave-fronts at oblique impact
("slow" failure wave case).

Let us consider a general 1D motion of an isotropic elastic material and denote as u and as v the horizontal and vertical components of the displacements which are assumed to depend on the spatial coordinate x and time t only: $u = u(x,t)$ and $v = v(x,t)$.

The elastic energy density $e(u_x, v_x)$, the longitudinal $\sigma = p_{xx}(u_x, v_x)$ and shear $\tau = p_{xz}(u_x, v_x)$ components of stress are the following

$$\rho^\circ e(u_x, v_x) = \frac{\lambda + 2\mu}{2} u_x^2 + \frac{\mu}{2} v_x^2, \quad \sigma(u_x, v_x) = (\lambda + 2\mu)u_x, \quad \tau = \mu v_x \tag{1}$$

The intact and damaged values of the elastic moduli should be used in (1) when dealing with each of the two states.

The bulk equations of motion within each of the states read

$$\left. \begin{array}{l} u_{tt} = a_{1\,\text{int}}^2 u_{xx} \\ v_{tt} = a_{2\,\text{int}}^2 v_{xx} \end{array} \right\} - \text{intact state}, \quad \left. \begin{array}{l} u_{tt} = a_{1dam}^2 u_{xx} \\ v_{tt} = a_{2dam}^2 v_{xx} \end{array} \right\} - \text{damaged state} \tag{2}$$

61

where $a_1 = \sqrt{(\lambda + 2\mu)/\rho^\circ}$ and $a_2 = \sqrt{\mu/\rho^\circ}$ are the speeds of the bulk longitudinal and shear shocks, respectively.

The displacements are assumed to be continuous across each shock and failure front

$$[u]_-^+ = [v]_-^+ = 0 \tag{3}$$

where $[A]_-^+$ is the jump of discontinuity of the field A across the corresponding interface. Across the "standard" shocks continuity of momentum implies

$$\rho^\circ c[u_t]_-^+ = [(\lambda + 2\mu)u_x]_-^+, \ \rho^\circ c[v_t]_-^+ = [\mu v_x]_-^+ \tag{4}$$

Across the failure front, which is considerably slower, continuity of momentum is accepted in the weaker form

$$[(\lambda + 2\mu)u_x]_-^+ = 0, \ [\mu v_x]_-^+ = 0 \tag{5}$$

It is not much harder to do the necessary calculations based on use of the equation (4_2) instead of (5_2); however, the corresponding central formula for the velocity of the failure wave becomes much more cumbersome, however).

In our model the failure is driven not by discontinuity of momentum, which is assumed continuous, in accordance with (5), but by the local jump in the energy density or more precisely, in the chemical potential. Namely, across the failure front we accept the kinetic condition of damage progress in the form

$$\frac{\rho^\circ}{K} c = \left[\frac{\lambda + 2\mu}{2} u_x^2 + \frac{\mu}{2} v_x^2 \right]_{dam}^{int} - q_b \tag{6}$$

where K is a positive function defining the rate of damage at the interface. Equation (6) is just a straightforward analogy to the standard condition $J = -K[\chi]_-^+$ for the mass flux J across phase interface. In the Lagrangean description the mass flux is given by the formula $J = \rho^\circ c$. For the solid substance in question, the analogy of the chemical potential χ is given by the enthalpy $\chi = e - u_x e_{u_x} - v_x e_{v_x}$. Combining (1) with the last relationship we arrive at the formula (6). We assume that K is a threshold-like function that vanishes for sufficiently small deformations and stresses.

Within each of the four domains in Fig. 2 we look for the solution of our initial/boundary value problem in the piece-wise linear form

$$\begin{cases} 1: & u(x,t) = G_1 x + H_1 t, \quad v(x,t) = P_1 x + Q_1 t \\ 2: & u(x,t) = G_2 x + H_2 t, \quad v(x,t) = P_2 x + Q_2 t \\ 3: & u(x,t) = G_3 x + H_3 t, \quad v(x,t) = P_3 x + Q_3 t \\ 4: & u(x,t) = G_4 x + H_4 t, \quad v(x,t) = P_4 x + Q_4 t \end{cases} \tag{7}$$

where G_i, H_i, P_i, Q_i are constants. Obviously, the functions (7) automatically satisfy the second order PDE (2) within the bulk.

The equations of the fronts in the $(x - t)$-plane are given by the equations

$$\begin{cases} \text{Longitudinal shock} : x = c_{long} t \\ \text{Transverse shock} : x = c_{trans} t \\ \text{Failure front} : x = c_{failure} t \end{cases} \tag{8}$$

where $c_{long}, c_{trans}, c_{failure}$ are unknown velocities of the longitudinal and transverse shocks and of the failure wave, respectively.

Values of the constants $G_i, H_i, P_i, Q_i, c_{long}, c_{trans}, c_{failure}$ should be found from the boundary conditions (3) – (6) and the initial conditions. Skipping lengthy routine calculations we show below the ultimate results. As expected (because of linearity in the bulk) velocities of the "standard" shock fronts coincide with their acoustic values

$$c_{long} = a_{1\,int} = \sqrt{(\lambda_{int} + 2\mu_{int}) / \rho^\circ}, \quad c_{trans} = a_{2\,int} = \sqrt{\mu_{int} / \rho^\circ} \tag{9}$$

The velocity of the failure wave is given by the formula

$$\frac{\rho^\circ}{K} c_{failure} = \frac{\sigma^2}{2} \left(\frac{1}{\lambda_{dam} + 2\mu_{dam}} - \frac{1}{\lambda_{int} + 2\mu_{int}} \right) + \frac{\tau^2}{2} \left(\frac{1}{\mu_{dam}} - \frac{1}{\mu_{int}} \right) - q_b \tag{10}$$

The remaining constants are described by the formulas which are too cumbersome to be presented here.

Further analysis depends on the function K. Assume that it is of the form

$$K = \begin{cases} 0 \; if \; |\tau| < \Theta(\sigma) \\ K_0 > 0 \; if \; |\tau| \geq \Theta(\sigma) \end{cases} \tag{11}$$

where the function $\Theta(\sigma)$ should be found from comparison with experiments. It was discovered in the experiments of Dandekar[27] with SiC that failure fronts in this material can appear only for loading with a sufficiently high tangential component of stress at the failure front. This sort of effect can be explained with the help of the threshold models for the rate of damage. Then, combining (10) and (11) we will get the following formula of the minimal velocity c_{min} of failure wave

$$\frac{\rho^\circ}{K} c_{\min} = \frac{\sigma^2}{2}\left(\frac{1}{\lambda_{dam} + 2\mu_{dam}} - \frac{1}{\lambda_{int} + 2\mu_{int}}\right) + \frac{\Theta^2(\sigma)}{2}\left(\frac{1}{\mu_{dam}} - \frac{1}{\mu_{int}}\right) - q_b \qquad (12)$$

To the best of our knowledge formulas (10) and (12) have never been subjected to any experimental verification. This is our plan for future studies.

CONCLUSIONS

In recent experiments at the Army Research Laboratory (Aberdeen Proving Ground) with SiC Dattatraya Dandekar[27] obtained evidence that failure waves in this ceramic material appear only using impact with a sufficiently high shear stresses. To model Dandekar's experiments we have analyzed a problem of oblique impact on a ceramic material undergoing intensive damage. The central result of our analysis is the simple formula (10) giving the velocity of a failure wave under oblique impact. The underlying theory is based on the simple model of a two-state solid substance. One of the states describes the intact modification, the other – the comminuted state. All the process of damage occurs at the interface interpreted as failure wave. Further experimental and theoretical efforts are in order to validate suggested model and approach. In particular, the effects of viscosity, dry friction, bulking etc. should be taken into account. Obviously, the damage occurs not only at the failure wave front but it is distributed in space and time. Our suggested approach does not allow exploration of the structure of the failure wave front when the most interesting physical phenomena take place. However, the authors believe that the established formula can be used to get useful insight for planning further experiments on failure fronts in ceramic materials, the very existence of which is still the subject of debate and controversy.

APPENDIX. A MODEL OF TWO-STATE SUBSTANCE

The damaged brittle material is actually a solid substance with numerous cracks, holes, voids, vacancies, etc. Of course, all these defects have different individual morphologies and sizes. Following Kachanov[28], for the damaged material we use an overall (homogenized) description that assumes we are dealing with phenomena on a large spatial scale that is much larger than the size of individual cracks and the distances between them. The level of damage is characterized by a single damage parameter κ. Therefore, for the damaged modification of our system we can use an "effective" elastic energy density $\rho^\circ e_-(u_{i,j}) = e^\circ(\kappa) + (1/2)c_-^{ijkl}(\kappa)u_{i,j}u_{k,l}$.

When describing the damaged modification, we can no longer ignore the "constant" $e^\circ(\kappa)$, and the damage has an effect on both elements: on $e^\circ(\kappa)$ and on the effective elasticity tensor $c_-^{ijkl}(\kappa)$. It is intuitively clear that with growth of damage the substance becomes weaker. It is somewhat less obvious that with growth of damage the constant $e^\circ(\kappa)$ becomes greater: this might even look counter-intuitive. Roughly speaking, we assume that $e^\circ(\kappa)$ represents the energy of broken bonds accompanying the development of macro-cracks. Schematically, "evolution" with damage of the elastic energy density is shown in Figure 3 for a 1D-case (where ε is the local deformation).

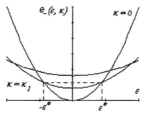

Figure 3. The specific energy density.

Figure 3 shows a parabola of the elastic energy density for different values of the damaged parameter κ. The parabola of the consolidated substance passes through the origin and has the fastest growth with ε. It is clear that for $|\varepsilon| < |\varepsilon^*|$ the elastic energy density of the consolidated phase is less than in the damaged phase with κ = $κ_I$ (for $|\varepsilon| > |\varepsilon^*|$ the elastic energy density of consolidated phase is greater than in the damaged phase with κ = $κ_I$). The trade-off between the opposite effects of κ on $e°(κ)$ and on $c_-^{ijkl}(κ)$ at various deformations is the central conflict under study with the help of our suggested model.

The simplest model of the above mentioned type, used in this paper, is the model of an isotropic two-state elastic substance. For this model, the energy densities for two modifications are given by the formula $\rho° e_\pm(u_{i|j}) = \kappa_\pm + \mu_\pm \left(\nu_\pm/(1-2\nu_\pm)u_{\cdot|i}^i u_{\cdot|j}^j + u_{(i|j)}u_{\cdot|}^{ij}\right)$, where $\mu_\pm$ are the shear Lame moduli and $\nu_\pm$ are the Poisson ratios.

REFERENCES

[1]Brar, N.S., Bless, S.J., and Rozenberg, Z., "Impact-induced failure waves in glass bars and plates", J. Appl. Phys., 1991, 59 (26), 3396-3398.

[2]Bless, S.J., Brar, N.S., Kanel', G., and Rozenberg, Z., "Failure waves in glass", J. Amer. Ceramic Soc., 1992, 75 (4), 1002-1004.

[3]Bourne, N., Rosenberg, Z., and Field, J.E., "High-speed photography of compressive failure waves in glasses", 1995, J. Appl. Phys., 78 (6), 3736-3739.

[4]D. P. Dandekar and P. A. Beaulieu, "Failure Wave Under Shock Wave Compression in Soda Lime Glass"; 211–18 in Metallurgical and Material Applications of Shock-Wave and High-Strain-Rate Phenomena. Edited by L. E. Murr, K. P. Staudhammer, and M. A. Meyers. Elsevier, Amsterdam, Netherlands, 1995

[5]Grigoryan, S.S., "Some problems of the mathematical theory of deformation and fracture of hard rocks", Prikl. Mat. Mekh., 1967, 31, 643-669.

[6]Slepyan, L.I., "Models in the theory of brittle fracture waves", Izv. Akad. Nauk SSSR, MTT, 1977, 1, 181-186.

[7]Clifton, R.J., "Analysis of failure waves in glasses", Applied Mechanics Reviews, 46, 1993, 540-546.

[8]Curran. D.R., Seaman, L., Cooper, T., and Shockey, D.A., "Micromechanical model for comminution and granular flow of brittle material under high-strain rate: application to penetration of ceramic target", Int. J. Impact Eng., 1993, 13 (1), 53-83.

[9]Chen, Z., and Schreyer, H.L., "On nonlocal damage models for interface problems", Int. J. Sol. Struct., 1994, 31(9), 1241-1261.

[10]Chaudhri, M. M., "Crack bifurcation in disintegrating Prince Rupert's drops", Phil. Mag. Lett., 78(2), 1998, 153-158.

[11]Brar, N.S., and Espinosa, H.D., "A review of micromechanics of failure waves in silicate glasses", Chemical Physics Reports, 1998, 17(1-2), 317-342.

[12]Grady, D.E., "Shock-wave compression of brittle solids", Mechanics of Materials, 1998, 29, 181-203.

[13]Kondaurov, V.I., "Features of failure waves in highly-homogeneous brittle materials", PMM, 1998, 62 (4), 657-663.

[14]Partom, Y., "Modeling failure waves in glass", International Journal of Impact Engineering, 1998, 21 (9), 791-799.

[15]Utkin, A.V., "Effect of a destruction wave on the structure of a compression pulse in brittle materials", Chemical Physics Reports, 2000, 18(10-11), 2061-2072.

[16]Feng, R., "Formation and propagation of failure in shocked glasses", J. Appl. Phys., 2000, 87(4): 1693-1700.

[17]Abeyaratne, R., and Knowles, J.K., "A phenomenological model for failure waves in glass", Shock Waves, 2000, 10, 301-305.

[18]Resnyansky, A.D., Romenski, E., Bourne, N.K., "Constitutive modelling of fracture waves", J. Appl. Phys., 2003, 93, 1537-1545.

[19]Courant, R., and Friedrichs, K.O., Supersonic Flow and Shock Waves, Springer-Verlag, 1948.

[20]Landau, L.D., and Lifshitz, E.M., Fluid Mechanics (Translation of: Mekhanika sploshnykh sred), Pergamon Press, 1987.

[21]Parshin, A.Ya., "Crystallization waves in He". In: Low Temperature Physics. Ed. Borovik-Romanov, A.S.. Mir, Moscow, 1985.

[22]Grinfeld, M.A., "Ramsey-like planets", Dokl. AN SSSR, 1982, 262, 1339-1344 (in Russian).

[23]Glimm, J., "The Continuous Structure of Discontinuities", in PDEs and Continuum Models of Phase Transitions", Eds. Rascle, M., Serre, D., and Slemrod, M., Lect. Notes in Phys., 1989, 344, 177-186, Springer-Verlag, New York-Heidelberg-Berlin.

[24]Grinfeld, M.A., Thermodynamic Methods in the Theory of Heterogeneous Systems, Sussex, Longman, 1991.

[25]Grinfeld, M.A., and Wright, T.W., "Thermodynamics of solids: recent progress with applications to brittle fracture and nanotechnology", 23 US Army Science Conference, 2002, Orlando.

[26]Grinfeld, M.A., and Wright, T.W., "Morphology of fractured domains in brittle fracture", Metall. Mater. Trans. A., 2004, 35A (9), 2651-2661.

[27]Dandekar, D.P., "Spall strength of Silicon Carbide under normal and simultaneous compression-shear shock wave loading", Int. J. Applied Ceramics Technology, 2004, 1(3), 2004.

[28]Kachanov, L.M., Introduction to Continuum Damage Mechanics, Dordrecht-Boston-Lancaster, Martinus Nijhoff Publishers, 1986.

FABRICATION AND SIMULATION OF RANDOM AND PERIODIC MACROSTRUCTURES

R. Mccuiston, E. Azriel, R. Sadangi, S. Danforth, R. Haber and D. Niesz
Rutgers University, Department of Ceramic Engineering
607 Taylor Road
Piscataway, NJ, 08854

J. Molinari
Johns Hopkins University, Department of Mechanical Engineering
3400 North Charles Street
Baltimore, MD, 21218

ABSTRACT

Shock wave propagation during an impact event can nucleate and grow strength limiting defects in an armor material before the penetration process. By limiting shock wave propagation before the penetration process an improvement in armor performance may be realized. Recent studies on phononic band gap structures have shown it is possible to create stop bands in which wave propagation is forbidden. This study will address the feasibility of applying concepts of phononic band gap structures to shock wave propagation during an impact event. Macrostructures comprised of an Al_2O_3 matrix containing discrete WC-Co inclusions have been fabricated using tape casting and lamination. The millimeter scale WC-Co inclusions were placed in the Al_2O_3 matrix in either a random or periodic fashion. After binder removal and sintering, samples were characterized using optical and electron microscopy. Vickers indentation was used to determine the hardness of, as well as the bond quality between the two materials. Through transmission acoustic characterization evaluated low amplitude wave propagation. Results show that the Al_2O_3 / WC-Co system is feasible for the fabrication of macrostructures. A series of 2D finite element simulations were used to study high amplitude shock wave propagation. The effect of WC-Co inclusion size, volume percentage and structure were studied under various loading conditions. Simulations indicate that shock wave propagation in the macrostructures can be significantly altered when compared with either of the monolithic parent materials.

INTRODUCTION

During a ballistic impact event between a ceramic target and a projectile, a shock wave precedes penetration and propagates through the target. The shockwave, though initially compressive in nature, can introduce and grow strength limiting defects in the target, weakening it. Upon arriving at free boundaries or discrete material interfaces, the shock wave is reflected and becomes tensile. This propagating tensile shock wave will further weaken the ceramic target substantially as ceramics typically perform poorly under tensile loads. Weakening of the target by the shock wave can happen before the projectile even begins to penetrate the target, essentially lowering the expected performance of the target. If the shock wave could be forbidden from propagating, then some increase in ceramic armor performance could be realized. The goal of this research is to determine whether it is feasible to purposefully design and fabricate a structure that is capable of limiting and or defeating shockwave propagation.

Work by Grady [1,2] has theorized wave scattering as a possible mechanism responsible for the creation of structured steady shock waves during propagation in a heterogeneous material. Scattering would lead to dispersion of the shock wave, which could directly counteract the tendency of the shock front to steepen as it propagates. Scattering would also create a limited spectrum of non-equilibrium acoustic phonon energy behind the shock front, which could be responsible for the excess stress between the Rayleigh line, the non-equilibrium path on which the shock wave travels, and the equilibrium thermodynamic path.

To verify the wave scattering mechanism theory of Grady, research on shock wave propagation in periodically layered composites was undertaken by Zhuang et al. [3-6]. The research studied the effects of acoustic impedance mismatch, interface density, and shock amplitude on shock wave propagation. The effect of increasing acoustic impedance mismatch was to decrease the slope of the shock front. An increase in both interface density and shock amplitude resulted in an increase in the slope of the shock front. It was found that the periodically layered composites could support steady structured shock waves and that the shock wave velocity could be lower than either of its components. It was concluded that wave scattering in the periodically layered composites was responsible for an increase in shock viscosity when compared with a homogenous material.

Hauver et al. [7,8] used periodically layered composites as shock wave attenuators to lessen the shock wave induced damage in ceramic targets during confined impact tests. To our knowledge Hauver et al. have not reported in open literature what materials the shock attenuators were made of. It could be assumed that the materials would have a high acoustic impedance mismatch and would be a combination of ceramic/plastic, ceramic/metal or a metal/plastic. Hauver et al. also did not address in open literature how the structure of the layered shock attenuator was devised, be it purposefully designed or randomly selected.

In semiconductor materials, there exist certain energy levels in which electrons are forbidden to occupy. These forbidden energy levels are referred to as an electron band gap. From this concept, Yablonovitch [9,10] developed the idea of the photonic band gap (PBG). Yablonovitch theorized that if semiconductors materials could contain an electron band gap for particular electron energies, and being that energy and frequency are directly related, then a structure could exist with band gap for other electromagnetic frequencies. After a dedicated research effort, a structure which blocked all electromagnetic frequencies from 13-16GHz was discovered.

Inspired in part by Yablonovitch, it did not take long for other researchers to realize that the photonic band gap concept could be applied to other waveforms besides electromagnetic, as the laws of physics apply to all other waveforms. Soon several research groups were at work to create structures containing acoustic or phononic band gaps (ABG, PBG), i.e. regions where no acoustic/elastic frequencies could be transmitted in a material [11-14]. This is not a new field of research though, as it has been extensively studied in the past [15].

It is the work of Grady, Zhuang et al. and Hauver et al., when viewed in light of recent work on phononic band gap structures that was the inspiration for this research. It is hypothesized that if a structure is fabricated properly using the design tools of phononic band gap research, a shock wave in part or in whole could be forbidden from propagating within a structure. Therefore in the event of a projectile impacting on such a structure, the reduction in shockwave propagation would provide some level of protection from shock wave induced damage to underlying materials. This research has focused on the design, fabrication, characterization and numerical simulation of such a structure.

Experiments were conducted using an alumina (Al_2O_3) / cobalt reinforced tungsten carbide (WC-Co) model system that was specifically selected for three reasons. First, Al_2O_3 and WC-Co have similar thermal expansions and sintering temperatures that make them amenable to co-processing. Second, both Al_2O_3 and WC-Co are recognized as ballistically important materials. Third, the pairing of Al_2O_3 and WC-Co results in a large acoustic impedance mismatch which favors the scattering of shock waves. In the model material system, the Al_2O_3 was selected as the host phase, while the WC-Co acted as a scattering phase in the form of discrete spherical inclusions.

Determination of design aspects of structures that would favor shock wave attenuation was done using the results of reported phononic band gap research. PBG research has determined the effects of structural dimensions, materials composition and materials placement with regards to wave length. A systematic study of the effects of WC-Co incorporation in Al_2O_3, specifically WC-Co diameter, percentage, and placement for various loading parameters was done using the finite element method.

RESULTS AND DISCUSSION
Experimental

The effect of sintering temperature on the model system was studied using die pressed Al_2O_3 (Malakoff RC-HP-MAR grade) samples containing randomly placed inclusions of WC-Co (Valenite). The inclusions were 300-500 μm in diameter. 3.8 cm diameter pellets were pressed containing 10, 20, 30, 40 and 50 volume percent WC-Co inclusions. The pellets were set in a BN powder bed in a graphite crucible and sintered under a dynamic vacuum of 100 mTorr in a graphite resistance furnace at 1350°C, 1450°C and 1500°C for 2 hours. The as-cut cross sections of samples sintered at 1350°C showed processing flaws as well as pull out of the WC-Co granules. As-cut cross sectioned of samples sintered at 1450°C showed a reduction of processing (due to improved processing) flaws as well as a reduction in pull out of the WC-Co granules. The cross sections of samples sintered at 1500°C show the alumina phase to be black, most likely due to cobalt contamination. An increase in weight loss after firing over the 1450°C samples would also indicate this. There was also an increase in WC-Co granule pull out during sectioning indicating a decrease in bond strength compared to samples sintered at 1450°C. The as-cut cross sections are shown in Figure 1.

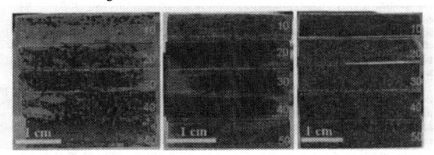

Figure 1. (l to r) As-cut cross sections of Set 1 sintered at 1350°C, Set 2 sintered at 1450°C and Set 3 sintered at 1500°C. The 10-50 labels on all sets denote the volume percent of WC-Co granules present.

To quantify the Al_2O_3/WC-Co bonding differences between the three sintering temperatures, Vickers hardness at the interface was measured. Cross sections of the 3 sets, containing 10 volume % WC-Co, were mounted in epoxy and polished to a 0.5 um surface finish with a combination of SiC grit paper and diamond slurry. Hardness was measured using a Leco M-400-G3 hardness tester using a 2 Kg load with a 10 second hold. 15 measurements were taken in the Al_2O_3, WC-Co and at the interface. The average hardness and deviation is shown in Table I.

Table I. Vickers hardness of 3 sets containing 10 vol % WC-Co. Units are Kg/mm^2.

Temp °C	Alumina Hv	WC-Co Hv	Interface Hv
1350	727 ± 25	867 ± 93	650 ± 143
1450	1866 ± 69	1726 ± 97	1685 ± 169
1500	1842 ± 113	1494 ± 180	1043 ± 254

The samples sintered at 1350°C have extremely low hardness in the Al_2O_3 and WC-Co, with an even lower hardness at the interface, but this is to be expected because 1350°C is below the sintering temperature for both materials. Samples sintered at 1450°C have high hardness in the Al_2O_3 and WC-Co, 1866 ± 69 and 1726 ± 97 respectively. The interface between the two has a composite hardness of 1685 ± 169 which is still high, indicating good bonding. Samples sintered at 1500°C show no statistical difference in hardness for Al_2O_3 and WC-Co, but the interfacial hardness has decreased drastically, indicating a decrease in bond strength.

Green Al_2O_3 samples containing periodic and random inclusions of WC-Co (Ceratizit CR24S grade) were fabricated by a tape casting and lamination technique. Al_2O_3 tapes of various thicknesses were cast based upon the diameter of the WC-Co inclusions. Periodic arrays of holes were punched in the Al_2O_3 tapes and the holes filled with the WC-Co inclusions. Alternating layers of tapes filled with WC-Co inclusions were laminated together. An Al_2O_3 sample containing no WC-Co was also fabricated as a baseline. The samples fabricated were approximately 50mm in diameter and 12-15mm thick. In the case of the periodic and random macrostructures, they contained 4 layers of WC-Co inclusions either 1mm or 2mm in diameter. A series of thirteen samples were acoustically tested using non-contact through transmission at a center frequency of 0.5MHz. The diameters of the transducer crystals were 12.5mm. Time of flight measurements were made at 5 different locations on the samples in an X shaped pattern and the results averaged in order to calculate the wave velocity. Unfortunately, four of the samples did not allow a strong enough signal to transmit and only resulted in one measurement each.

The baseline alumina monolith returned the fastest sonic velocity of 1688 m/s. The slowest sonic velocity of 1169 m/s was returned by a sample containing 20 areal % of 2mm diameter inclusions in body centered tetragonal arrangement. (Structures containing WC-Co inclusions will be referred to by the areal % of the WC-Co, diameter of the WC-Co and stacking arrangement) Velocity scans for these two samples are shown in Figure 2. The 20% 2mm body centered tetragonal sample is equivalent to a 20% 2mm hexagonal structure in cross section. From 2D FEM wave propagation simulations, the 20% 2mm hexagonal structure has been shown to behave similarly to a 40% 2mm hexagonal structure. The 40% 2mm hexagonal structure has been judged to have the best simulated performance to date.

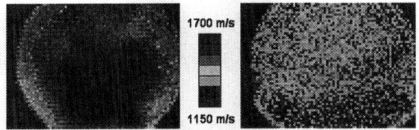

Figure 2. (left) Through transmission of green (right) baseline Al₂O₃ monolith and (left) 20% 2mm body centered tetragonal composite.

Numerical Simulation

Two dimensional elastic FEM simulations were performed on 77 mm wide x 13 mm thick cross sections. The FEM code was provided by Professor J. Molinari of Johns Hopkins University. Structures were meshed containing 20 and 40 areal percent of 2mm, 1mm and 0.5mm diameter WC-Co inclusions. They were arrayed in simple cubic, hexagonal and random arrangements. A 5 GPa load was applied in a trapezoidal manner, with rise and fall times of 1.0 nanosecond and three plateau durations of 20, 60 and 100 nanoseconds. The load was applied to the center of the top surface over a 5 mm width.

Figure 3 shows two equivalent stress contour plots for the baseline Al₂O₃ monolith and the 40% 2mm hexagonal composite, both loaded with 20 nanosecond plateaus. The plots are shown at 1.1 µs which is just as the shock pulse reaches the rear surface for both. It is seen in Figure 3 (left) for the Al₂O₃ monolith, that the majority of the high stress is located near the rear face in the longitudinal component. The shear wave component is also clearly visible approximately halfway through the monolith. In Figure 3 (right) the opposite is apparent. The majority of the high stress is still located near the top surface. The shear wave and longitudinal wave are no longer distinctly separate components. It is apparent that the 40 % 2mm hexagonal composite is a having a large influence on the shock pulse propagation.

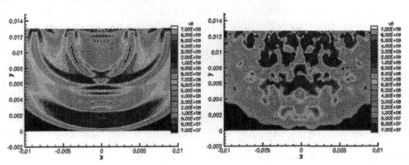

Figure 3. Equivalent stress contour plot for 20 nanosecond load plateau for (left) baseline Al₂O₃ monolith (right) 40% 2mm hexagonal composite. The stress scale is in GPa and the x y scales are in meters. Both plots were captured at 1.1 µs.

71

The location of the kinetic energy (KE) and strain energy (SE) in the structure during the loading event was monitored using a partitioning scheme. A schematic of the partitioning scheme is shown in Figure 4. The partitioning scheme first divides the model axisymmetrically along the vertical axis. The right half of the model is then partitioned into six zones. The zones do not extend to the horizontal edge of the model to avoid any potential problems arising from the boundary conditions.

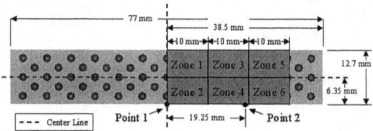

Figure 4. Schematic of the partitioning scheme. The partitions are labeled as Zones 1 through 6.

Results for partitioned KE and SE for the baseline Al₂O₃ monolith and the 40% 2mm hexagonal composite are shown in Figure 5. The load plateau was 20 nanoseconds. The plot of KE for the Al₂O₃ monolith, Figure 5 (top left), initially shows the level of KE in zone 1 increase and reach a maximum for a short duration. The KE in zone 1 then decreases as it is transferred into zone 2. The KE level in zone 2 then increases and also reaches a maximum at which point the shock pulse is reflected from the rear face and travels back towards zone 1. The process of KE transfer between zone 1 and zone 2 continues for the duration of the simulation with the maximum KE levels in zone 1 and 2 decaying as the KE is spread laterally into the remaining four zones. From the plot of SE for the Al₂O₃ monolith, Figure 5 (bottom left), it is clear that there are two distinct groupings of SE. The upper zones 1, 3, and 5 contain the majority of SE. The bottom zones 2, 4, 6 contain the least amount of SE. As the loading occurs on the top surface, it will experience the most elastic deformation and should have the bulk of the SE. Zones 5 and 6 exhibit spikes in the SE, which are likely due to wave interaction with the sides.

Plots of KE and SE for the 40% 2mm hexagonal composite are shown in Figure 5 as well. It is immediately seen, Figure 5 (top right), that the shape of KE curves for the composite are vastly different from the Al₂O₃ monolith. Just as in the Al₂O₃ monolith, the KE in zone 1 increases to a maximum, but rather than transfer immediately into zone 2, it instead begins a gradual decay. There are no large amplitude oscillations of the KE as in the Al₂O₃ monolith, which corresponded to shock pulse reflections from the top and bottom surfaces, but instead a series of smaller amplitude oscillations superimposed on the decaying curve. These smaller oscillations are most likely attributed to reflections from the WC-Co inclusions. There is still KE transfer to the remaining four zones. The plot of SE for the composite, Figure 5 (bottom right), also shows a vastly different behavior when compared with the Al₂O₃ monolith. There is an immediate increase of SE in all six zones, attributed to rigid body motion, but rather than segregating into two distinct SE groupings as with Al₂O₃ monolith, the SE curves all converge towards a single cluster. There are no large amplitude SE oscillations, like in zones 5-6 of the baseline, but again a series of smaller amplitude oscillations most likely attributed to shock pulse reflections for the WC-Co inclusions.

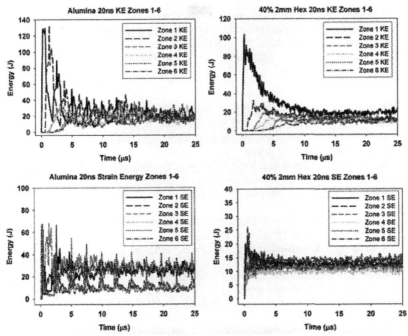

Figure 5. Partitioned kinetic (KE) and strain energy (SE) for zones 1 through 6 of the Al₂O₃ monolith and 40% 2mm hexagonal macrostructure. (top left) monolith KE zones 1-6, (bottom left) monolith SE zones 1-6, (top right) 40% 2mm hexagonal KE zones 1-6, (bottom right) 40% 2mm hexagonal SE zones 1-6.

SUMMARY AND CONCLUSION

A model system of Al_2O_3 containing inclusions of WC-Co has been developed. The optimal sintering temperate was determined to be at 1450°C, by means of an interface indentation technique. Acoustic characterization of green samples shows a large difference in sonic velocity between the baseline monolith and the 20% 2mm BC tetragonal composite. Two dimensional elastic FEM simulations show that shock pulse propagation is for the 40% 2mm hexagonal composite is significantly altered compared with the Al_2O_3 monolith. Plots of partitioned KE and SE for the Al_2O_3 monolith and the 40% 2mm hexagonal structure also show vastly different behavior. The differences in shock pulse propagation and partitioned KE and SE are attributed to incorporation of the WC-Co inclusions.

ACKNOWLEDGEMENTS

This work was sponsored by the Army Research Laboratory (ARMAC-RTP) and was accomplished under the ARMAC-RTP Cooperative Agreement Number DAAD19-01-2-0004. The views and conclusions contained in this document are those of the authors and should not be interpreted as representing the official policies, either expressed or implied, of the Army

Research Laboratory or the U.S. Government. The Government is authorized to reproduce and distribute reprints for Government purposes notwithstanding any copyright notation hereon. We would also like to thank Dr. J. McCauley, Dr. J. Adams, and Dr. E. Chin of the ARL for technical input.

REFERENCES

[1]D.E. Grady, "Physics and Modeling of Shock-Wave Dispersion in Heterogeneous Composites", *J. de Phys. IV*, Colloque C3, 669-674 (1997).

[2]D.E. Grady, "Scattering as a Mechanism for Structured Shock Waves in Metals", *J. Mech. Phys. Sol.*, **46**, 2017-2032 (1998).

[3]D.E. Grady, G. Ravichandran and S. Zhuang, "Continuum and Subscale Modeling of Heterogeneous Media in the Dynamic Environment", Science and Technology of High Pressure – Proceedings of AIRAPT-17, editors M.H. Manghnani, W.J. Nellis, and M.F. Nicol, 189-193 (2000).

[4]S. Zhuang, "Shock Wave Propagation in Periodically Layered Composites", Ph.D. Thesis, California Institute of Technology, (2002).

[5]S. Zhuang, G. Ravichandran and D.E. Grady, "Influence of Interface Scattering on Shock Waves in Heterogeneous Solids", Shock Compression of Condensed Matter, Edited by M.D. Furnish, N.N. Thadhani and Y. Horie, 709-712 (2001).

[6]S. Zhuang, G. Ravichandran and D.E. Grady, "An Experimental Investigation of Shock Wave Propagation in Periodically Layered Composites", *J. Mech. Phys. Sol.*, **51**, 245-265 (2003).

[7]G. Hauver and J. Dehn, "Interface Defeat Mechanisms in Delayed Penetration", Proceedings of the 1995 US-GE Armor/Anti-Armor Workshop 13-16 June, 217-229 (1995).

[8]G. Hauver, P. Netherwood, R. Benck and L. Kecskes, "Ballistic Performance of Ceramic Targets", Proceedings of the13th U.S. Army Symposium on Mechanics, Plymouth, MA, 17-19 August (1993).

[9]E. Yablonovitch, "Inhibited Spontaneous Emission in Solid-State Physics and Electronics", *Phys. Rev. Let.*, **58**, 2059-2062 (1987).

[10]E. Yablonovitch, "Photonic Crystals: Semiconductors of Light", *Sci. Amer.*, **285**, 46-55, (2001).

[11]M.S. Kushwaha, P. Halevi, L. Dobrzynski and B. Djafari-Rouhani, "Acoustic Band Structure of Periodic Elastic Composites", *Phys. Rev. Let.*, **71**, 2022-2025 (1993).

[12]J.O. Vasseur, B. Djafari-Rouhani, L. Dobrzynski, M.S. Kushwaha, and P. Halevi, "Complete Acoustic Band Gaps in Periodic Fibre Reinforced Composite Materials: The Carbon/Epoxy Composite and Some Metallic Systems", *J. Phys. Condens. Matter*, **6**, 8759-8770 (1994).

[13]M.M. Sigalas, and C.M. Soukoulis, "Elastic-Wave Propagation through Disordered and/or Absorptive Layered Systems", *Phys. Rev. B*, **51**, 2780-9 (1995).

[14]M. Kafesaki, M.M. Sigalas, and E.N. Economou, "Elastic wave bad gaps in 3-D periodic polymer matrix composites", *Sol. State. Comm.*, **96**, 285-289 (1995).

[15]L. Brillouin, *Wave Propagation in Periodic Structures*, Dover Publications, Inc., New York, (1946).

Dynamic and Static Testing to Predict Performance

THE CORRELATION OF MICROSTRUCTURAL AND MECHANICAL CHARACTERISTICS OF SILICON CARBIDE WITH BALLISTIC PERFORMANCE

Ian Pickup
Dstl
Porton Down
Salisbury
Wiltshire SP4 0JQ, UK

ABSTRACT
The ballistic response to tungsten long rod penetrator attack of two silicon carbide variants has been characterised. The variants, (SiC B, Cercom Inc. and SiC 100, Ceramique et Composite) which have nominally similar densities and quasi-static mechanical properties, exhibit significant differences in ballistic performance. This is shown to be a result of the different dwell response of the variants, i.e. the propensity to erode the rods on the surface without significant penetration.

The paper reviews methods developed to dynamically and statically characterise the two materials to determine deviatoric strength and a measure of the damage kinetics; the former from Plate Impact experiments and the latter from Split Hopkinson Pressure Bar techniques. Microstructural and fractographic studies are employed to elucidate the relative mechanical behavioural characteristics which are in turn related to the ballistic performance.

INTRODUCTION
Determining the role of ceramic microstructure and mechanical characteristics on ballistic performance is a difficult proposition. A plethora of interrelated properties has to be correlated with, often ill-defined and complex, ballistic phenomenology. The efficacy of high performance ceramic armour materials like silicon carbide is dependent on the ability to significantly disrupt the ballistic threat on the impact surface. In the case of a long rod tungsten alloy penetrator this will be partial or total erosion of the rod prior to penetration [1-3]. This so-called dwell phenomenon is very sensitive to target geometry and boundary conditions in addition to the intrinsic properties of the ceramic. Correlation of ceramic properties with ballistic performance requires reasonably careful ballistic test analysis to decouple the contribution from material effects and geometry.

This paper compares two nominally high performance armour grade silicon carbide variants in terms of their relative ballistic performance against long rod tungsten alloy penetrators. Previously, the author has published some aspects of their shock characterisation [4,5] and separately some aspects of their ballistic behaviour [6]. Aspects of microstructure have been alluded to in the various publications yet the relationship between microstructure, mechanical properties and consequential ballistic behaviour has not been brought together in one publication. This paper seeks to address this need for coalescence.

The paper will be presented in an order which does not necessarily represent the sequence of the research programme but it seems logical to start with observed differences in ballistic behaviour and further to analyse the ballistic response to separate the penetrator dwell phase from steady penetration. Having observed a significantly different ballistic response from the two materials, identification of the relevant material characteristics that may be responsible for the observed behaviour is addressed. Plate impact techniques are described which allowed the dynamic deviatoric strength to be determined as are techniques which allowed assessment of relative damage kinetics of the two ceramics. Finally fractographic and microstructural analysis is used to propose reasons for the respective mechanical responses.

MATERIALS

The two silicon carbides, which are compared in this study, are SiC 100 and SiC B manufactured by Ceramic et Composite (France) and Cercom Inc (US), respectively. Table I lists their principal mechanical characteristics. The elastic properties, Young's modulus (E), Shear modulus (G) and Poisson's ratio (v) were determined from the longitudinal and transverse wave speeds measured using a Panametrics 5052PR ultrasonic pulse generator / receiver in conjunction with a 5 MHz quartz transducer. The hardness was determined using a Knoop indentor (ASTM C 1326) with a 2 kg load. The grain size was measured according to the ASTM E112 and E1382 method, using a linear intercept method via a scanning electron microscope and thermally etching the grains by heating in a vacuum at 1300°C.

Table I. The physical properties of silicon carbide

	Density kgm^{-3}	E GPa	G GPa	v	Hardness HK$_2$	Mean grain diameter (μm)
SiC B	3216	453	195	0.16	2100±69	2.9±2.4
SiC 100	3163	423	183	0.16	2102±68	4.5±3.5

SiC 100 is manufactured using a pressureless sintering technique on α SiC powder with a boron sintering aid material and a carbon-rich organic binder. To ensure full conversion of surface oxide excess carbon is added. This tends to graphitise at the high sintering temperatures (2150-2200°C) employed. Highly defected graphitic particles were left in the structure with sizes ranging from 1-4 μm. No other second phases are present leaving 'clean' grain boundaries.

SiC B is also formed from an α SiC powder but is pressure densified at approximately 2000°C. The sintering aid in this case is aluminium based. The processing ensures that there is no continuous glassy phase in the matrix. All impurities tend to segregate into small well dispersed pockets along the SiC grain boundaries.

Both variants are primarily α SiC with 6H polytypes. Table 2 lists the mixture of polytypes as a percentage.

TABLE II. Percentage of polytypes for the silicon carbides

Polytype	6H	4H	3C	15R
SiC B	85	2	4	9
SiC 100	71	19	3	7

BALLISTIC TESTING

Ceramic tiles of each of the variants (30 mm thick x100x100 mm) were laterally confined with 75-mm thick steel frames clamped with 1.5mm thick annealed brass inserts between the

78

ceramic and clamp to minimize impedance mismatch and improve mechanical contact. The surface ground tiles were clamped to a 100x100x75 mm thick Rolled Homogeous Armour (RHA) steel backing. The targets were impacted at velocities ranging from 1150 to 1850 ms^{-1} with Densimet D176 FNC (manufactured by Plansee Metals) tungsten alloy rods. These had a 5mm diameter and 20:1 aspect ratio. The rod, including an aluminium flare stabilizer, had a mass of 38g was fired from a 40mm smooth bore gun using a base pushed, 3 part, discarding sabot. The mean yaw angle measured using flash X-ray, at impact was 0.8°.

The principal measurand was the residual depth of penetration into the steel backing, Fig. 1.

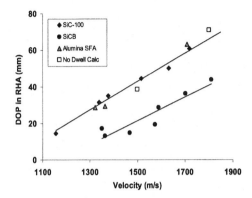

Figure 1: Residual depth of penetration into the RHA backblock for a range of rod impact velocities

This simple ballistic experiment, which was part of a more complex programme [6,7] serves to demonstrate differences in the ballistic response of the two materials. The residual DOP for SiC B is significantly lower than SiC 100; between 14 – 23 mm less, over the velocity range 1350 to 1811 m/s. If a correlation between the ceramic's mechanical properties and ballistic behaviour is to be identified, it is important to differentiate between the ceramics ballistic resistance to steady penetration and differences in rod erosion on the surface i.e., dwell.

Previously it was demonstrated [6,8] that the steady penetration rate and the erosion rate of tungsten rods, Fig. 2, through SiC 100 and SiC B was almost identical. The penetration rate of the rods in the ceramics was measured using the SWARF flash x-ray facility at Foulness UK, operated by the Atomic Weapons Establishment [8]. SWARF is a 3.5 MV machine with a 50 ns pulse width. This allows good imaging of rods penetrating through relatively thick tiles (150mm) of ceramic. Tip and tail positions were monitored for a series of impacts. Only one tip and tail position was measured per shot. Typically, four shots would be performed per target configuration. The rod in this case was an STA tungsten rod (manufactured by Royal Ordinance Specialty Metals) of 8.6mm-diameter and 20:1 aspect ratio at an impact velocity of 1800M/S. The rod penetration rate and erosion rate in the silicon carbides were almost identical, and these were the same as those measured for an 97% purity alumina (Morgan Matroc D975). Since the penetration rates and the rod erosion rates are almost identical, the difference in ballistic response observed in Fig. 1, is primarily due to the different dwell times exhibited by SiC B and SiC 100. The residual depth of penetration in RHA after penetrating 30mm of SiC without any surface dwell was calculated[6], based on the experimentally determined erosion rates and numerical simulation of the eroded rod's penetration and arrest in steel. These calculated zero dwell DOPs are shown to lie on the SiC 100 data (open, square data points) in Fig. 1 suggesting that SiC 100, tested under these prescribed conditions has minimal or no rod dwell propensity. A number of alumina (Morgan Matroc, 96% purity Sintox FA) targets were tested under identical conditions; these are known to have minimal dwell capability (unpublished work by the author), and these too lay on the SiC 100 data curve, Fig 1.

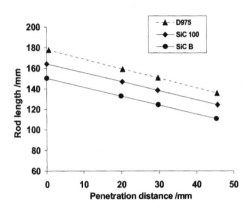

Figure 2: Rod erosion with respect to ceramic penetration distance. The rod erosion lengths were derived from the difference of the linear regression fits to the tip and tail position data.

Having attributed the difference in ballistic behaviour to be due to no-dwell or partial dwell, it is possible to estimate the amount of rod eroded on the surface. It is noteworthy that SiC B caused partial dwell of the rod over the full range of velocities, with ~24% of the rod eroded on the surface at an impact velocity of 1350 ms^{-1} whilst ~21 % was eroded at 1811 ms^{-1}. This is in contrast to the well defined dwell/no dwell transition at ~ 1600 ms^{-1} that was evident in the heavily confined reverse ballistic experiments of Lundberg et al [2,3]. Work on the effect of axial confinement[6] using coverplates on the impact surface has confirmed SiC B exhibits partial dwell even at the highest velocities but also identified a velocity dependent, step-like, change in dwell performance for axially confined targets at ~ 1550 ms^{-1} which is around the transition velocity identified by Lundberg.

What then are the specific material characteristics of the two silicon carbides which are responsible for the observed differences in ballistic behaviour?

MECHANICAL CHARACTERISATION

It is difficult to correlate a single material property of a ceramic to its ballistic behaviour. Researchers have considered hardness, fracture toughness, Weibull modulus and combinations of these to generate ceramic failure models [2,9,10]. The difficulty is compounded for ceramics where there is a dwell transition which brings into play many material and geometrical factors. In this section two material characteristics are examined, dynamic shear strength and relative damage kinetics.

The dynamic shear strength of the ceramics was measured using plate impact experiments conducted at Dstl's facility based at the Cavendish Laboratory, University of Cambridge. The experimental methods are detailed elsewhere[5,11] but summarised here. Manganin stress gauges were mounted to sense the longitudinal stress (σ_x) parallel to the shock direction and also in the lateral direction, normal to the shock (σ_y). The shear strength (τ) was calculated using the well known relationship, Equation 1. The technique employed allowed the measurement of the shear strength ahead of and also behind a failure front that trails the shock front, Fig. 3. The line in the figure is the shear stress calculated assuming elastic behaviour, Equation 2.

$$\tau = (\sigma_x - \sigma_y)/2 \tag{1}$$

$$\tau = \frac{1-2v}{2(1-v)}\sigma_x \quad \text{Where;} \quad v = \left(\frac{E}{2G}\right) - 1 \tag{2}$$

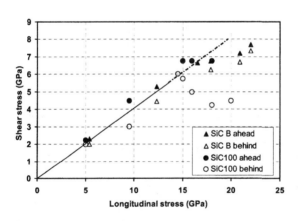

Figure 3: Shear stress variation with impact stress. Closed data points are values measured ahead of the failure front. Open points are measured behind the front.

Clear differences exist between SiC B and SiC 100. The former follows the elastic relationship up to a longitudinal stress of ~ 15GPa. Beyond 15 GPa the strength of the ceramic, both pre and post failure-front, falls below that of the elastic value, indicating both instantaneous

81

and delayed damage. SiC 100 undergoes a loss in strength at virtually the same longitudinal stress of 15 GPa as SiC B. The pre-failure front strength values follow those of SiC B, but the post failure front, in contrast to SiC B, undergoes a significant drop in shear strength, to approximately 60% of the value exhibited by SiC B. Since the shear strength indicates resistance to penetration it is apparent that on initial impact, the strength of the ceramic is significantly greater than that of the rod (for SiC B by a factor of approximately 6). It is clear that initially the rod will yield in preference to the ceramic giving rise to the dwell effect. However, on impact, the ceramic will inevitably start to damage and at some point the strength of the rod will exceed that of the ceramic. This transition which has been the subject of significant analytical activity [2,9,12] will be strongly affected by the confining pressure in the zone around the penetrator, the initial deviatoric strength of the ceramic and the rate of damage accumulation of the ceramic. SiC 100 has a significantly lower initial strength than SiC B; it is more comparable to a high density armour grade alumina, which has little propensity to demonstrate dwell. Consequently, if all other factors remain constant it may be expected to take less time for a critical level of damage to be attained in SiC 100 than SiC B.

The relative damage kinetics of these ceramics (using a consistent experimental procedure) have been determined by the authors previously[4]. This was achieved by measuring the difference in compressive strength in standardized, 1-D stress experiments between quasi-static rates and high strain rates measured using a Split Hopkinson Pressure Bar (SHPB). This represents a strain rate range of $\sim 10^6$ s^{-1} (10^{-3} to 10^3 s^{-1}). The rationale behind this approach is based on work, for example, by Grady [13] and Janach [14] in which apparent strain rate effects of compressive strength are analysed in terms of the kinetics of localized failure coalescing to allow macroscopic failure. In Grady's analysis, if a threshold strain rate exceeds 10^3 to 10^4 s^{-1}, the tensile failure stress should have a cube root relationship with strain rate, Equation 3. Localized tensile stresses may be generated in a compressive stress field which coalesces flaws leading to macroscopic failure[15].

$$\sigma_t = \sqrt[3]{\rho C_0 K_{1C}^2} \, \dot{\varepsilon}^{1/3} \tag{3}$$

Janach's rate dependent model was based on the physical separation of failed particles, i.e. the particles must move radially from the axis of the specimen thus propagating an ejecta front to the axis of the specimen prior to macroscopic failure. The velocity of this ejecta front (C_f), and consequently, time to catastrophically fail is inversely proportional to the associated volumetric strain ($\Delta V/V$) associated with bulking of the damaged material, Equation 4.

$$C_f = \sqrt{\eta} \, \frac{\sigma_1}{\rho C \, \Delta v / v} \tag{4}$$

Where η is the fraction of strain energy converted into kinetic energy and ρ and C are the density and elastic wave speed respectively. The compressive stress histories at both quasi-static and SHPB rates were measured using dumbbell shaped specimens with steel radial constraint rings, thermally shrunk on to the specimen ends, Fig 4. The specimen geometry was designed to minimize the effect of the interaction between the specimen load bearing ends and the testing platens which may induce axial splitting and hence premature failure.

82

Figure 4: Ceramic compression specimens (left). Quasi-static compression rig (right)

The quasi-static specimens were tested on a load partition rig, Fig. 4, to ensure even loading of the specimen and minimize negative effects of strain energy release from either sub-critical or catastrophic failure. In this method three circular section, high yield strength steel columns were positioned at 120° intervals around self-levelling platens. The ceramic specimen was placed on a steel plinth placed centrally on the compression platen with tungsten carbide discs on the top and bottom faces. The columns were engineered so that the strain in the ceramic specimen would exceed failure limits before the steel columns exceeded their elastic limit. Each column was strain gauged and calibrated so that the load it was supporting could be monitored throughout the test. The load in the specimen was determined by subtracting the sum of the three column loads from the load registered by the testing machine's load cell.

The high strain rate compressive strength was determined using 16mm diameter silver steel Hopkinson pressure bars with a consistent stress being applied for all specimens via the input bar. A tungsten carbide disc (2 mm thick) was placed between the specimen ends and the Hopkinson bar end surfaces to prevent damage to the bar. Investigations were conducted to verify that the impedance difference of these end-discs had minimal effect on the measured output bar stress. An Ultranac digital high speed camera was used to observe the specimens during loading. Eight frames were taken at various intervals throughout the loading and failure cycle which would typically be 50-70 µs. The compressive strengths for the two materials are given in Table III.

Table III. Compressive characteristics

	QS Strength (Strain rate 10^{-3} s^{-1}) GPa	SHPB strength (Strain rate 10^{3} s^{-1}) GPa	Characteristic failure time (T_f) µs
SiC B	5.15±0.35	8.17±0.16	30±5
SiC 100	5.21±0.50	7.47±0.32	20±2

The relative difference in damage kinetics is assessed by comparing a characteristic failure time, T_f, for a specific test configuration. This is the time taken during a Hopkinson bar test from attaining an average QS failure stress, as determined using the method described above, to catastrophic failure indicated by a sudden reduction in load, Fig. 5.

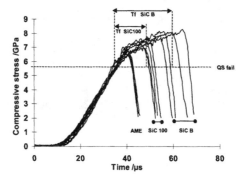

Figure 5: SHPB Stress histories for SiC 100 and SiC B at strain rates of ~10^3 s^{-1}.

The stress histories of three silicon carbides are presented; in addition to SiC B and 100, a reaction bonded material is compared. This material, AME has high defect content with areas of unreacted silicon. In each case the load increases linearly and exceeds the QS value, ~5.5 GPa. At stresses in excess of 6.5 GPa there is a significant change in slope of the loading curve; the AME material catastrophically fails at this point but for SiC 100 and SiC B the load monotonically increases until failure. It is believed that this apparent inelastic behaviour is due to incipient axial splitting making the specimen behave as a weaker elastic body. It is evident that the time taken for a critical degree of damage accumulation is significantly greater for SiC B than SiC100 with a characteristic failure time of 30μs compared to 20 μs for SiC 100, Table III.

The high speed photographs were correlated with their respective stress history by monitoring a fiducial pulse each time an exposure was taken. For SiC 100 as soon as the mean QS strength was attained in the SHPB tests damage was observed in the form of particles being ejected from the axis of the specimen across the gauge length, Fig 6a. This continued until catastrophic failure. No apparent damage was observed in SiC B after exceeding the QS strength limit until the peak SHPB load was attained, whereupon the specimen disintegrated explosively into particulate ejecta across the entire gauge length, Fig. 6b.

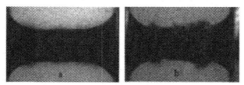

Figure 6: Damage ejecta in SiC 100 immediately after exceeding the mean QS strength (a). The moment of catastrophic failure in SHPB compression tests for SiC B (b).

Unpublished work of the author indicates the increased propensity to initiate flaws in SiC 100 c.f. SiC B using ultrasonic attenuation methods to assess flaw activation. Various pre-loads were applied to the QS compression specimens at standard QS rates. The specimen was monitored after each loading for ultrasonic attenuation whilst ramping hydrostatic pressure from atmospheric pressure to 300 MPa. Increased ultrasonic attenuation was attributed to flaw activation. The two materials had similar attenuation levels at zero pre-load. Little effect on flaw activation was noted until pre-load levels exceeded 70% of the QS failure stress. At a pre-load value of 85% of the mean QS failure stress, SiC 100 had double the attenuation of SiC B indicating increased flaw activation.

The difference in characteristic failure times observed above lead to the proposition that a shorter characteristic failure time may result, not only from the microcrack inter-linking more rapidly, but also from an ability of the comminutia particles formed prior to catastrophic failure to move past each other more easily. If the grains cleave under the applied stress, then the comminutia may move more easily under hydrostatic pressure than if the grains interlock and resist deformation.

MICROSTRUCTURAL EXAMINATION

The ejecta from the SiC100 and SiC B captured during SHPB testing was fractographically examined [16]. There are clear differences in the fragment distribution size with SiC 100 having significantly finer particles than SiC B, Fig 7.

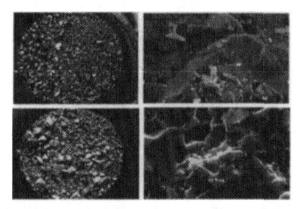

Figure 7: Fragments collected from the SHPB tests for SiC 100 (top left) and SiC B (bottom left). The SEM platen that the particles are on is 17 mm diameter. The trans-granular fracture of SiC 100 is shown (top right) and the inter-granular fracture of SiC B (bottom right).

From scanning electron microscopy (SEM), it is apparent that failure in SiC 100 is predominantly trans-granular, Fig. 7 (top right). The dark regions in the fracture surface are where graphitic regions have been pulled out. In SiC B failure is predominantly inter-granular, Fig 7 (lower right). X-ray photelectron spectroscopy was used to analyse the surface compositions of the grains. Aluminium was found on the surface of the SiC B grains. Quantitative electron probe micro-analysis (EPMA) was used to determine that the aluminium content on the surface was approximately 1.5% by weight[17]. The aluminium from the sintering aid has diffused to form an aluminium rich shell around each grain. It is also believed that the aluminium promotes wetting of the ceramic and leads to a structurally sound dispersion of metal silicide inclusions at the boundaries.

DISCUSSION

It was proposed above, that the fracture mode may strongly influence the damage kinetics. SiC 100 fails in a trans-granular manner and thus may allow co-operative movement of particulate matter more easily than the SiC B which failed inter-granularly. Even though SiC 100 had a significant population of graphitic regions which act as flaws in triple point positions, the grains failed in preference to the boundary in contrast to SiC B, where the grains stayed intact. No quantitative measurement is presented here, but it is believed that the aluminium shell contributes to localised plasticity in the grains, reducing the probability of activating flaws within the grain and favouring failure along grain boundaries. Under impact stress, the grains of SiC B interlock leading to the toughening mechanism that Flinders[10] refers to as crack-bridging. This explains SiC B's increased characteristic failure time in the SHPB tests.

The mode of failure not only affects the way in which the particles move relative to each other but affects flaw activation as indicated by the ultrasonic attenuation studies. The way in which the microstructure may influence fracture mode, and hence damage kinetics, is apparent but it is not clear why SiC 100 should undergo such a relatively strong reduction in shear strength on impact as indicated in Fig.3. It is possible that SiC 100's increased propensity for grain cleavage may allow a population of disparate flaws to be generated modifying the bulk elastic behaviour. If the flaw distribution generated modified the shear modulus relative to the Youngs modulus, then Poison's ratio of the modified elastic body would be altered, From Equation 2, it can be seen that a reduction in shear modulus, G, with respect to E would reduce Poison's ratio and consequently the shear strength relative to the longitudinal stress. It would only take a 15% relative reduction in G to account for the reduction in shear strength observed.

The ability to accommodate particle movement by shear will enhance the termination of rod dwell by allowing damage in the zone around the impacting rod to reach a free surface and reduce the hydrostatic pressure. The strength of ceramic is dependent on the confining pressure; if the pressure drops, then the strength drops and the rod may penetrate. If dwell cannot be maintained SiC performs no better against long rod threats than alumina at many times the cost.

CONCLUSIONS

Penetrator dwell has been demonstrated to be of paramount importance in realizing the possible ballistic efficiency of silicon carbides against long rod threats. Without dwell, silicon carbides perform little better than alumina which may cost 7 times less. Significant differences in material characteristics which directly affect dwell have been observed in the two silicon carbides SiC B and SiC 100. These are dynamic shear strength and relative damage kinetics. SiC 100's reduced shear strength on impact (as measured by plate impact experiments) compared to SiC B means that the strength reduction required before a long rod may penetrate is significantly lowered. In addition the relative rate of damage accumulation in SiC 100 is higher than SiC B.

A key difference in mechanical response between the two materials which governs the dynamic shear strength and damage kinetics is the mode of failure; SiC 100 is prone to grain cleavage whereas SiC B grains are resistant to fracture and the ceramic fails inter-granularly leading to crack-bridging and resistance of co-operative movement resulting in slower damage rates and consequently a tougher structure. It is hypothesised that the resistance to grain cleavage in SiC B may result from the internal microstructure of the grain, specifically the aluminium rich shell which SiC B grains exhibit which may promote locallised plasticity in preference to flaw activation in the grain.

ACKNOWLEDGEMENTS

Funding support for Dstl from UK MoD, WPE Domain is gratefully acknowledged. The author would also like to acknowledge the work of his Dstl colleagues in the ceramic programme, in particular; Bryn James, Andrew Baxter, Antony Barker, Ian Elgy, Reza Chenari and Doug Imeson. Roger Morrell from the National Physical Laboratory (UK) is thanked for hardness testing the ceramics as are Neil Bourne, Jeremy Millett and Natalie Murray for their work on plate impact experiments.

REFERENCES

[1] G. E. Hauver, P. H. Netherwood, R.F. Benck and L. J. Kecskes, "Ballistic Performance of Ceramic Targets," in Army Symposium on Solid Mechanics, Plymouth, Massachusetts US, (1993).

[2] P. Lundberg, R. Renstrom and L Holmberg, An experimental investigation of interface defeat at extended interaction time, 19th International Symposium of Ballistics, Interlaken, Switzerland, pp1463-1469, (2001).

[3] P. Lundberg, R. Renstrom and B. Lundberg, "Impact of metallic projectiles on ceramic targets: transition between interface defeat and penetration," Int. J. Impact Engng., 24, 259-275, (2000).

[4] I.M. Pickup, and A.K. Barker, "Damage Kinetics in Silicon Carbide," Proceedings of the Conference on Shock Compression of Condensed Matter-1997, edited by S. C. Schmidt et al., AIP Press, 513-516 (1998).

[5] I. M. Pickup and A. K. Barker, "The Deviatoric Strength Of Silicon Carbide Subject To Shock," Proceedings of the Conference on Shock Compression of Condensed matter, Snowbird USA, (1999).

[6] I.M. Pickup, A. K. Barker, I.D. Elgy, G.J.J.M. Peskes and M van de Voorde, "The Effect Of Coverplates On The Dwell Characteristics Of Silicon Carbide Subject To KE Impact," Proceedings of the 21st International Symposium on Ballistics, Adelaide, South Australia, 207-213 (2004).

[7] I.M. Pickup, A. K. Barker, B. J. James, V. Hohler, K. Weber and R. Tham, "Aspects Of Geometry Affecting The Ballistic Performance Of Ceramic," Proceedings of the Symposium on Ceramic Armor Materials, ed. J.W. McCauley PAC RIM IV, Maui, H.I., (2001).

[8] B. J. James and A G Baxter, Dstl unpublished data, (1994).

[9] J.C. LaSalvia, "A physically-based model for the effect of microstructure and mechanical properties on ballistic performance", Ceram. Sci. and Eng. Proceedings, 23[3], 213-20, (2002).

[10] M. Flinders, D. Ray and R. Cutler, "Toughness-Hardness Trade-Off In Advanced Sic Armor," Ceramic Armor and Armor Systems, 105th Annual Meeting of the Am. Cer. Soc., Nashville, Tennessee, USA (2003).

[11] N.K.Bourne, J.C.F. Millett and I.M.Pickup, J.Appl. Phys. **81**, 6019-6023 (1997).

[12] J.C. La Salvia, E. J. Howarth, E.J. Rapacki, C. J. Shih and M. Meyers, "Microstructural and Micromechanical Aspects of ceramic/Long-Rod Projectile Interactions: Dwell/Penetration Transitions" pp437-446 in *Fundamental Issues and Applications of Shock-Wave and High Strain-rate Phenomena*, ed. K.P. Staudhammer, L.E. Murr and M.A. Meyers, Elsevier Science, New York (2001).

[13] D.E. Grady, "The Spall Strength of Condensed Matter," *J. Mech. Phys. Solids,* Vol. 36. No. 3, 353-384, (1988).

[14] W. Janach, "The Role of Bulking in Brittle failure of Rocks Under Compression," *J. Rock Mech. Min. Sci. & Geomech. Abstr.*, Vol. 13, pp177-186, Pergammon Press (1976).

[15] H. Horii and S. Nemat-Nasser, "Brittle Failure in Compression: Splitting Faulting and Brittle-Ductile Transition," *Phil. Trans. R. Soc.* London A, 319, 337-374(1986).

[16] D. Imeson, "Fractography of Silicon Carbide Samples," DERA technical Memorandum, FMC-98-069, (1998).

[17] N. D. Harrison, D. Imeson, M. A. Barnett and A. R. Bhatti, "Examination and analysis of silicon carbide for ceramic armour applications, DERA Technical Report, DERA/MSS2/WP980048/1.0, (1998).

HIGH STRAIN RATE COMPRESSION TESTING OF CERAMICS AND CERAMIC COMPOSITES

William R. Blumenthal
Los Alamos National Laboratory
Mail Stop G-755
Los Alamos, NM 87545

ABSTRACT
The compressive deformation and failure behavior of ceramics and ceramic-metal composites for armor applications has been studied as a function of strain rate at Los Alamos National Laboratory since the late 1980s. High strain rate ($\sim 10^3$ s^{-1}) uniaxial compression loading can be achieved using the Kolsky-split-Hopkinson pressure bar (Kolsky-SHPB) technique, but special methods must be used to obtain valid strength results. This paper reviews these methods and the limitations of the Kolsky-SHPB technique for this class of materials.

INTRODUCTION
The Kolsky-split-Hopkinson pressure bar (Kolsky-SHPB) technique was originally developed to characterize the mechanical behavior of ductile materials such as metals and polymers where the results can be used to develop strain-rate- and temperature-dependent constitutive behavior models that empirically describe macroscopic plastic flow.[1] The flow behavior of metals and polymers is generally controlled by thermally-activated and strain-rate-dependent dislocation motion or polymer chain motion in response to shear stresses.

Conversely, the macroscopic mechanical behavior of dense, brittle, ceramic-based materials is dominated by elastic deformation terminated by rapid failure associated with the propagation of defects in the material in response to resolved tensile stresses. This behavior is usually characterized by a distribution of macroscopically measured failure strengths and strains. The basis for any strain-rate dependence observed in the failure strength must originate from rate-dependence in the damage and fracture process, since uniform, uniaxial elastic behavior is rate-independent (e.g. inertial effects on crack growth).[2] The study of microscopic damage and fracture processes and their rate-dependence under dynamic loading conditions is a difficult experimental challenge that is not addressed in this paper.

The purpose of this paper is to review the methods that have been developed at the Los Alamos National Laboratory to perform valid, uniaxial, dynamic compression experiments on brittle materials using the Kolsky-SHPB technique and to emphasize the limitations of this technique. Kolsky-SHPB results for several ceramic and ceramic-metal (cermet) materials of interest for armor applications have been measured and show little or no strain rate sensitivity compared to quasi-static compression results.

KOLSKY-SHPB TEST LIMITATIONS FOR HIGH-STRENGTH, BRITTLE MATERIALS
The Kolsky-SHPB technique consists of a pair of long "pressure bars" used to rapidly load a specimen sandwiched between them by a loading pulse applied to one of the pressure bars by a smaller "striker bar". The classic technique is thoroughly described elsewhere and only highlights or modifications applicable to high strength, brittle materials will be addressed below.[1] A general-purpose Kolsky-SHPB system was used for this study and consisted of 12 mm diameter by 122 cm long pressure bars made of high-strength maraging steel with a nominal

yield strength of ~2 GPa; however, this paper also applies to systems with different bar dimensions (or even different bar materials) that are optimized for testing high-strength, low-failure-strain brittle specimens. Load and displacement are conventionally measured using strain gages mounted on the two pressure bars and used to calculate the applied strain and strain rate and the resulting stress response of the specimen material. It is essential that the pressure bars remain completely elastic during the test because elastic analysis is used in the data reduction and because the pressure bars can be damaged by plastic deformation (i.e. indentation). In particular, the stress generated at the pressure bar-specimen interface must remain below the yield strength of the pressure bars at all times.

A universal condition for valid Kolsky-SHPB testing is that stress-state equilibrium must occur by "ringing up" the specimen stress during the test so that uniaxial stress conditions are achieved throughout the specimen. Early in the test (with respect to time or strain), the stress and strain within the specimen are not uniform and can vary significantly from one end to the other. Hence, prior to achieving stress-state equilibrium, the gage-measured specimen stress and strain are, at best, average values and cannot be considered uniaxial stress conditions. Therefore, these measurements are invalid until equilibrium has been achieved. Because brittle materials fail at extremely low strains, extraordinary care must be taken to design a proper specimen geometry, precisely align the pressure bars with the specimen for axial loading, and choose strain rates that ensure stress-state equilibrium can be achieved during the test. A minimum "ring-up time" is required that is approximately equal to three times the specimen length divided by the effective sound speed of the specimen material (a "ring-up strain" is similarly defined as the product of the ring-up time and the time-averaged strain rate).[1-3]

Stress-State Equilibrium Considerations

To perform a valid uniaxial compression experiment on brittle materials using the Kolsky-SHPB technique, the specimen length and the applied strain rate must be restricted relative to the strain-to-failure and the sound speed of the specimen material through the ring-up time. Fortunately, the sound speeds of ceramics are very high which aids the rapid attainment of ring-up. Assuming a sound speed of 9.5 km/s (the minimum for materials in this study), a strain-to-failure of 0.8% under compression (conservative for both ceramics and cermets[4,5]), and a target strain rate of 2000 s[-1], then the maximum specimen length that ensures stress-state equilibrium is 12.7 mm (with a corresponding ring-up time of four microseconds).

Our Kolsky-SHPB specimen geometry was therefore designed around a total length of 12.7 mm. Shorter lengths are obviously desirable at high strain rates; however, additional geometry constraints discussed in the next section require reduced diametral dimensions and difficult machining which becomes impractical with shorter specimen lengths. For B_4C, with a very high sound speed of 14 km/s, the proposed specimen length allows testing at strain rates up to ~ 3000 s[-1]. However, strain gage size and data acquisition time resolution impose additional constraints on the maximum strain rate that will be discussed later. Therefore, the maximum strain rate for valid Kolsky-SHPB testing of brittle ceramics with current techniques can be no more than ~ 5000 s[-1] (see similar arguments elsewhere[2]).

Pressure Bar Maximum Stress Considerations

In order to test armor-grade ceramics with compressive strengths as great as 8 GPa, the stress at the pressure bar-specimen interface must be reduced by at least a factor of four so that the yield strength of the pressure bars (nominally 2 GPa) is never exceeded. Otherwise plastic indentation of the pressure bars will occur that invalidates the test results. Either very high-strength loading platens and/or a reduced-gage-diameter specimen (dumb-bell-shaped) have been used as stress reducers and these constrain the ratio of the maximum-to-minimum diameters of the specimen or the platen-specimen combination to be ≥ 2.

The geometry and material for loading platens must be carefully chosen. First, their lengths must generally be included in the total ring-up length (unless they are perfectly impedance-matched to the bars) and therefore they must be relatively short to allow rapid stress equilibrium to be achieved within the specimen. However if the platens are too thin, then they can undesirably flex and fracture. Next, a non-uniform stress-state will result within a constant-diameter cylindrical specimen loaded between larger diameter loading platens instead of the desired uniaxial stress-state. Tapered loading platens can minimize these stress concentrations, but because armor ceramics are extremely strong in compression the platen material must have a yield strength greater than the ceramic being tested. Otherwise, the platens will fail before the ceramic specimen (Note: platen failure is a good indication that they are inadequate). In general, platens are also susceptible to damage and should be discarded after each test to ensure reliable results, which is expensive. Finally, loading platens create multiple, undesirable interfaces in the load train. Each interface surface must be machined to very precise flatness and parallelism (within a few micrometers), which is again very costly. Loading platens are also difficult to align precisely in a horizontal load train and each interface must remain in good contact during the test. Overall, platens create unacceptable sources of test variability and should be avoided.

The Kolsky-SHPB Compression Specimen Design

Instead, a one-piece, dumb-bell-shaped specimen with a diameter ratio of two is a superior design for achieving the desired uniform, uniaxial, high stress state in the gage-section, plus it is relatively easy to manipulate and align in either a Kolsky-SHPB or a quasi-static load frame (Note that the same specimen design is used for both tests). The dumb-bell-shaped specimen designed and developed by Tracy[4] for quasi-static compression testing was modified by the author[5] by proportionally scaling the dimensions down to the desired overall length of 12.7 mm necessary for dynamic stress-state equilibrium. Downscaling results in a specimen gage diameter of 2.2 mm and a nominal gage length of 3.3 mm. A gage length-to-diameter ratio of 1.5 allows for compression failure by either resolved shear or tensile stresses while avoiding buckling failure. The smooth, gradual contours of this specimen are necessary to minimize stress gradients within the gage-section.

Specimen Machining Considerations

The small, intricate geometry and precise dimensional specifications of the dumb-bell-shaped specimen are challenging and obviously more costly to fabricate than a simple right-circular cylinder, especially for ceramics that require abrasive grinding. However, this specimen geometry remains large enough for precise machining of the required tolerances, without being prohibitively expensive. In general, ceramic materials are also susceptible to fracture initiation from surface damage caused by improper machining, so that careful specimen fabrication is a critical task for obtaining accurate, high-quality strength data. However, the need to polish specimen surfaces is not considered to be necessary with proper machining techniques.

Specimens were fabricated by first precision surface grinding a sheet of material to the final specimen thickness from which multiple specimens were created. Specimen blanks were then roughed-out using electro-discharge machining (if the material was sufficiently conductive) or by slicing the sheet into parallelepipeds. The blanks were then mounted onto a lathe by pressure-padding the precision-ground ends and then the final contours were abrasively machined using a template. Grinding perpendicular to the specimen axis minimizes axially-aligned damage and reduces the probability of failure from machining-induced surface flaws.

KOLSKY-SHPB COMPRESSION TEST METHODOLOGY FOR BRITTLE MATERIALS

Specimen Strain Measurements

The major shortcoming of the dumb-bell-shaped specimen (or specimens loaded using platens for that matter) under both quasi-static and dynamic loading conditions is that the uniaxial strain in the reduced gage-section cannot be accurately determined using "far field" displacement measurements (i.e. from Kolsky-SHPB reflected bar strain gage measurements) because strain from the entire specimen is averaged and is therefore not representative of the gage section alone. Furthermore, strain calculated using the Kolsky-SHPB reflected bar strain gage signal is highly uncertain at low strains due to the large amplitude oscillations associated with the leading edge of the reflected wave. Therefore, in situ strain gages mounted within the gage section are necessary to obtain the necessary strain data. The use of specimen strain gages has drawbacks for dynamic testing: they are time-consuming and difficult to apply, faulty mounting can compromise their accuracy, and they require expensive high-speed signal conditioning and data acquisition equipment. Therefore, the use of strain gages is a significant, but indispensable, time and cost factor associated with this test methodology.

The selection, mounting, and application of strain gages is a substantial subject area and only key aspects will be discussed below. The strain gages selected for this application were stock 120-ohm resistance, general-purpose, miniature gages (designated EA-06-050AH-120) and were manufactured by Measurement Group, Inc. (Raleigh, NC). These gages consist of a constantan foil grid with a polyimide backing that are capable, if properly mounted, of measuring maximum strains of 3-5% (far greater than the largest ceramic or cermet failure strains measured to date). The active grid length and width are 0.05" (1.27 mm) and 0.04" (1.02 mm), respectively. This miniature size allows the strain gage to be centered and fully contained within the specimen gage length. The length of the active grid also governs the time resolution of the strain gage signal. Assuming no influence from the backing or adhesive, the time resolution can be defined as the time required for a stress wave to traverse the entire length of the active portion of the strain gage grid (i.e. the grid length divided by the sound speed of the specimen material). Therefore, the maximum time resolution of the strain signals in this study is calculated to be about 0.1 microseconds (for boron carbide) and is equal to the sampling time used for data acquisition of the strain gage signals (discussed below). Furthermore, the maximum strain resolution (defined as the product of the strain rate and the time resolution) is about 0.00025 (or 0.025%), assuming a strain rate of 2500 s^{-1}. This level of strain resolution is considered adequate and any improvement would require shorter strain gages and higher data sampling rates. Again we are faced with a practical high strain rate limit of ~ 5000 s^{-1}.

Three axial strain gages per specimen were generally used for several reasons. First, multiple gages allow the strain signals to be averaged for higher measurement confidence compared to a single measurement. Second, three gages provide substantial redundancy in case one or two gages prove to be faulty. Finally, three gages placed uniformly (at 120 degrees) around the gage section circumference allow loading eccentricity to be evaluated such as due to bending. Bending is an undesirable condition because it causes a non-uniform stress state in the specimen and can invalidate the test if it is greater than the measurement uncertainty. For example, during a failure test one strain gage signal will often rapidly drop to near zero output first and act as a "break" gage signifying the precise time and strain of failure in the specimen. Both axial and transverse strain gages were used in some cases to explore the volumetric damage and failure behavior of cermet materials. However, only one axial gage can be mounted if transverse gages are used, due to space limitations in the gage section.

The relative output of a strain gage is defined by the gage factor (GF) and specified by the manufacturer for each lot of strain gages, including an accuracy range. The gage factor of constantan gages is nominally 2.05 $\pm$ 2%, so the measurement accuracy of the gages used in this study without consideration of various aspects of the actual test method and the instrumentation is only considered "moderate" by the manufacturer at a level of $\pm$ 2-4%. To maximize reproducibility for a given test, all three gages were selected from the same manufactured lot; however the total uncertainty associated with the dynamic strain measurements was estimated to be about $\pm$5%.

The small strain gages used in this study were difficult to mount with precision and reproducibility onto the dumb-bell-shaped specimens (Note that the polyimide backing of these strain gages is several times larger than the active gage and the excess is trimmed off prior to mounting). Curvature is high in the specimen gage section due to the small diameter, so that thumb pressure was used to achieve a thin, uniform adhesive thickness over the entire gage surface and to accelerate curing of the adhesive. Precise alignment of the strain gages to the specimen axis is critical for obtaining accurate axial strain measurements. Strain gages were first precisely aligned on a strip of transparent tape and then the tape was carefully aligned with the specimen axis to within several degrees during mounting. The reduction in the measured versus actual axial strain associated with a two degrees gage misalignment angle is less than 0.2%. If a gage was found to be misaligned by more than two degrees after mounting, then it was removed, the specimen surface re-prepared, and a new strain gage was applied (a difficult and tedious process when applying multiple gages close together on the specimen). M-Bond 200 (Measurement Group, Inc. Raleigh, NC) is a special cyanoacrylate ("superglue") certified by the manufacturer for use in bonding strain gages. It was chosen to mount the specimen strain gages because it is very thin, fast-setting, and strongly bonding. The recommended curing schedule is one minute of thumb pressure, followed by a minimum two-minute delay before tape removal. These schedules were tripled in practice to ensure good bonding. Each mounted gage was resistance tested and evaluated for bonding defects and alignment using optical microscopy. Unfortunately, proper bonding of the gages to the specimen (i.e. high bond strength and uniform bond thickness) cannot be entirely verified prior to testing. The output of strain gages mounted to a specimen can be partially checked prior to testing by carefully loading specimens quasi-statically to very low stress levels; however, this was not routinely done for several reasons. First, pre-loading specimens may introduce undesirable changes in the specimen, such as damage or crack growth. Second, this procedure requires high-precision loading equipment and is labor-intensive and therefore costly. Finally, satisfactory strain gage performance at low strain levels did not always ensure proper gage function at higher strain levels.

Each of the three strain gage signals on a specimen was recorded and evaluated separately for consistency. Occasionally a gage record was disregarded when it was clearly faulty. Variations in stress within the gage section of the specimen due to eccentric loading were calculated when three strain gage signals were available. Normally, these indicated the presence of very small bending moments; however, the strain variations were also usually within the uncertainty of the strain measurements, so that these could be artifacts of the uncertainty. Note that strength measurements alone are of limited use for characterizing material behavior.

High Speed Signal Conditioning and Data Acquisition

The measurement of Kolsky-SHPB pressure bar and in situ specimen strain gage signals requires specialized signal conditioning and data acquisition equipment capable of high bandwidth ($\geq$ 3 MHz), high precision ($\geq$ 10 bits), and high sampling rates ($\leq$ 0.1 microseconds per point as discussed above). Ectron Model 778 whetstone bridge and wideband amplifiers (Ectron Corp. San Diego, CA.) were used to condition and amplify the strain signals. The amplified signals were simultaneously recorded with Tektronix (Beaverton, OR) high-speed digital storage oscilloscopes (models used are now obsolete) and then transferred to a personal computer for analysis. (Note that modern high-speed data acquisition boards are now capable of recording Kolsky-SHPB strain gage signals directly to a personal computer without an oscilloscope interface).

Kolsky-SHPB Striker Conditions and Stress-Strain Analysis

Because ceramics fail at very low strain levels (~1%), it is important to select the proper striker length in order to limit the pulse duration and thereby the amount of displacement generated during a Kolsky-SHPB test. Striker bars that are too long forces excess energy into the specimen after failure that undesirably comminutes the specimen. Ideally, the original fracture surfaces can be preserved for post-test observations. For the present study where a nominal strain rate of 2000 s^{-1} is applied, a striker bar of only 13 mm in length is calculated to generate the required 1% strain in the dumb-bell-shaped specimen. However, to allow for lower applied strain rates and slightly over-drive the applied displacement, various striker bar lengths between 19 and 40 mm were evaluated. It was observed that the 19 mm striker bar produced a "triangular-shaped" pulse with a ramp-up time of about 7-8 microseconds. The ramp time of longer strikers was the same, but was followed by a flat peak lasting about 7 microseconds in the case of the 40 mm length. Consequently, for applied strain rates $\geq$ 2000 s^{-1}, constant loading conditions could not be established and lower strain rates generally provided greater fidelity in the strength measurement. "Over-driving" the specimen was minimized by progressively adjusting the striker velocity (and therefore the peak stress) until specimens just fractured. Each specimen was loaded only once so that some specimens were observed to fully load and unload elastically which validated both the dynamic modulus and strength results.

Specimen strain and strain rate were calculated using the in-situ strain gages and the specimen stress was calculated using the transmitted pressure bar strain gage (i.e. the 1-wave stress analysis[1]). Vibrations due to radial ringing in the Kolsky-SHPB strain signals are largest during the initial, low-strain regime and unfortunately cannot be eliminated. These oscillations effectively determine the resolution of the stress measurements and can vary greatly from a few percent up to $\pm$ 15% depending on the test conditions. The amplitude of these leading-edge oscillations can be reduced by modifying the striker impact conditions. One method is to slightly round the impact end of the striker bar to a radius of curvature of about 15 cm and another is to insert a thin sheet of metal or "tip" material (e.g. 0.4 mm stainless steel) between the incident bar and the striker bar that will deform slightly upon impact.[6] Both of these methods reduce the amplitude of radial ringing in the incident bar (especially the leading edge oscillations), but also tend to increase the ramp-up time of the pulse. A slow ramping pulse can also be generated by using thick (~2 mm), ductile "tip" materials.[1,7] However, the maximum strain rate is limited to about 300 s[-1] and varies significantly during the test so that only the value at failure can be cited.

RESULTS

The results of Kolsky-SHPB compression tests using the methods described above are shown along with quasi-static data using the same specimen geometry for selected ceramics and cermets.[5,8,9] Commercial-grade TiB_2 and B_4C are shown in Fig. 1. Dynamic strengths are about the same as quasi-static results considering the large scatter and uncertainty in the measurements, but the strains-to-failure are consistently higher in the dynamic data. Nevertheless, the stress-strain paths are in good agreement with the ultrasonically-derived Young's modulus and, other than differences in the failure strains, there is little indication of strain rate effects.

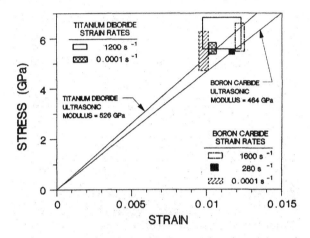

Fig. 1. Compression strengths for TiB_2 and B_4C at quasi-static and dynamic strain rates.

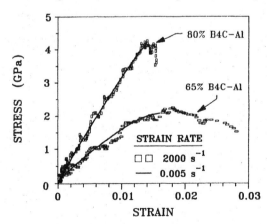

Fig. 2. Compression stress-strain curves for cermets at quasi-static and dynamic strain rates.

95

Figure 2 shows results for 65% B$_4$C-Al and 80% B$_4$C-Al cermets produced by metal infiltration. For these materials, there is very little difference between the dynamic and quasi-static stress-strain behavior, indicating the absence of strain rate sensitivity. The most significant effect is the appearance of a peak followed by a decay in the dynamic stress with further strain indicating damage-tolerance in the 65% B$_4$C-Al cermet. This effect is likely a result of the superior time resolution of the Kolsky-SHPB diagnostics such that the load decay associated with damage evolution (i.e., microcracking and Al plasticity) can be resolved.

CONCLUSIONS

The use of the Kolsky-SHPB compression technique for testing very high-strength, low-ductility materials, such as ceramics, at high strain rates ($\sim 10^3$ s^{-1}) requires special methods and controls to ensure valid test results. The establishment of uniaxial stress-state equilibrium within the specimen and the reduction of stress at the pressure bars-specimen interface below the yield strength of the bars necessitate the use of a specially designed dumb-bell-shaped specimen in addition to limiting the applied strain rate and pulse duration of the tests. In situ strain gages were used to obtain strain-to-failure data and to ensure proper loading of the specimen and specimen stress was calculated via strain measured remotely in the transmitted pressure bar. Unfortunately, the low precision of the stress measurement combined with the inherent distribution of strength of brittle ceramic materials severely limits quantitative comparisons of the ceramic strengths.

REFERENCES

[1]G.T. Gray III, "Classic Split-Hopkinson Pressure Bar Testing," *ASM Handbook Vol. 8: Mechanical Testing and Evaluation,* H. Kuhn and D. Medlin, Eds., ASM International, Materials Park, Ohio, 462-76 (2000).

[2]G. Subhash and G. Ravichandran, "Split-Hopkinson Pressure Bar Testing of Ceramics," *ASM Handbook Vol. 8: Mechanical Testing and Evaluation,* H. Kuhn and D. Medlin, Eds., ASM International, Materials Park, Ohio, 497-504 (2000).

[3]E.D.H. Davies and S.C. Hunter, "The Dynamic Compression Testing of Solids by the Method of the Split Hopkinson Pressure Bar (SHPB)," *J. Mech. Phys. Solids,* 11, 155-79 (1963).

[4]C.A. Tracy, "A Compression Test for High Strength Ceramics," *J. Test. Eval.,* 15, 14-19 (1987).

[5]W.R. Blumenthal and G.T. Gray III, "Structure-Property Characterization of a Shock-Loaded B$_4$C-Al Cermet," *Inst. Phys. Conf. Ser.,* 102, Bristol, 363–70 (1989).

[6]C.E. Frantz, P.S. Follansbee, and W.T. Wright, "Experimental Techniques with the SHPB," *High Energy Rate Fabrication-1984*, I. Berman and J.W. Schroeder, Eds., American Society of Mechanical Engineers, 229–36 (1984).

[7]K.T. Ramesh and G. Ravichandran, "Dynamic Behavior of a Boron Carbide-aluminum Cermet: Experiments and Observations," *Mech. of Mater.,* 10, 19-29 (1990).

[8]W.R. Blumenthal, "High-Strain-Rate Compression and Fracture of B$_4$C-Aluminum Cermets," *Shock-Wave and High-Strain-Rate Phenomena in Material,* M.A. Meyers, L.E. Murr, and K.P. Staudhammer, Eds., Marcel Dekker, NY, 1093-1100 (1992).

[9]W.R. Blumenthal, G.T. Gray, and T.N. Claytor, "Response of Aluminum-Infiltrated Boron-Carbide Cermets to Shock-Wave Loading," *J. Matls. Sci.,* 29, 4567-76 (1994).

RECENT ADVANCEMENTS IN SPLIT HOPKINSON PRESSURE BAR (SHPB) TECHNIQUE FOR SMALL STRAIN MEASUREMENTS

Bazle A. Gama[*], Sergey L. Lopatnikov and John W. Gillespie Jr.
Center for Composite Materials (UD-CCM)
University of Delaware, Newark, DE 19716, USA

ABSTRACT

Recent advances in SHPB technique are discussed. A compliance calibration technique together with an exact 1-D Hopkinson bar theory has been developed to accurately measure the rate dependent elastic modulus in the small strain range. A right circular cylinder (RCC) specimen with chamfered edges is found to minimize the stress concentration at the edges and thus provide a more uniform uni-axial stress state in the specimen. A new SHPB experimental methodology using an equal diameter RCC specimen and a small diameter RCC specimen is developed in determining the dynamic stress-strain behavior of materials.

INTRODUCTION

A critical review by the authors[1] has identified several fundamental questions associated with the split Hopkinson pressure bar (SHPB) experimental technique (Figure 1), the validity of the assumptions of its classic 1-D theory, and the limits of its application. Several important issues have been considered in the present research.

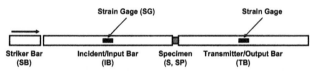

Figure 1: Schematic Diagram of SHPB.

In the fundamental assumptions of classic 1-D Hopkinson bar theory[2], the stress wave propagation in the bars is assumed one-dimensional, which is not true in the vicinity of the bar-specimen (B-SP) interfaces. The physical presence of material discontinuity at the incident bar-specimen (IB-S) and specimen-transmitter bar (S-TB) interfaces produces wave reflections and wave scattering which in turn induces higher order vibration modes. The net effect is that small strain measurement using Hopkinson bar experiment becomes challenging. A compliance calibration methodology for SHPB has been proposed to minimize these errors and provide higher accuracy for small strain measurements.

Stress equilibrium is assumed at the IB-S and S-TB interfaces in the classic theory, and thus the stress wave propagation effect in the specimen (SP) is neglected. Since this assumption is not true in the first few reverberations in the specimen, it has been mentioned[3] that the data in this time window should be used with caution. A significant amount of data in the linear elastic zone of the stress-strain plot comes from the early time when stress is not in equilibrium. The lack of stress equilibrium also contributes to the fact that elastic strain measurement in SHPB is inaccurate. It has been identified that the compliance of SHPB is the root cause for inaccuracy in small strain measurement. Since the stress wave equilibrium in the specimen is a function of the

---

[*] Corresponding author. Tel: (302) 831-8352. E-mail: gama@ccm.udel.edu, gamab@asme.org

---

material properties of the specimen, this issue is resolved by developing an exact 1-D theory considering stress wave propagation in the specimen and denoting the 'time-averaged' Hopkinson bar data as 'time-averaged non-equilibrium dynamic behavior of the material.' The exact 1-D theory provides a strain correction factor, which in conjunction with compliance calibration technique provides the correct elastic modulus of an equal diameter specimen. An experimental methodology is developed for the determination of rate dependent elastic modulus of materials.

The B-SP interfaces are assumed planar at all time, and thus the stress in the specimen is assumed uni-axial, which is true for thin soft materials, and equal diameter thin specimens (Figure 2). In the case of small diameter hard specimens, required to load the specimens up to failure, the B-SP interfaces are non-planar (Figure 2), and the stress in the specimen is not uni-axial. The non-planar deformation of the B-SP interfaces is found to overestimate the strain in the specimen, and induce stress concentration at the edges of a right circular cylinder (RCC) specimen at the B-SP interfaces. A chamfered specimen geometry in this case is found to reduce the stress concentration, improve the uni-axial state of stress in the specimen, and thus minimize undesirable damage modes.

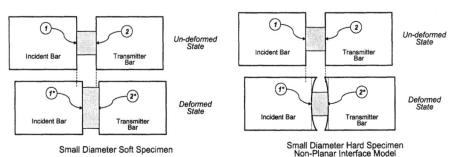

Figure 2: Bar-Specimen (B-SP) Interface Deformation Hypothesis[1].

The diameter of a RCC specimen used in a compression SHPB test is usually smaller than the diameter of the bars. The diameter ratio (D_S / D_B) is chosen such that the stress in the specimen is more than its failure stress, and the stress in the bars is less than its yield stress. Larson[4] (2003) studied the effect of specimen diameter on the stress-strain behavior of aluminum specimens and found that the elastic modulus is a function of specimen diameter (Figure 3). The calculated elastic modulus for small diameter specimen is found to be smaller than the actual modulus. Since the stress in the specimen calculated by '3-wave' analysis is exact[3], it is clear that the strain of the specimen is over-predicted for small diameter specimens due to the non-planar deformation of the bar-specimen interfaces, and thus the linear-elastic portion of the stress-strain curve is not acceptable. Since error in strain measurement is minimal in case of equal diameter specimens, a meaningful stress-strain plot can be constructed if the specimen actually fails. However, in most cases a small diameter specimen is needed to reach the failure limit of the material.

An experimental methodology combining an equal diameter RCC specimen (i.e. elastic modulus) and a small diameter RCC specimen with chamfered edges (i.e. strength) is proposed to construct the complete dynamic stress-strain curve of a material. Since the experimental data is time-averaged, the stress-strain behavior can be termed as 'time-averaged non-equilibrium dynamic stress-strain behavior.' The present research effort is an important step towards standardizing the SHPB test methods for measurement of small strain behavior, elastic moduli and strength of brittle materials as a function of strain rate.

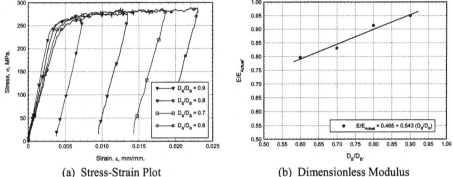

(a) Stress-Strain Plot (b) Dimensionless Modulus

Figure 3: Effect of D_S / D_B on Elastic Modulus, Aluminum SP with Inconel Bars[4].

COMPLIANCE CALIBRATION OF SHPB

The review of Hopkinson bar experimental technique[1] has identified that equal diameter specimens can be used to minimize the error in displacements at the B-SP interfaces. Experiments with rate independent equal diameter aluminum specimens have identified that the initial modulus calculated from the classic theory is less than the known value. This raises questions about the validity of the stress-strain plot as obtained from a classic Hopkinson bar experiment. In order to address this issue, a compliance calibration methodology for small strain measurement using compression split Hopkinson pressure bar (SHPB) is developed and presented.

The main components of a compression SHPB are an incident bar (IB), a transmitter bar (TB) and a striker bar (SB) as presented in Figure 1. In a 'bars-together' calibration experiment, the IB and TB are coupled together without a specimen (SP) and are impacted at the impact end of the IB. Figure 3 shows a 'bars-together' calibration experiment of a 12.72-mm diameter inconel bar using a copper pulse shaper (PS). Ideally, the reflected pulse from this experiment should be zero and the transmitted pulse should be equal to the incident pulse. However, the reflected pulse obtained from the experiment is not zero (Figure 3, Reflected Pulse × 10).

For a non-zero reflected pulse, the difference in displacements at the IB-TB interface is also a non-zero quantity, which can be expressed as:

$$\Delta = -\int \{c_{0IB}[\varepsilon_I(t) - \varepsilon_R(t)] - c_{0TB}\varepsilon_T(t)\}dt \qquad (1)$$

where, Δ is the difference between IB and TB ends at IB-TB interface, c_{0IB} and c_{0TB} are the corrected bar velocities of IB and TB determined following 'bars-apart' calibration experiment and dispersion correction methodology[1, 5]. Following the analogy of a '3-wave' analysis[3], the stress at this interface can be expressed as:

$$\Sigma = \left[E_{IB}^{C}\left\{\varepsilon_{I}(t)+\varepsilon_{R}(t)\right\}+E_{TB}^{C}\left\{\varepsilon_{T}(t)\right\}\right]/2 \tag{2}$$

where, E_{IB}^{C} and E_{TB}^{C} are the corrected elastic modulus of the IB and TB calculated from the corrected bar velocities.

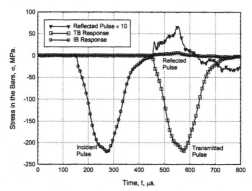

Figure 3: 'Bars-Together' Calibration Experiment, Inconel Bars, Copper PS, $D_B = 12.72$-mm.

The IB-TB interface stress is plotted as a function of the difference in displacements in Figure 4 for two different 'bars-together' calibration experiments. Two different experiments are conducted using copper and rubber pulse shapers, respectively. However, the $\Sigma - \Delta$ plot for these two different experiments is very similar in shape. Therefore, the $\Sigma - \Delta$ behavior of a 'bars-together' calibration experiment is assumed independent of the pulse shapes. Additionally, the presence of an imperfect interface between the IB and TB acts as a defect and produces 3-D wave scattering at the sharp edges of the cylindrical bars, which appears as a reflected pulse form in this experiment. If it is assumed that for a particular set of bars, the initial linear loading portion of the $\Sigma - \Delta$ behavior is a function of stress in the bars only, then the slope of the initial region can be used in calculating the error in small strain measurements using SHPB, which is exactly analogous to procedures developed for quasi-static compliance calibration of universal testing machines. For this specific set of bars the slope of the $\Sigma - \Delta$ curve is found to be $m_{\Delta}^{\Sigma} = \Sigma / \Delta = 14,067$ MPa/mm.

If it is assumed that the reflected pulse from a real test with an equal diameter specimen of length, H_S, is an algebraic sum of reflection from the specimen and the reflection from wave scattering at the IB-S interface, the error in the strain measurement can then be expressed as:

$$\varepsilon_{Error} = \frac{\Delta}{H_S} = \frac{\sigma_S}{H_S \cdot m_{\Delta}^{\Sigma}} \tag{3}$$

where the stress in the specimen is equivalent to the IB-TB interface stress, i.e., $\sigma_S = \Sigma$.
The corrected specimen strain can then be calculated by using the following relationship:

$$\varepsilon_{Correct} = \varepsilon_{Experimental} - \varepsilon_{Error} \qquad (4)$$

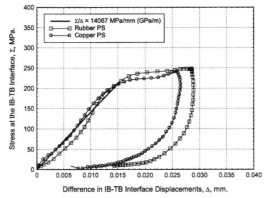

Figure 4: Interface Stress vs. Difference in Displacements, Inconel Bars, , $D_B = 12.72$-mm.

An equal diameter aluminum specimen ($H_S = 10.69$-mm, $D_S = 12.74$-mm) is tested twice under identical loading conditions using a copper PS. The experimental data is reduced following '3-wave' analysis procedure utilizing corrected bar velocities and elastic modulus and is presented in Figure 5a. The data obtained from '3-wave' analysis is then corrected for strain using Equations (3) and (4) and is also presented in Figure 5a. The elastic modulus is calculated in the stress range 50 MPa $< \sigma_S < 150$ MPa, and is presented in Figure 5b and is summarized in Table 1. The modulus calculated using classic '3-wave' analysis is found to be about 25% less, and the corrected modulus is found to be about 15% higher than the known value. Further data reduction utilizes the exact 1-D Hopkinson bar theory described next.

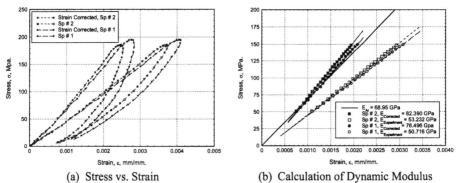

(a) Stress vs. Strain   (b) Calculation of Dynamic Modulus

Figure 5: Equal Diameter Aluminum SP and Inconel Bars, Copper PS, $D_B = 12.72$-mm, $D_S = 12.74$-mm, $H_S = 10.69$-mm, $c_{0IB} = 4878.9$ m/s, $c_{0TB} = 4868.6$ m/s, $E_{IB}^C = 191.16$ GPa $E_{TB}^C = 190.36$ GPa, $\rho_{IB} = \rho_{TB} = 8031$ kg/m^3, $\rho_S = 2700$ kg/m^3

101

AN EXACT 1-D HOPKINSON BAR THEORY FOR EQUAL DIAMETER LINEAR SPECIMENS

An exact 1-D Hopkinson bar theory for an equal diameter linear specimen has been developed[5, 6] that considers the stress wave propagation in the specimen. According to this exact 1-D theory, the strain and stress in the specimen is given by:

$$\bar{\varepsilon}_S(t) \approx \left[-\frac{(1-r)^2}{4r}\right] \cdot \left[-\frac{2}{\tau} \cdot \int^t \varepsilon_R(t,0) \cdot dt\right] + \left[\frac{(1+5r) \cdot (1-r)^3}{4r(1+r)^2}\right] \cdot \left[\frac{\rho_S}{\rho_B} \cdot \varepsilon_R(t,0)\right]$$

$$= K_S \cdot \bar{\varepsilon}_S^{cl}(t) + K_S^{dyn} \cdot \frac{\rho_S}{\rho_B} \cdot \varepsilon_R(t,0)$$

(5)

$$\bar{\Sigma}_S(t) \approx E\varepsilon_T(t,H) + E\frac{\rho_S}{2\rho_B} \cdot \tau \cdot \frac{d}{dt}\varepsilon_T(t,H)$$

(6)

where, ρ_B and c_{0B} are the density and wave velocity of the bar material, ρ_S and c_S are the density and wave velocity of the specimen, $r = (\rho_S c_S - \rho c)/(\rho_S c_S + \rho c)$, K_S is the strain correction factor applied to the classic 1-D strain in the specimen, $\bar{\varepsilon}_S^{cl} = -(2/\tau)\int\varepsilon_R(t)dt$, and K_S^{dyn} can be neglected[5]. The strain correction factor, K_S, corrects some unique problems of classic 1-D theory and is discussed elsewhere[4]. The stress in the specimen appears to be the classic 1-D stress with some additional terms originating from the consideration of stress wave propagation in the specimen.

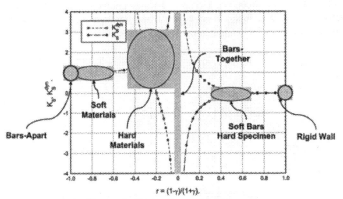

Figure 6: Strain Correction Factors

Figure 6 shows K_S and K_S^{dyn} as a function of r. For acoustically soft materials (or post-yield regime), $K_S \rightarrow 1$, and is equivalent to the classic 1-D strain expression. However, for acoustically hard materials in comparison to the bar materials, $K_S > 1$, and needs to be considered to accurately predict the strain in the specimen (and hence the elastic modulus). The values of reflection coefficient, r, and the strain correction factor, K_S (see Table 1) are calculated by trial and error since the wave speed of the specimen is not known a priori. After the strain correction factor is applied to the compliance corrected strain of the specimen, the 'exact 1-D theory' predicts the modulus of the aluminum specimens with ±4% accuracy. This experimental methodology can be used in determining the rate dependent elastic modulus (and stress-strain response in the small strain regime) of any materials of interest which utilizes an equal diameter specimen, the compliance calibration technique, and the '1-D exact' strain correction factor. However, the complete stress-strain curve up to failure may not be achievable unless a smaller diameter specimen is tested.

Table 1: Dynamic Elastic Modulus of Aluminum

Properties	Specimen # 1	Specimen # 2
$E_{Experiment}$, GPa	50.72	53.23
$E_{Corrected}$, GPa	76.50	82.39
$r = \dfrac{\rho_S c_S - \rho_B c_{0B}}{\rho_S c_S + \rho_B c_{0B}}$	-0.48735	-0.47482
$K_S = -\dfrac{(1-r)^2}{4r}$	1.13482	1.14522
$E_{Exact}^{1-D} = E_{Corrected} / K_S$, GPa	67.41	71.95
$E_{Book\ Value}$, GPa	68.95	68.95
Error, %	-2.23	+4.35

STRESS ANALYSIS OF A SMALL DIAMETER SPECIMEN

An in-depth numerical analysis[5] of small diameter specimens revealed that the bar-specimen interfaces are non-planar for small diameter specimens[1] and the stress in the specimen is not uni-axial. Figure 7 shows the dimensionless difference in displacements of the IB edge (u_Z^E) and the IB center (u_C^E) as a function of dimensionless time for different D_S / D_B ratio (SBD) under a ramp load of duration T. Obviously, the difference in displacement is zero for equal diameter specimens ($SBD = 1.0$), however, non-zero for small diameter specimens ($SBD < 1.0$). This numerical experiment is clear proof of the non-planar B-SP interface deformation hypothesis presented earlier in Figure 2. The net effect is the stress concentration at the edges of the SP at the B-SP interfaces. Figure 7b shows the dimensionless axial stress on the surface of a RCC specimen without and with chamfered edges along the length. In case of a RCC specimen, the axial stress distribution is not uniform and the stress concentration at the SP edges is about 2. It is well known that smaller diameter specimens are usually used in SHPB technique to load the SP up to failure. In such cases, premature failure can occur at the edges of the SP due to the high stress concentration and the non-uniform axial stress distribution in the SP violates the uni-axial stress assumption of the SHPB theory. It is also understood that this problem is not critical for an acoustically soft material where the B-SP interfaces are planar.

This problem can be resolved by making the edges of the specimens contoured at the B-SP contact interfaces such that the non-planar B-SP interface deformation matches with the contour of the SP. For a small diameter SP, the specimen to bar diameter ratio can be determined using the relation, $D_S/D_B = \sqrt{\sigma_{max}^B / \sigma_F^S}$, where σ_{max}^B is the maximum allowable elastic

103

stress in the bars and σ_F^S is the estimated dynamic failure strength of the specimen. The contour dimensions, L_R and L_H, and the shape of the curve to which the specimen loading surfaces should be machined is a function of the ratio of specimen to bar diameter, the impedance of the specimen and the bar, and the stress amplitude of the incident pulse.

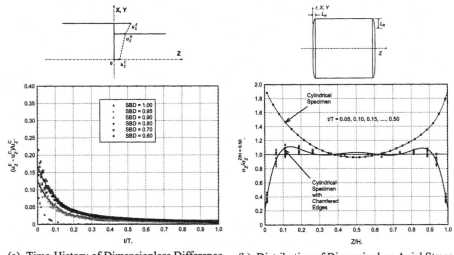

(a) Time-History of Dimensionless Difference between Displacements at IB Edge and Center, as a Function of Specimen to Bar Diameter Ratio, $SBD = D_S / D_B$

(b) Distribution of Dimensionless Axial Stress along the Length of a Cylindrical Specimen with and without Chamfered Edges, $SBD = 0.60$

Figure 7: Analysis of a Small Diameter Specimen (Aluminum SP, Steel Bars).

The difference in displacements between the IB edge and IB center ($\Delta = u_E - u_C$) at IB-S interface is calculated at the maximum stress in the IB for $\sigma_{IB} = 500\,\text{MPa}$ for two different impedance ratios, $\gamma = \rho c / \rho_S c_S = 1.0$ (steel bars – steel specimens) and $\gamma = \rho c / \rho_S c_S = 0.0005$ (steel bars – rigid specimens) and is presented in Figure 8a. Behavior of any acoustically hard material ($0.0005 < \gamma < 1.0$) will fall between the two trend lines shown in Figure 8a, while the soft materials ($\gamma > 1.0$) will fall below the trend line designated by $\gamma = 1.0$. The upper limit of the contour dimension L_H is given by, $L_H = \Delta = u_E - u_C$. Since the shape of the contour is also unknown, a linear profile (chamfer) can be assumed to define the contour angle, $\alpha \approx L_H / L_R = 2(u_E - u_C)/(D_B - D_S)$ and is presented in Figure 8b. The contour/chamfer angle shows similar behavior as the difference in displacements, $\Delta = u_E - u_C$. In all practical cases, the upper bound of the chamfer angle is less than one degree. The determination of the exact contour shape is a complex problem, and this is why a RCC specimen (RCCS) with chamfered edges has been investigated. Results show that the axial stress distribution is much more uniform than the RCC specimen without chamfered edges, and the most critical fact is that the stress concentration factor has reduced from 2 to about 1.1 (Figure 7b). Thus, the use of a RCCS with chamfered edges will provide better uni-axial stress in the SP, and will also provide desired damage modes by preventing any premature failure of the SP.

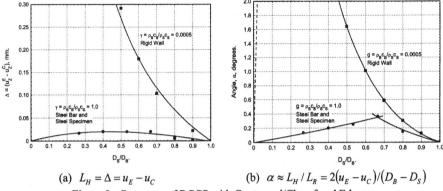

(a) $L_H = \Delta = u_E - u_C$ (b) $\alpha \approx L_H / L_R = 2(u_E - u_C)/(D_B - D_S)$

Figure 8: Geometry of RCCS with Contoured/Chamfered Edges.

A NEW SHPB EXPERIMENTAL METHODOLOGY IN DETERMINING THE DYNAMIC STRESS-STRAIN BEHAVIOR OF MATERIALS

The dynamic stress-strain behavior of a material should provide correct dynamic modulus and correct dynamic large deformation behavior (non-linear damage behavior). In earlier sections, it has been shown that the compliance calibration technique and the 'exact 1-D' analysis with an equal diameter RCC specimen can be used to calculate the rate dependent elastic modulus of any material up to a maximum stress amplitude that can be generated in the SHPB. Ideally, if an equal diameter specimen fails without too much radial deformation at a time much larger than its characteristic time (condition for stress equilibrium), the failure mode is acceptable, and the stress-strain plot is the correct dynamic stress-strain behavior of the material at that specific loading rate/ strain rate. Practically, most material expands under large compressive deformation, and may not reach the failure stress if equal diameter SP is used. This is why, a small diameter specimen is used in real tests, and the stress-strain behavior in the linear-elastic region is inaccurate.

It has also been shown, that a small diameter RCCS with chamfered edges can be used to test a material suppressing premature failure and not violating the uni-axial stress assumption of the SHPB theory. Assuming that the linear dynamic stress-strain behavior of an equal diameter specimen can be extrapolated up to the maximum stress that can be obtained from a small diameter RCC specimen with chamfered edges, it is plausible to combine the results of the two tests to construct the correct and complete dynamic stress-strain plot of the SP.

It has been shown in Table 1 that the dynamic elastic modulus of equal diameter rate independent aluminum specimen can be measured from SHPB experiments following the compliance calibration methodology and using the strain correction factor as obtained from the '1-D exact' theory. This value is plotted in Figure 9 as a straight line (with solid triangle symbols, experiments were done at stress level less than 200 MPa as shown in Figure 5). A small diameter RCC aluminum specimen with chamfered edges is tested up to a stress level 300 MPa and is presented in Figure 9 with legend "'3-wave' Data". The apparent elastic modulus

105

obtained from this experiment is found to be $E_{Apparent} = 49.05$ GPa, much lower than reality as expected. Knowing the fact that the small strain measurement is incorrect, the total strain of the specimen is decomposed into linear and non-linear strain at a constant stress.

$$\varepsilon_S(\sigma,t) \approx \varepsilon_S^E(\sigma,t) + \varepsilon_S^{NL}(\sigma,t) \qquad (7)$$

The stress vs. non-linear strain behavior of the small diameter RCCS with chamfered edges is presented in Figure 9 with the legend "Stress vs. Non-Linear Strain." The tangent modulus of the stress vs. non-linear strain is very small and the material behaves as a soft material. Since the SHPB measurement is correct for soft materials, the 'Stress vs. Non-Linear Strain' of the small diameter aluminum specimen can be considered as accurate. At this point, it is obvious that one can independently measure the dynamic elastic behavior and the dynamic non-linear behavior of any material from two independent SHPB experiments. Assuming that the error in the strain during elastic to non-linear transition is negligible, one can numerically add the elastic and non-linear behavior by replacing the elastic strain of a small diameter RCCS specimen with chamfered edges with the correct elastic strain as obtained from the equal diameter RCCS.

$$\varepsilon_S(\sigma,t)\Big|_{Exact} \approx \frac{\sigma(t)}{E_{Equal\ Diameter}^{Exact}} + \varepsilon_S^{NL}(\sigma,t) \qquad (8)$$

The reconstructed stress-strain behavior of aluminum is presented in Figure 9 with the legend 'Corrected Stress-Strain.' Thus, the present experimental methodology combines the accurate elastic and non-linear large deformation response to represent the complete and accurate dynamic stress-strain response of a material under uni-axial stress of loading.

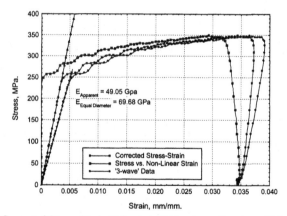

Figure 9: Corrected Stress-Strain Behavior of Aluminum Specimen Using the New Experimental Methodology.

106

CONCLUSIONS

Significant contributions to the SHPB experimental technique are reported. These advancements include the development of:

- a compliance calibration experimental methodology for compression SHPB,
- an 'exact 1-D' Hopkinson bar analysis for accurate calculation of rate dependent elastic modulus, and
- an experimental methodology to determine the correct dynamic stress-strain behavior of materials.

ACKNOWLEDGMENTS

The funding for this research provided by the Composite Materials Technology (CMT, DAAD19-01-2-0005) program is gratefully acknowledged.

REFERENCES

[1] B. Gama, S. Lopatnikov, J. Gillespie Jr., "Hopkinson Bar Experimental Technique: A Critical Review," *App. Mech. Rev.*, **57(4)**, 223-250 (2004).

[2] *ASM Handbook, Volume 8, Mechanical Testing and Evaluation*, ASM International, Materials Park, Ohio 44073-0002, 462-476 (2000).

[3] G. Gray III, "Classic Split-Hopkinson Pressure Bar Testing," *ASM Handbook, Volume 8, Mechanical Testing and Evaluation*, ASM International, Materials Park, Ohio 44073-0002, 462-476, (2000).

[4] M. Larson, "An Evaluation of Methodologies Applied to the Split Hopkinson Pressure Bar," *MS Thesis*, University of Delaware, Fall 2003.

[5] B. Gama, "Split Hopkinson Pressure Bar Technique: Experiments, Analyses and Applications," *Ph. D. Dissertation*, University of Delaware, Spring 2004.

[6] S. Lopatnikov, B. Gama, C. Krauthauser, J. Gillespie, "Applicability of the Classical Analysis of Experiments with Split Hopkinson Pressure Bar," *Tech. Phys. Letters*, **30(2)**, 102-105, (2004).

[7] H. Kolsky, "An Investigation of the Mechanical Properties of Materials at very High Rates of Loading," *Proc. Phys. Soc., London*, **62**(II-B), 676-700, (1949).

COMPRESSION TESTING AND RESPONSE OF SIC-N CERAMICS: INTACT, DAMAGED AND POWDER

Kathryn A. Dannemann, Arthur E. Nicholls, Sidney Chocron, James D. Walker,
Charles E. Anderson, Jr.
Southwest Research Institute
P.O. Drawer 28510
San Antonio, Texas 78228-0510

ABSTRACT

The objective of this work was to determine the fundamental compression response of SiC-N ceramics and obtain an improved understanding of the transition from intact to damaged material to aid in ceramics modeling efforts. Existing models [1,2] for evaluating the performance of ceramics under ballistic impact do not adequately address this behavior owing to the lack of experimental data in this regime. Compression experiments were conducted on SiC-N material in various forms: intact, damaged, and powder. A technique was developed for obtaining "comminuted" SiC-N material from intact material by in-situ failure of pre-damaged samples. A thermal shock procedure was devised to pre-damage the SiC-N material. The pre-damaged samples were inserted in a high-strength steel confining sleeve, and then loaded and re-loaded at quasistatic strain rates to fail the material in-situ. Several load/reload cycles were conducted at successively increasing loads. Strain gages mounted on the outer diameter of the confinement sleeve allowed the measurement of hoop strain. A high-strength steel confining ring was also used for testing SiC-N powder. The powder samples were compacted to approximately 50% of theoretical density prior to testing. The experimental findings are presented for the three material forms. The "comminuted" material exhibited a decline in performance versus the intact material. The results are compared and discussed in light of existing damage models. Numerical and analytical modeling of the experiments for damaged and powder materials was performed to enhance understanding; the modeling results are presented in another paper by Chocron, et al.[3] at this conference.

INTRODUCTION

The compressive strengths of various intact and powder ceramic materials (e.g., Al_2O_3, AlN, B_4C) have been characterized previously using innovative experimental techniques.[4,5,6,7] Similar characterization studies for SiC-N ceramics are the focus of the current work. Characterization of the compressive strength of in-situ comminuted material is of most interest, though the performance of SiC-N powder was also evaluated as a lower bound for comparisons.

Existing models [1,2] of ceramic performance under impact do not adequately address the transition from intact to damaged material behavior owing to the lack of experimental data in this regime. The present work focused on obtaining a better understanding of this transition to aid in modeling efforts for SiC-N ceramics. Initial test plans were to obtain comminuted material by loading intact SiC-N samples at quasistatic rates in an autofrettage apparatus[8], failing the material in-situ. Samples would then be reloaded while maintaining the confining pressure to provide useful information on the behavior of comminuted SiC-N ceramic. However, the very high strength of the solid SiC-N material presented a challenge in selecting platen materials that could survive such high loads (> 6 GPa). Testing of solid SiC-N samples at high confining pressures was limited by the failure strength of the tungsten carbide (WC) platens. Alternatively, a thermal shock procedure was used to create pre-damaged SiC-N material for testing comminuted material. Compression experiments were conducted to determine the effects of initial condition of the SiC-N ceramic (intact, powder, *in-situ* comminuted) as a function of strain rate and confining pressures.

MATERIALS

A SiC-N plate, measuring approximately 15-cm x 15-cm x 1.6-cm, was obtained from Cercom, Inc. (Vista, CA). The material was produced using a pressure assisted diffusion (PAD) process. The grain size distribution in the PAD SiC-N is that of the starting material. SiC-N powder was also obtained from Cercom for comparison testing with intact and comminuted material.

Cylindrical test samples, measuring 1.27-cm long by 0.635-cm diameter, were machined from the SiC-N plate. Machining was performed by a skilled vendor that consistently met specifications for surface finish and tight machining tolerances. Flatness and parallelism of the sample ends, especially critical for the high strain rate tests, was maintained to within 0.0025-mm. WC platens and anvils were also machined for testing; the shapes of these parts differed depending on the material type (i.e., intact, pre-damaged or powder). Tapered anvils were used for tests with confining sleeves. The WC material grade used was selected for its high impact and shock resistance, and is similar to that used previously in autofrettage testing [8] at Southwest Research Institute (SwRI®). The confining sleeves for tests on the SiC-N powder and pre-damaged materials were fabricated from maraging steel, Vascomax C350. The inner diameter of each sleeve was honed to ensure the appropriate fit between each sample and confining sleeve.

EXPERIMENTAL PROCEDURE

Compression tests were conducted on different SiC-N material forms, including: intact material (with and without confinement), comminuted material obtained by pre-damaging intact samples using a thermal shock procedure, and SiC-N powder. Low strain rate ($\sim10^{-5}$ to 10^{-1} s^{-1}) tests were conducted using an MTS servohydraulic machine. The high strain rate ($\sim10^3$ s^{-1}) tests were conducted in the SwRI split Hopkinson pressure bar (SHPB) using 2.54-cm diameter maraging steel bars.

Intact

Cylindrical WC platens were utilized in low strain rate tests on intact SiC-N. Confinement tests were conducted at low pressures (i.e., 100 MPa) using a hydraulic pressure device used in previous studies.[9] The pressure limit of this device is approximately 350 MPa. The 100 MPa confining pressure was maintained almost constant during testing. For the confined and unconfined tests on the intact material, axial strains were measured using an extensometer attached to the Vascomax loading platens.

Pre-Damaged

A thermal shock procedure was devised to pre-damage the ceramic. Samples were exposed for two 1-h cycles at 750°C in a resistance tube furnace. The 1-h cycles allowed temperature equilibration, and were each followed by an ice water quench. Stereomicroscopy evaluation of thermally shocked SiC-N samples revealed a microcrack pattern on the sample ends without loss of sample integrity; the samples remained relatively intact and could be readily tested. Several samples were sectioned, polished and evaluated to quantify the extent of damage and determine the consistency of the damage pattern throughout the entire sample length (i.e., 1.27-cm). For the thermal shock procedure chosen, a consistent damage pattern was observed from sample to sample.

Initial compression tests on the pre-damaged samples were conducted without a confining sleeve to determine the effect of pre-damage on material performance. Upon confirming that the thermal shock procedure reduced the strength relative to the intact SiC-N material, the remaining tests on the pre-damaged material were performed using a 6.35-mm-thick steel confining ring with tapered WC anvils at the specimen ends. "Comminuted" SiC-N material was obtained by quasistatically loading the pre-damaged samples, contained in a steel confining sleeve, and "failing" the material in-situ. To prevent failure of the WC platens, each sample was loaded and and re-loaded at quasistatic (QS) strain rates. Several load/reload cycles were conducted at successively increasing loads; maximum loads were less than the failure strength of the damaged material. Strain gages mounted on the outer diameter of the confinement sleeve allowed the measurement of hoop strain. Axial strain was measured with an extensometer. A schematic of the test assembly is shown in Figure 1. For high strain rate tests, the comminuted samples, still contained in the steel confining sleeve, were transferred to the SHPB system for testing.

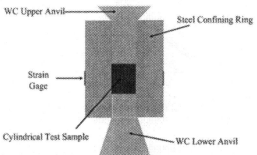

Figure 1. Schematic of the test assembly used for testing powder and pre-damaged SiC-N.

Powder

The 6.35-mm thick (22.2-mm long) steel confining sleeves were also used to test SiC-N powder. Tapered WC platens were positioned at both ends of the confining sleeve to contain the powder in the chamber during testing. The geometry of the upper anvil was modified slightly vs. the corresponding anvil for the tests on comminuted material to ensure containment of the powder. Hoop strain measurements were accomplished with strain gages mounted to the external surface of the confining sleeve; axial strain was measured with an extensometer.

The powder was packed into the chamber and loaded to approximately 1 MPa pressure to achieve compaction prior to testing. The powder samples were compacted to approximately 50% of theoretical density prior to testing. Compression tests were performed at quasistatic strain rates. Additional tests were performed using a compression load/release/reload sequence at low strain rates. The load was released at displacement intervals of approximately 1.27-mm. High strain rate tests were also conducted on powder using the SHPB. Following load/reload cycling, the entire confining fixture with platens was transferred to the SHPB for testing.

RESULTS

Intact

Unconfined compression test results served as a baseline for comparison. For strain rates less than 10 s^{-1}, the results indicate a maximum compressive strength in excess of 5 GPa and the absence of a significant strain rate effect. The recent data correlate very well with earlier SiC data when cross-plotted on the strength vs. pressure plot for SiC in Ref. 1. The engineering stress-strain curves for confined and unconfined tests on the intact SiC-N are compared in Figure 2. A 100 MPa confinement pressure was achieved with a hydraulic pressure device. The confined test results show lower failure strengths than expected - comparable or less than the unconfined strengths. This is related to premature failure of the WC platens.

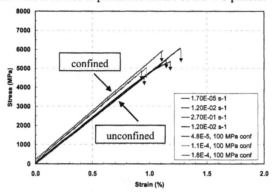

Figure 2. Comparison of engineering stress-strain plots obtained for unconfined and confined compression tests on intact SiC-N material.

Pre-Damaged

The effectiveness of the thermal shock procedure for damaging the SiC-N material was determined from comparison of the microstructures, as well as measurement of the strength decline. Stereomicroscopy evaluation of polished sample ends was used to determine the extent of damage incurred prior to sample loading. Representative images of the damage observed are shown in Figure 3 for one of the pre-damaged samples prior to loading. The thermal shock procedure produced a crack pattern without causing a loss of sample integrity.

A 40 to 60 % strength decline was demonstrated for the pre-damaged samples relative to intact samples tested at similar conditions. The strength decrease was sufficient to confirm the feasibility of conducting room temperature compression tests on the pre-damaged samples with confinement. Since pre-damaging the SiC-N samples decreases the strength, the likelihood of WC platen failure is also less.

Pre-damaged samples were subsequently loaded in compression while contained in a steel confining sleeve. The purpose of the compression loading was to obtain comminuted SiC-N material by "failing" the material *in-situ*. Each test sample was loaded and re-loaded several times at successively increasing loads. Load cycling was performed to prevent failure of the WC platens. The number of load/reload cycles was determined based on the hoop and axial strains observed during quasistatic loading. The results of a representative test are shown in Figure 4 where the effect of load cycling on the axial and hoop strains is evident. Eight load/reload cycles were conducted for this sample. Note the decrease in modulus following initial loading; subsequent loading does not result in further modulus changes. The hoop strain increased with loading; the peak stress obtained was less than the intact strength. These changes are indicative of further damage to the sample due to load cycling. This was confirmed upon removal of the confining sleeve for several pre-damaged samples following load cycling. The findings indicate significant damage to the samples. The photo in Figure 5 is representative of the damage that was observed. The results indicate that comminuted SiC-N material can be obtained by *in-situ* failure of thermally shocked, pre-damaged samples. Similar peak stresses were measured in dynamic compression for the comminuted samples.

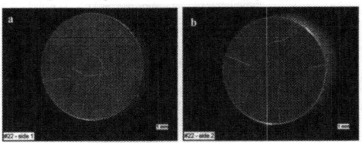

Figure 3. A thermal shock procedure was used to pre-damage the SiC-N test samples prior to load/reload testing and SHPB testing. (a) and (b) are views of both polished ends of the test sample before testing.

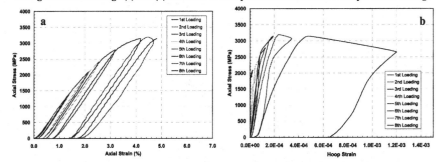

Figure 4. Axial stress versus strain for pre-damaged SiC-N samples during quasistatic compression load/reload cycling to "fail" the material *in-situ* and obtain comminuted SiC-N material. Note the effect of eight load/reload cycles on: (a) axial strain and (b) hoop strain.

Figure 5. A representative pre-damaged SiC-N test sample following quasistatic loading/reloading. Load/reload cycling was effective in "failing" the material *in-situ* to create comminuted material.

113

Powder

Compression test results for low strain rate, monotonic loading of powder samples contained in the steel sleeve were very consistent. Representative axial stress-strain curves are shown in Figure 6. The compressive load was increased until failure of the WC platens occurred. Analysis of the pressure load on the powder sample during testing gave a loading of approximately 2 GPa. This provides a reasonable lower bound for the slope of the equivalent stress vs. pressure curve in existing ceramic models (e.g. JH-1).

Additional tests were also performed using a compression load/release/reload sequence. The load was released at approximate displacement intervals of 1.27-mm. Representative results are shown in Figure 7. The results revealed a similar maximum axial stress vs. the continuous loading case. Some hysteresis was observed, though the shape of the axial stress-strain curve is similar to the monotonic loading curve; compare Figure 6 and Figure 7(a). There was some permanent set as evidenced by the hysteresis loops in the axial stress vs. hoop strain plot in Figure 7(b). Repeat testing of powders using the indicated loading cycles gave consistent results. SHPB testing of the SiC-N powder, following cyclic load/reloading, resulted in similar behavior to the quasistatic results at axial strains exceeding ~5%. The difference in peak stress for the high strain rate results correlates with increasing impact velocity.

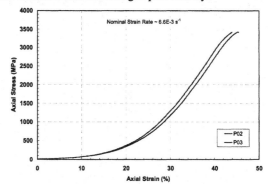

Figure 6. Axial stress –strain curves for two separate tests on SiC-N powder during quasistatic compressive loading. The results are quite consistent.

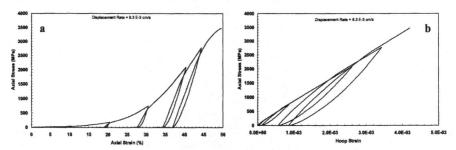

Figure 7. Representative results for load/reload cycling of SiC-N powder at quasistatic strain rate: (a) axial stress versus axial strain, (b) axial stress versus hoop strain.

114

Comparisons

Overlay plots of the test data are beneficial in highlighting differences among the intact, comminuted and powder SiC-N materials. The axial stress vs. hoop strain plot in Figure 8(a) shows a significant difference in the quasistatic (QS) compression behavior for comminuted vs. powder SiC-N. The extent of compaction of the powders is evident: the powder material endures a much higher hoop strain for the same stress level vs. comminuted SiC-N. Higher hoop strains for the powder material indicate the higher confinement needed for powder material to achieve the same level of axial stress as the comminuted (pre-damaged) material. The difference between the powder and the comminuted materials is somewhat less upon comparison of axial stress vs. axial strain, as shown in Figure 8(b). QS results for intact SiC-N are also included in Figure 8(b). The extent of the degradation in strength and modulus is evident for the comminuted and powder materials versus intact SiC-N. Some of the QS curves for the powders show slightly higher maximum stresses vs. the comminuted material. This difference in maximum stress is attributed to the premature termination of many of the comminuted material tests, following several QS loading cycles, to prevent failure of the tungsten carbide platens prior to transferring the confined sample to the SHPB apparatus for high strain rate testing. The maximum stress obtained for both the comminuted and powder materials is limited by the failure strength of the WC anvils (i.e., approximately 3600 MPa).

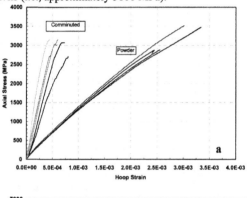

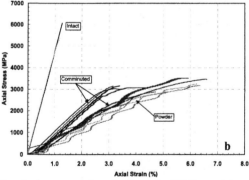

Figure 8. Comparison plot s for intact, comminuted and powder SiC-N materials at quasistatic strain rates: (a) axial stress vs. hoop strain, (b) axial stress vs. axial strain.

115

CONCLUSIONS

Materials experiments (i.e., non-ballistic tests) were conducted on Sic-N material in various forms: intact, damaged and powder. The purpose of the experiments was to determine the fundamental compression response of Sic-N ceramics, and to aid in deriving constants for ceramics models (e.g., Johnson-Holmquist) independent of ballistics experiments. A procedure was developed for obtaining "comminuted" Sic-N material by *in-situ* failure of thermally shocked, pre-damaged samples. Experiments on the comminuted material enhanced understanding of the transition from intact to damaged material. A significant degradation in performance was demonstrated for comminuted and powder materials versus intact Sic-N.

ACKNOWLEDGMENTS

The authors gratefully acknowledge the financial support of the US Army Research Laboratory. Technical assistance and insights from Dr. Michael Normandia (ARL) and Dr. Doug Templeton (TARDEC) is also gratefully acknowledged.

REFERENCES

1. T. Holmquist and G.R. Johnson, "Response of Silicon Carbide to High Velocity Impact", *J. Appl. Phys.*, **91** (9), 5858-5866 (2002).

2. G.R. Johnson, T. Holmquist, "Response of Boron Carbide Subjected to Large Strains, High Strain Rates, and High Pressures", *Journal of Applied Physics*, **85** (12), 8060-8073, (1999).

3. S. Chocron, K.A. Dannemann, A.E. Nicholls, J.D. Walker, C.E. Anderson, Jr., "A Constitutive Model for Damaged and Powder Silicon Carbide", American Ceramic Society 29th International Conference on Advanced Ceramics and Composites, Cocoa Beach, FL (2005).

4. J. Lankford, "Compressive Strength and Microplasticity in Polycrystalline Alumina", *J. Mat. Sci.*, **12**, 791-796 (1977).

5. J. Lankford, "Mechanisms Responsible for Strain-Rate Dependent Compressive Strength in Ceramics Materials", *J. Am. Ceram. Soc.*, **64** (2) C33-C34 (1981).

6. J. Lankford, C.E. Anderson, Jr., A.J. Nagy, J.D. Walker, A.E. Nicholls, and R.A. Page, "Inelastic Response of Confined Aluminum Oxide under Dynamic Loading Conditions", *J. Mat. Sci*, **33**, 1619-1626 (1998).

7. J. Lankford, W.W. Predebon, J.M. Staehler, G. Subhash, B.J. Pletka, C.E. Anderson, "The Role of Plasticity as a Limiting Factor in the Compressive Failure of High Strength Ceramics", *Mechanics of Materials*, **29**, 205-218 (1998).

8. J.D. Walker, A. Nagy, C.E. Anderson, Jr., J. Lankford, Jr., A.E. Nicholls, "Large Confinement High Strain Rate Test Apparatus for Ceramics", *Metallurgical and Materials Applications of Shock-Wave and High Strain Rate Phenomena*, L.E. Murr, K.P. Staudhammer and M.A. Meyers (editors), Elsevier Science (1995).

9. J. Lankford, "Dynamic Compressive Failure of Brittle Materials under Hydrostatic Confinement", AMD-Vol. **165**, *Experimental Techniques in the Dynamics of Deformable Solids*, ASME (1993).

DAMAGE EFFECTS ON THE DYNAMIC RESPONSE OF HOT-PRESSED SIC-N

H. Luo
Dept. of Aerospace and Mechanical Engineering
The University of Arizona
1130N. Mountain Ave.
Tucson, AZ 85721

W. Chen
Schools of Aero/Astro. and Materials Engineering
Purdue University
315 N. Grant St.
West Lafayette, IN 47907

ABSTRACT
    The dynamic compressive responses of a hot-pressed silicon carbide, SiC-N, have been determined at various damage levels. We employed a novel dynamic compressive experimental technique modified from a split Hopkinson pressure bar (SHPB) to determine the dynamic properties of ceramics under loading conditions simulating those encountered in ceramic armors subjected to impact, in which a ceramic specimen was loaded by two consecutive stress pulses. The first pulse determines the dynamic response of the intact ceramic material and then crushes the specimen to a desired damage level. The second pulse then determines the dynamic compressive constitutive behavior of the damaged but still interlocked ceramic specimen. The first pulses were slightly varied to control the damage levels in the ceramic specimen while the second pulse was maintained identical. The results show that the compressive strengths of damaged ceramics depend on a critical level of damage, below which the specimen retains its load-bearing capacity.

INTRODUCTION
    For ceramics in armor-system applications, the lack of reliable dynamic material constitutive models remains to be a significant design challenge [1,2]. The local tensile stresses around the grain boundaries, defects, and other inhomogenities, exceed the tensile strength of the ceramic under even moderate compressive loading from a long-rod projectile striking a ceramic armor [3], which causes the ceramic material to be eventually pulverized. Many cracks will propagate/interact simultaneously, forming a comminuted zone around and ahead the tip of the penetrator [4,5,6,7]. The fine but interlocked fragments in ceramic target ahead of the penetrator flow radially around the penetrator's nose and are then ejected backwards along the shank. This process erodes the penetrator until it vanishes or the ceramic is perforated [8]. It is therefore critical to determine the dynamic compressive response of the damaged or pulverized but still interlocked ceramic at high strain rates. The behavior of the damaged AD995 alumina ceramic as a function of strain rate and damage has been explored [9,10,11,12].
    In this paper, we present the dynamic compressive responses of a damaged hot-pressed silicon carbide, SiC-N. The results were obtained using a dynamic experimental technique recently modified from a SHPB that sends two consecutive stress pulses to a ceramic specimen. The first pulse determines the dynamic compressive response of the intact ceramic and then damages the specimen to a desired level through proper pulse shaping. The second pulse

determines the dynamic compressive stress-strain behavior of the damaged ceramic. The damage in the specimen was found to be very sensitive to the loading conditions. Therefore the first loading pulses were controlled to have slightly different profiles to study the damage-induced effects on the compressive response of the ceramic specimen.

EXPERIMENTS

In order to dynamically load the damaged but still interlocked ceramic specimens, we recently modified the SHPB technique such that two strikers with pulse shapers in series generated two consecutive loading pulses. This method has been found to be an effective approach to obtain the dynamic compressive response of damaged ceramics [9,10,11,12]. The first pulse crushes the intact specimen into different damaged rubble after characterizing the intact material. The profiles of loading pulses may be varied to study the damage-induced effects on the compressive response of the ceramic specimen, as well as effects from other parameters such as strain rates. Besides the double-loading-pulse feature in a single SHPB experiment, we also incorporated a second modification to the SHPB in this study. A spherical joint along the loading axis was employed to ensure the proper alignment of the brittle specimen.

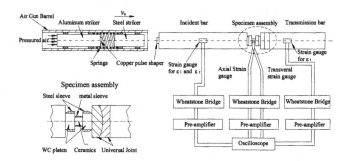

Fig. 1. A schematic illustration of the dynamic experimental setup for loading-reloading.

A schematic of the modified SHPB setup for the loading-reloading dynamic experiments on ceramics is shown in Fig. 1. A spring and a pulse shaper separated the two strikers launched from the gas gun barrel. The first striker was a steel bar. It impacted the incident bar through a B260 annealed brass pulse shaper attached to the impact end of the incident bar. This impact generated a nearly linear ramp pulse in the incident bar, which would load and deform the intact ceramic specimen at a near constant strain rate. The second striker was an aluminum bar. This second striker impacted the incident bar through the first striker and another pulse shaper, which generated the loading pulse to characterize, at a nearly constant strain rate, the dynamic compressive behavior of the ceramic specimen damaged by the first loading pulse. The cross-sectional area of the aluminum bar was varied depending on the desired strain rate in the damaged specimen. A combination of the geometry of the first and second strikers, the striking velocity, and the geometry of two pulse shapers controls the desired strain rates during both loading stages, as well as the damage level in the specimen [13,14]. A spring between the two strikers separated the second loading pulse from the first one at desired intervals. Variations in the different springs, gap between two strikes and a C11000 annealing copper shaping provided various separation time and profile of the second pulse, and ensured the damaged samples to deform under dynamic stress equilibrium at constant strain rates. The dimensions of pulse shapers at various strain rates were selected through trial experiments guided by analytical models [13].

To minimize the stress concentrations in the brittle ceramics due to mismatches in material rigidity [10,15,16,17], a pair of stainless steel-confined tungsten carbide platens were placed between the specimen and the bars, preventing the specimen from indenting into the bars. A simplified spherical universal joint was placed between the tungsten carbide platen and the transmission bar to for a self-adjusting alignment [15]. To prevent the specimen from shattering completely during the first loading, the cylindrical ceramic specimen was slightly confined by a thin-walled metal sleeve on the lateral surface, following an established heated-up procedure [17]. The contribution to the transmitted signal by the thin sleeve was subtracted from the total transmitted signal in the stress calculation. The contribution from all the contacting surfaces was also subtracted from the reflected signal [9,10] in data reduction.

The dimensions of the cylindrical specimen used in this research were 6.35 mm in diameter and 6.35 mm in length. The material studied was Cercom SiC-N with density 3200 kg/m^3, Young's modulus 460 GPa, Poisson's ratio 0.16, and compressive strength 4.9-7 GPa. The thin metal sleeve on the cylindrical surface of the sample was made of a stainless steel with an inner diameter of 6.345 mm and an outer diameter of 7.94 mm, with the same length as the specimens. During an experiment in this study, a digital oscilloscope recorded the strain pulses in the incident and transmission bars. After using the 1-wave, 2-wave method to check the dynamic equilibrium, the data was reduced to obtain the dynamic stress and strain histories in the specimens. The level of damage in the ceramic specimen is an important parameter that affects the mechanical response of the specimen. To examine the damage effects, dynamic experiments were conducted under four different testing conditions, differing in the first pulse profile that controls the "initial damages" before the second pulse arrives. The second pulses were kept approximately identical in the experiments in an effort to load the damaged specimens under identical conditions. The second pulse arrived 60 μs after the first pulse was completely unloaded.

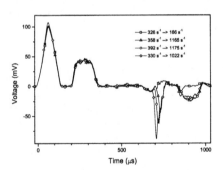

Fig. 2. Incident bar signals of designed to determine the damage effects on SiC-N

Figure 2 shows the incident-bar pulses to generate the four testing conditions. It is clear that the second pulses in Fig. 2 are nearly identical. It is noted that the differences in the first pulses are not significant either. This is because the damage state in the specimen after the first loading is very sensitive to the profile of the first loading. A slight variation in the first pulse can result in significant differences in the damage level in the specimen, as presented later. Figure 3 shows the corresponding dynamic compressive stress-strain curves obtained on SiC-N specimens. During the deformation of the ceramic specimen, the confining pressure from the steel sleeve on the ceramic can initially be neglected when the specimen is under elastic deformation or is damaged only slightly due to the small deformation involved. After substantial damage, the sleeve confines the pulverized specimen at an estimated pressure of 104 MPa through the plastic deformation of the metal sleeve according to the thin-wall pressure vessel theory and a yield strength of 450 MPa for the stainless steel material, which is directly measured by a SHPB experiment on the sleeve material for the correction of the transmitted signals.

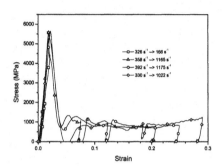

Fig. 3. Dynamic stress-strain curves of damaged SiC-N

As shown in Fig. 3, a specimen initially behaves as a typical brittle material with a linear stress-strain response. It deviates from linear response at a stress of ~ 4.5 GPa and then fails catastrophically near the peak stress of ~5.6 GPa. After failure, the stress sharply decreases to near zero. The slight differences in the first loading pulse mentioned earlier generated different damage level in the ceramic specimens after first loading. One of the four specimens was not completely crushed, corresponding to less damage in the specimen. When the second loading was applied to the damaged specimen, the specimen behaved nearly as if undamaged. The stress-strain curve from the second loading overlapped with that from the first loading. This is shown in Fig. 3, although not very clearly due to the number of curves presented in the figure. This specimen withstood a peak stress of ~2000 MPa due to the second pulse loading. The specimens under other three loading conditions were crushed after first loading. As the axial strains in the crushed samples increase under the compression from the second load pulse, the lateral confinement from the thin metal sleeve maintained a nearly constant axial "flow" stress, as shown in Fig. 3. At large strains beyond 20%, the initially constant flow stress increased with increasing strain, which is considered to be the result of an increased effective cross-sectional area of the specimen at large strains. The results in Fig.3 indicate that when the ceramic specimen is damaged below a critical level, the specimen remains nearly elastic to the second loading along the axial direction. When the specimen is damaged beyond the critical level, the specimen flows under the second loading pulse. No clear variation was observed in the amplitude of the "flow stress" for the completely damaged specimens, which indicates that the remaining strength of the critically damaged ceramic is insensitive to the damage levels anymore.

After the double pulse loading on the ceramics, the specimens were observed to have two post-impact types. In the first type, the comminuted SiC-N ceramic was contained within the severely deformed steel sleeve. In the second type, the damaged ceramics were separated from their confining sleeves. Figure 4 shows a typical post-impact SiC-N specimen with contained rubbles. Figure 5 shows the ceramic rubbles that were separated from its sleeve after dynamic loading. The rubble was collected from a small chamber mounted around the test section of the modified SHPB.

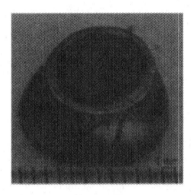

Fig 4. A post-impact SiC-N specimen with contained rubble.

Fig 5. A shattered SiC-N specimen after the double pulse loading.

CONCLUSIONS

Dynamic compressive mechanical responses of damaged SiC-N were determined with a novel dynamic experimental technique modified from a SHPB that loaded the ceramic specimen by two successive stress pulses. The results suggest that there exists a critical damage level. Before the ceramic material is damaged to this level, the ceramic specimen remains its load-baring capacity. When the material is damaged beyond the critical level, the specimen behaves like plastic flow and is insensitive to further damage.

ACKNOWLEDGMENTS
This research is supported by the U.S. Army Research Office (ARO) through a Grant to The University of Arizona (DAAD 19-02-1-0257).

REFERENCES

[1] T. Holmquist and G. Johnson, "Modeling Ceramic Dwell and. Interface Defeat," Ceramic Armor Materials by Design, Ceram. Trans. 134, 309-316 (2002).

[2] T. Holmquist, D. Templeton and K. Bishnoi, "Constitutive Modeling of Aluminum Nitride for Large Strain, High-Strain Rate, and High-Pressure Applications", Int. J. Impact Engng., 25, 211-231 (2001).

[3] M. Ashby and C. Sammis, "The Damage Mechanics of Brittle Solids in Compression," PAGEOPH, 133, 489-521(1990).

[4] D. Shockey, A. Marchand, S. Skaggs, G. Cort and M. Burkett, "Failure Phenomenology of Confined Ceramic Targets and Impacting Rods," Int. J. Impact Engng., 9, 263-275 (1990).

[5] R. Klopp and D. Shockey, "The Strength Behavior of Granulated Silicon Carbide at high Strain Rate and Confining Pressure," J. Appl. Phys., 70, 7318-7326 (1991).

[6] A. Rajendran, "Modeling the Impact Behavior of AD85 Ceramic under Multiaxial Loading," Int. J. Impact Engng., 15, 749-768 (1994).

[7] R. Clifton, "Response of Materials under Dynamic Loading", Int. J. Solids Struct., 37, 105-113 (2000).

[8] S. Sairam and R. Clifton, "Pressure-Shear Impact Investigation of Dynamic Fragmentation and Flow of Ceramics," in Mechanical Testing of Ceramics and Ceramic Composites, ASME, AMD197, 23-40 (1994).

[9] W. Chen and H. Luo, (2004), "Dynamic Compressive Responses of Intact and Damaged Ceramics from a Single Split Hopkinson Pressure Bar Experiment," Exp. Mech., 44, 295-299 (2004).

[10] W. Chen and H. Luo, "Dynamic compressive testing of intact and damaged ceramics," Ceram. Engng. Sci. Proc., 24, 411-416 (2003).

[11] H. Luo and W. Chen, "Dynamic compressive response of intact and damaged AD995 alumina", Int. J. Appl. Ceram. Tech., 1, 254-260 (2004).

[12] H. Luo and W. Chen, "Dilatation of AD995 alumina impacted by two consecutive stress pulse during a SHPB experiment," Proc. of the 2004 Int. Conf. on Comp. Exp. Engng. Sci., Madeira, Portugal, 69-74 (2004).

[13] D. Frew, M. Forrestal and W. Chen, "Pulse-shaping techniques for testing brittle materials with a split Hopkinson pressure bar", Exp. Mech., 42, 93-106 (2002).

[14] D. Frew, M. Forrestal, W. Chen, "A split Hopkinson bar technique to determine compressive stress-strain data for rock materials," Exp. Mech., 41, 40-46. (2001).

[15] Y. Meng and S. Hu, "Some Improvements on Stress Homogeneity for Concrete Test under Impact Compressive Loading," (in Chinese) J. Exp. Mech., 18, 108-112 (2003).

[16] W. Chen and G. Ravichandran, "An Experimental Technique for Impact Dynamic Multiaxial-Compression with Mechanical Confinement," Exp. Mech. 36, 155-158 (1996).

[17] W. Chen and G. Ravichandran, "Dynamic Compressive Behavior of a Glass Ceramic under Lateral Confinement," J. Mech. Phys. Solids, 45, 1303-1328 (1997).

122

EFFECTS OF POROSITY DISTRIBUTION ON THE DYNAMIC BEHAVIOR OF SiC

Samuel R. Martin
US Army Research Laboratory
AMSRD-ARL-WM-TA
APG, MD 21005

Min Zhou
G.W. Woodruff School of Mechanical Engineering
Georgia Institute of Technology
Atlanta, GA 30332-0405

ABSTRACT

In order to understand the microstructural reasons behind variations in ballistic performances, plate impact tests were conducted on two sintered silicon carbides with slightly different microstructures. The materials are referred to as Regular Hexoloy (RH) and Enhanced Hexoloy (EH) here. The porosity distribution in the EH samples had fewer large pores, leading to an 18% increase in flexural strength over that for RH samples. Plate impact experiments were conducted utilizing a VISAR to measure free surface velocities. The Hugoniot Elastic Limit (HEL) and spall strength for each material were determined. The spall strength was measured as a function of impact stress and pulse duration. Results show that the difference in porosity distribution between the EH and RH samples leads to no discernable difference in their HEL values and spall strengths. Both materials demonstrated finite spall strengths under loading above the HEL. Furthermore, the spall strengths were independent of the pulse width and showed a trend similar to that found in other studies on SiC.

INTRODUCTION AND BACKGROUND

For years, the military has sought to use ceramics for armor applications. Hauver et al.[1,2] demonstrated a new mechanism in the defeat of long rod penetrators using ceramics. Under certain impact conditions, instead of penetrating the target, a long rod projectile can be defeated at the surface of a ceramic in a process termed "interface defeat." Lundberg et al.[2] looked further into this phenomenon in a study that examined the velocity range where interface defeat of a long rod penetrator transitioned to a more traditional penetration behavior. He called this the transition velocity interval. A higher transition velocity correlates to increased resistance to penetration and greater erosion of the projectile. In a separate study, Orphal et al.[3-5] performed tests in which penetration depth versus impact velocity were examined using a semi-infinite ceramic target configuration. From these studies, it has been determined that under similar impact conditions and configurations, SiC demonstrated a high transition velocity interval and less total penetration than other ceramics. Overall, SiC has proven to be one of the more attractive options for armor applications; not only due to its good penetration behavior, but also because of its low density and lower relative cost.

In addition to performance variations from ceramic to ceramic, it has also been observed that there can be significant variations even within the same type of ceramic. Minor differences in microstructure drastically effect ceramic properties and ballistic performance. Porosity, density, impurities, glassy phases, and grain morphology are examples of these microstructural attributes. They are directly related to the processing techniques used to take the initial powders to the final product. The interest of this study is to: (i) quantitatively characterize two sintered silicon carbides with slightly different microstructures, (ii) study their response to dynamic loads, and (iii) develop a better fundamental understanding of microstructural influences on dynamic material behavior.

MICROSTRUCTURAL CHARACTERIZATION

The materials used in this study are two variations of a silicon carbide with the trade name Hexoloy SA, which is commercially available through Saint Gobain. Regular Hexoloy (RH) and Enhanced Hexoloy (EH) samples are pressureless sintered and have exactly the same chemistries. The EH samples went through additional powder processing prior to sintering, producing a final product with a slightly different morphology than the RH samples. The samples of each material were characterized microstructurally, including phase morphology, density, elastic wavespeeds, microhardness, fracture toughness, and flexure strength.

Density was measured using Archimedes principle and elastic wavespeeds were obtained through ultrasonic methods. For grain sizes and distributions, micrographs of polished and etched samples were analyzed using ImagePro. This same process was followed on unetched samples to obtain pore sizes and distributions. Fracture toughness was determined according to ASTM Standard 1421 using single edged pre-cracked beam specimens. Flexure strength measurements were conducted following ASTM standard C1161. Hardness measurements followed ASTM Standard C1327 using a 1 kg load for 20 seconds. The results are presented in Table 1.

Table 1: Material Properties

	RH	EH
Density (g/cm^3)	3.152±0.006	3.156±0.005
$C_{longitudinal}$ (km/s)	12.03±0.04	12.13±0.01
C_{shear} (km/s)	7.62±0.03	7.64±0.02
C_{bulk} (km/s)	8.21±0.08	8.33±0.03
Fracture Toughness (MPa√m)	2.61±0.05 (5)	2.53±0.05 (5)
Flexure Strength (MPa)	380±30 (25)	450±40 (22)
Vickers Hardness (GPa)	23.4±0.9 (9)	23.3±1.6 (10)
Avg. Grain Size (μm)	3.44±0.23	3.79±0.11

Figure 1 and Figure 2 show polished micrographs of RH and EH samples, respectively. RH samples had a higher number of large sized pores as can be seen from the figures and quantified by Figure 3, which shows the total pore size frequency distribution of each material. The graph to the right magnifies the larger pore sizes. RH samples had a greater number of both larger and smaller sized pores. The larger pore sizes, above 9 microns, are only seen in samples of RH. In proportion to the total number of pores analyzed, the larger pores are small in number, but they have a disproportionate and adverse effect on flexure strength. Cracks initiate from these sites and, due to the brittle nature of the material, propagate and cause the material to fail.

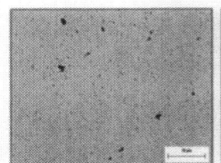
Figure 1. Regular Hexoloy

Figure 2. Enhanced Hexoloy

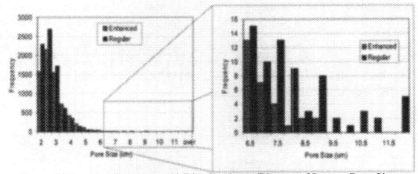

Figure 3. Pore Size Frequency % Distribution w/Blowup of Larger Pore Sizes

For the flexure strength tests, type B four-point bend specimens with loading at quarter points were used. Roughly 20 specimens of each material were loaded to failure. This failure load was entered into the beam equation for simply supported beams and the failure stress was calculated. RH samples had a flexure strength of 380±30 MPa compared to 450±40 MPa for EH samples, an 18% higher value for the latter.

EXPERIMENTAL CONFIGURATIONS
Plate impact experiments were conducted at the Army Research Laboratory (ARL) using a gas gun with a 102 mm bore diameter and 8 meter long barrel. A VISAR (Velocity Interferometer System for Any Reflector [6]) was utilized to measure free surface velocities. Tests were performed on each material to determine the Hugoniot Elastic Limit (HEL) and spall strength. The spall strength was measured as a function of impact stress and pulse duration. The impact stresses ranged from 1-16 GPa.

The spall strength is a measure of dynamic material tensile strength. It is calculated from the pullback velocity drop seen in the free surface velocity profile. The free surface velocity will drop from the Hugoniot state at the end of the compressive pulse, the velocity will then rebound and return to the Hugoniot state (assuming a PMMA or LiF window is not used behind the target) from a reflected compressive wave generated by the newly formed spall plane. Spall strength is calculated as $\sigma_{spall} = 1/2 * \Delta v * Z_{tar}$, where Δv is the pullback velocity and Z_{tar} is the elastic impedance of target.

The spall strength can vary with impact stress and duration of the compressive pulse. The former is believed to be a result of the magnitude of compressive damage; the latter provides information on the time-dependent damage experienced by the material throughout the initial compressive wave. Pulse width is calculated as $pw = 2 * th_{flyer}/c_{L\_flyer}$, where pw is pulse width, th_{flyer} is the thickness of flyer, and c_{L_flyer} is the longitudinal wave speed of the flyer material. Free surface velocity profiles of spall strength experiments for three different impact stress levels are shown in Figure 4.

Grady[7] defines the HEL as "the axial stress at which a solid, loaded in compression under constraint of uniaxial strain, can no longer support elastic distortion and begins to flow through plastic or cataclastic [i.e., crushing] fracture processes." HEL can be seen on free surface velocity graphs and is represented by a slope change on the rise to the Hugoniot state. The free surface velocity at the break in elastic rise is used to calculate the HEL and is calculated as $HEL = 1/2 * v * Z_{tar}$. Free surface velocity profiles of HEL experiments can be seen in Figure 5.

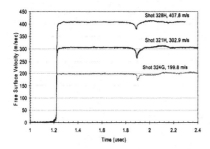

Figure 4. Profiles of EH   Figure 5. HEL Profiles for RH

The plate impact specimens were cylindrical and machined to three different thicknesses: 2.047±0.003 mm, 4.047±0.003 mm, and 6.397±0.003 mm. Each face was flat to within 3 light bands, parallel to within 5 microns of the opposite face, and polished to a mirror finish. All specimens were 31.8±0.03 mm in diameter. These l/d ratios were such that the cylindrical release waves would not interfere with the measurement of spall strength. The diameters also allowed two samples to be impacted side by side in one shot. For the paired shots, two probes and two VISARs were used; one for each sample.

Testing this way had the increased benefit of allowing two tests to be run at exactly the same conditions (i.e. identical striking velocity).

Spall strength experiments were conducted at four different impact stresses (4 GPa, 6 GPa, 8 GPa, and 15 GPa). See Table 2 for a shot summary. Experiments at the first three stresses were conducted using similar flyer and target plate materials. Results were obtained for two different flyer thicknesses, which resulted in two different pulse widths. The pulse width was approximately 340 ns for 2 mm thick flyers and approximately 670 ns for the 4 mm flyers. The tests with impact stresses of approximately 15 GPa were achieved using higher impedance flyer plate materials. K68 was used for two of these tests and WC was used for the other two. K68 is a tungsten carbide with a cobalt binder produced by Kennametal. The thickness of each K68 flyer created pulse widths within 36 ns of the 2 mm SiC pulse width. The WC flyer plates were manufactured by Cercom. The two experiments using WC were for HEL measurements only.

Table 2. Shot Summary

Material	Shot #	Flyer T (mm)	Targ. T (mm)	Impact Vel. (km/sec)	Impact Stress (GPa)	Pullback (m/sec)	Spall (GPa)	HEL (GPa)
EH	323G	2.048	6.392	302.9	5.80	43.3	0.830	-
EH	328G	2.050	6.401	407.8	7.81	34.1	0.653	-
EH	411H	2.050	4.045	200.8	3.84	43.7	0.836	-
EH	321H	4.044	6.398	302.9	5.80	42.2	0.807	-
EH	324G	4.048	6.400	199.8	3.82	26.8	0.513	-
EH	328H	4.045	6.393	407.8	7.81	28.2	0.539	-
EH	408H*	1.289	6.398	601.9	15.30	14.4	0.276	13.9
EH	412H**	3.999	9.854	607.0	15.30	-	-	13.1
RH	323H	2.045	6.397	302.9	5.74	34.6	0.657	-
RH	325G	2.051	6.396	203.4	3.86	32.5	0.615	-
RH	327H	2.043	6.393	406.7	7.71	35.0	0.664	-
RH	403G	2.050	6.402	302.6	5.74	37.6	0.712	-
RH	403H	2.048	6.397	302.6	5.74	35.3	0.669	-
RH	322H	4.049	6.401	302.3	5.73	37.4	0.709	-
RH	325H	4.052	6.397	203.4	3.86	28.6	0.542	-
RH	327G	4.049	6.398	406.7	7.71	32.5	0.616	-
RH	405H*	1.288	6.396	593.7	15.10	15.2	0.288	13.6
RH	410H**	4.002	9.929	598.6	15.10	-	-	13.9
	* K68 flyer							
	** WC flyer							

RESULTS

Spall strengths were plotted as a function of impact stress in Figure 6. The square symbols represent experiments conducted with 2 mm thickness flyers, while the circles were conducted with 4 mm flyers. The solid symbols represent experiments of EH samples and the hollow symbols represent RH samples. The triangular symbols represent the spall tests conducted using K68 flyers. The asterisks in the legend are to point out that the K68 flyers had pulse widths equivalent to that of 2 mm of SiC.

127

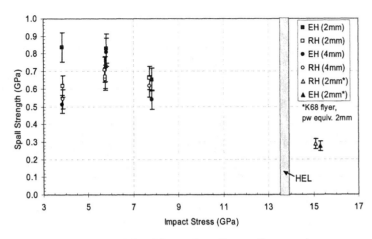

Figure 6. Spall Strength vs. Impact Stress

Within the variability of the tests, there appears to be no difference in spall strength between the two materials. For the most part, the longer pulse width experiments show slightly lower spall strengths. However, within the variability of the tests, the spall strength could be considered pulse width independent.

Experiment 411H appears to be an anomaly. All the other experiments follow a similar trend with little pulse width dependence and little difference between the RH and EH samples. The EH target in experiment 411H has a much higher spall strength than the similar test with the longer pulse width (324G) and than those of RH samples (325G and 325H). The only reasonable explanation is that the result is one of material variability. Dandekar[8] had similar results in his study.

The spall strength results, with the exception of experiment 411H, show similar trends to those found in Dandekar and Bartkowski [9, 10]. Spall strengths increase up to a threshold in impact stress and then show a subsequent decline, as seen in Figure 6. Recently, Dandekar[8] performed simultaneous compression-shear plate impact experiments which supported his original assumption that plasticity dominates deformation below the threshold [8]. He found that above the threshold, deformation mechanisms become much more complex. Crack dynamics become more significant and he detected the presence of a failure wave under induced shear.

Shots 408H* and 405H* were conducted to determine both the HEL and spall strength. The striking velocities were chosen to result in impact stresses above 15 GPa, well above normal values for HEL in SiC (usually 12-14 GPa). These experiments revealed that, at impact stresses above the HEL, the materials retained finite spall strengths.

Shots 403G and 403H replicated the conditions of shot 323H to analyze the repeatability in the spall strength measurements. The results fell within the error bars of the original, see Figure 6. With this said, Shot 0411H seems to be an anomaly in the data since the same test with a shorter pulse width has a lower spall strength. The only reasonable explanation is that the result is one of material variability. Dandekar had similar results in his study [8, 10].

At an impact stress of 15 GPa, RH and EH samples had HEL values of 13.6 GPa and 13.9 GPa, respectively, with a target thickness of 6.4 mm. At similar impact stresses with a target thickness of 9.9 mm, they had HELs of 13.9 GPa and 13.1 GPa, respectively. The decreasing HEL with increased target thickness for the EH sample may appear to be precursor decay. However, it has been found that there is little precursor decay in ceramics. The trend in the RH HEL with thickness cannot be explained. Either it was material variability or the error associated with testing. If averages were taken of the HEL values, the HEL for RH samples would be 13.8 GPa and the HEL for EH samples would be 13.5 GPa. There is no difference in HEL values between the RH and EH samples within the repeatability of the test data.

CONCLUSIONS

It was determined that the difference in porosity distribution between Regular Hexoloy (RH) and Enhanced Hexoloy (EH) samples led to an increased flexure strength in the latter by 18%. RH samples had larger and a higher number of larger sized pores. These acted as large flaws and effectively lowered the flexure strength of RH samples. Plate impact experiments were conducted to examine the influence of porosity distributions on the HEL and spall strengths of these two materials.

Results of these plate impact experiments showed that within the variability of the experimental results:

- There was no discernable difference between the spall strength values for RH and EH samples within the range of impact stresses studied.
- Spall strength was independent of pulse width.
- Experiment 411H appears to be an anomaly, see discussion in Results section.
- Spall strengths, with the exception of 411H, increase to a threshold value between 5-7 GPa and then decline with impact stress.
- Spall strengths showed a similar trend as observed by other researchers for SiC with respect to impact stress.
- Both materials show finite spall strengths above their respective HEL.
- The HEL values for RH samples and EH samples were similar, at 13.8±0.2 GPa and 13.5±0.6 GPa, respectively.

Despite the effect the large pores had on the quasi-static flexure strength levels between RH and EH samples, they had little influence on the material behavior under dynamic loading conditions. The triaxial stresses associated with uniaxial strain conditions seen in the plate impact tests did not allow the bigger pores in RH samples to affect spall strength or HEL. The stresses confine the formation of large cracks from larger pores. Without the stress release from the large cracks, a greater number of cracks may form from smaller pores and flaws. In contrast, the uniaxial stress conditions imposed by the four-point bend strength tests allow the initiation and propagation of large cracks from the large pores in RH samples.

Another possible reason for the lack of difference between the spall strength values of RH and EH samples could be the result of the compressive wave that passes through the target prior to spallation. The damage caused by the compressive loading could be more severe than the damage caused by the large pores. This would explain the similar spall strength values between the two material varieties.

Acknowledgements
The authors would like to thank Dr. Dattatraya Dandekar for reviewing this manuscript and for all his advice and help with the plate impact experiments. The authors would like to thank Dr. Jerry LaSalvia for obtaining the material samples and for his advice on the materials characterization. The authors would also like to thank Tim Cline and Mike Blount for their assistance with the gas gun.

REFERENCES
[1]Hauver, G.E., P.H. Netherwood, B. R.F., and L.J. Kecskes. "Enhanced ballistic performance of ceramic targets." in *Proceedings of the 19th Army Science Conference, USA*. 1994.
[2]Lundberg, P., R. Renstrom, and B. Lundberg, "Impact of metallic projectiles on ceramic targets: transition between interface defeat and penetration," *International Journal of Impact Engineering*, **24**: p. 259-75 (2000).
[3]Orphal, D.L. and R.R. Franzen, "Penetration of Confined Silicon Carbide Targets by Tungsten Long Rods from 1.5 to 4.6 km/s," *International Journal of Impact Engineering*, **19**(1): p. 1-13 (1997).
[4]Orphal, D.L., et al., "Penetration of Confined Boron Carbide Targets by Tungsten Long Rods at Impact Velocities from 1.5 to 5.0 km/s," *International Journal of Impact Engineering*, **19**(1): p. 15-29 (1997).
[5]Orphal, D.L., R.R. Franzen, A.J. Piekutowski, and M.J. Forrestal, "Penetration of Confined Aluminum Nitride Targets by Tungsten Long Rods at 1.5-4.5 km/s," *International Journal of Impact Engineering*, **18**(4): p. 355-68 (1996).
[6]Barker, L.M. and R.E. Hollenbach, "Laser interferometer for measuring high velocities of any reflecting surface," *Journal of Applied Physics*, **43**(11): p. 4669-75 (1972).
[7]Grady, D.E. "Shock-Wave Properties of High-Strength Ceramics." in *Shock Compression of Condensed Matter-1991*. 1991: Elsevier Science.
[8]Dandekar, D.P., "Spall Strength of Silicon Carbide Under Normal and Simultaneous Compression-Shear Shock Wave Loading," *International Journal of Applied Ceramic Technology*, **1**(3): p. 261-68 (2004).
[9]Dandekar, D.P. and P. Bartkowski, *Spall strengths of silcon carbides under shock loading*, in *Fundamental Issues and Applications of Shock-Wave and High Strain-Rate Phenomena*, K.P. Staudhammer, L.E. Murr, and M.A. Meyers, Editors. 2001, Elsevier Science Ltd. p. 71-77.
[10]Dandekar, D.P. and P. Bartkowski, "Tensile Strengths of Silicon Carbide (SiC) Under Shock Loading," ARL-TR-2430: p. 38 (2001).

EFFECT OF ROOM-TEMPERATURE HARDNESS AND TOUGHNESS ON THE BALLISTIC PERFORMANCE OF SiC-BASED CERAMICS

Darin Ray, R. Marc Flinders, Angela Anderson, and Raymond A. Cutler
Ceramatec, Inc.
2425 South 900 West
Salt Lake City, Utah, 84119

William Rafaniello
Consultant
1715 Sylvan Lane
Midland, Michigan

ABSTRACT
Depth of penetration (DOP) testing using 7.62 mm x 51 mm M993 (WC core) simulants shot at a velocity of 907±4 m/s at 6.35 mm thick ceramic targets backed with 130 mm thick 5083 Al was used to evaluate six SiC materials having a wide range of room temperature properties. DOP values correlated well with hardness, with Knoop measurements being the best indicator of depth of penetration for the six materials tested. The DOP of solid-state sintered SiC was comparable to the commercial SiC-B material despite having more porosity and lower flexure strength and fracture toughness. The fracture mode of the two materials was very different in static testing, yet both materials produced similar diameter holes in the Al during penetration. V_{50} testing of selected samples using 14.5 mm BS-41 simulants showed that DOP results were a poor predictor of performance for 12.5 mm thick SiC adhered to 25.4 mm 5083 Al. The V_{50} of a SiC-YAG composition was similar to the reported SiC-B value, despite having three times the depth of penetration.

INTRODUCTION
Ceramic armor has been under development for the past forty years[1], with early recognition that low specific gravity ceramics[2,3] could potentially be used to defend against armor piercing bullets and long-rod penetrators.[4] Early work showed that hardness and the Hugoniot elastic limit[2,5] were important parameters for identifying lightweight materials with desirable properties. Due to the amorphization of B_4C at high pressures[6,7], nearly fully-dense SiC with intergranular failure[8,9] is the material of choice for heavy threats. The current state-of-the-art ceramic is produced by Cercom under the trade name of SiC-N. It has lower hardness than solid-state SiC sintered with B and C, but has improved fracture toughness.[10,11] LaSalvia has suggested that further improvements can be made if higher short-crack toughness can be obtained without decreasing hardness.[12]

Liquid-phase sintered SiC, using either yttrium aluminates[13,14] or Al, B, and C[15] as additives are two alternatives for obtaining high toughness in SiC-based ceramics. Single-edged precracked beam (SEPB) toughness[16] values in excess of 6 MPa-m$^{1/2}$ can be obtained in either the YAG-SiC or ABC-SiC systems, though this comes at the cost of decreasing the hardness of the material.

The purpose of this paper is to report on ballistic testing of six SiC-based ceramics with different hardness–toughness relationships using depth-of-penetration (DOP) testing[17] as a guide to assess the more realistic protection provided by armor on thin backing[18]. The V_{50} of two high-toughness SiC-based ceramics were assessed using conditions identical to those used for SiC-B, a predecessor of SiC-N, and solid-state sintered SiC.[19]

EXPERIMENTAL PROCEDURES

The six materials chosen for this study are listed in Table I. SiC-B, a predecessor of SiC-N, was supplied by the Army Research Laboratory (ARL). The solid state composition (SiC-0.6 wt. % B-2 wt. % C) and two ABC-SiC compositions (SiC- x wt. % Al-0.6 wt. % B-2 wt. % C, where x was either 1 or 6) were prepared in accordance with the method described by Cao, et al.[15] Al (Valimet grade H-3), B (Starck grade S-432), C (Apiezon grade W assuming a 50 % yield after pyrolysis), and β-SiC (Superior Graphite grade HSC-059) were ball milled for 16 hours in reagent grade toluene using Y-TZP spherical media. The powders were stir dried, pyrolyzed by heating in N_2 to 600°C, and then screened through a 44 µm screen prior to hot pressing at 28 MPa inside graphite dies in stagnant Ar as described in Table I.

Alpha-SiC (Starck grade UF-15) with 1.7 wt. % Al and 2.0 wt. % Y_2O_3 (Molycorp grade 5600) additives were prepared by milling using WC-Co media, with a ball to charge ratio of 10:1, in isopropanol for 24 hours. The powders were sieved through a 44 µm screen before hot pressing at 28 MPa in stagnant Ar under conditions as described in Table I.

The hot pressed billets were ground with a 320 grit diamond wheel to make 3 mm x 4 mm x 45 mm bars as specified by ASTM C-1421-99.[2] Toughness was measured using the single-edge precracked beam (SEPB) technique[16] with details as given previously.[10] All crack planes were parallel to the hot pressing direction. Each data point is the mean of 4-7 bars tested, with error bars representing one standard deviation.

A microhardness machine (Leco model LM-100) was used to obtain Vickers and Knoop hardness data on polished SEPB bars. Data were taken at a load of 9.8 N. Each data point represents the mean of ten measurements, with error bars representing the standard deviation.

Rietveld analysis[20,21] was used to determine SiC polytypes present in the densified samples with X-ray diffraction patterns collected from 20-80° 2-theta, with a step size of 0.02°/step and a counting time of 4 sec/step.

Polished samples of SiC densified with a liquid phase were plasma-etched by evacuating and back-filling with 400 millitorr of CF_4-10% O_2 and etching for 20-40 minutes. Solid state sintered materials were etched in molten KOH at 550°C for 10-15 seconds. Grain size was determined by the line-intercept method, where the multiplication constant ranged between 1.5 (equiaxed grains) and 2.0 (elongated, plate-shaped grains).[22] Typically, 200-300 grains were

Table I

Materials Evaluated

Designation	Composition (wt. %)	Processing Conditions Temp (°C)	Time (h)
Solid State	SiC-0.6 % B-2 % C	2150	0.75
Low-Toughness ABC	SiC-1.0 % Al-0.6 % B-2 % C	2100	1.0
High-Toughness ABC	SiC-6.0 % Al-0.6 % B-2 % C	1900	1.0
SiC-B	Proprietary (Cercom)	Proprietary (Cercom)	
Low-Toughness YAG	SiC-2.0 % Y_2O_3-1.7 % Al	1900	1.0
High-Toughness YAG	SiC-2.0 % Y_2O_3-1.7 % Al	2150	3.0

132

measured for each composition in order to get a mean grain size. The aspect ratios of the five most acicular grains in each of three micrographs were used to estimate a comparative aspect ratio.

The fracture mode was determined from broken SEPB bars. The precracked SEPB bars were subsequently etched to get a quantitative estimate of the fracture mode by viewing the crack path over a distance of 100-500 μm, depending on grain size.

Flexural strength was measured on 15 bars (3 mm x 4 mm x 45 mm) using a 40 mm support span and a 20 mm loading span, with the crosshead speed at 0.5 mm/min. A two-parameter Weibull analysis was used to calculate the characteristic strength. Young's modulus was measured in flexure using strain gages.

Ballistic testing was performed at H. P. White Co. (Street, Maryland). Samples for armor testing were mounted on armor-grade Al (alloy 5083-H131 supplied by Clifton Steel Co.) using a urethane adhesive (Uralite 3548 from H. B. Fuller Co.) with a nominal thickness of the elastomer of 25 μm. The Al was milled flat and parallel prior to mounting the ceramic targets.

Depth-of-penetration (DOP) testing was conducted on all six materials, in accordance with the methodology proposed by Moynihan, et al.[17] using 7.62 mm x 51 mm M993 (WC-Co core) simulant shot at a velocity of 907±4 m/s (2975±14 ft/s) at ceramic targets 100 mm round (five experimental materials supplied by Ceramatec, Inc.) or 100 mm square (SiC-B supplied by ARL). The SiC samples were 6.35 mm thick, backed with 130 mm thick 5083 Al (150 mm x 150 mm x 130 mm). The depth of penetration of the projectile into the Al is compared with and without the ceramic target. Depth gauges and radiography were used to measure the depth of penetration into the Al backing plate. Each data point represents between 3 and 12 measurements.

V_{50} testing was performed as described by Lillo, et al.[19] on two of the six materials (high-toughness ABC (SiC-6 wt. % Al-2 wt. % C-0.6 wt. % B) and high toughness YAG (SiC-2 wt. % Y_2O_3-1.7 wt. % Al) compositions). A 14.5 mm diameter BS-41 (WC-Co core) simulant was shot at velocities ranging from 640 to 752 m/s at ceramic targets (100 mm diameter by 12.7 mm thick) backed with 25.4 mm thick 5083 Al (150 mm x 150 mm x 25.4 mm). The V_{50}, in theory, is the velocity of the bullet at which the probability of the projectile penetrating through the aluminum backing plate is 50%. In practice, this value is taken as mean of the two highest velocity tests at which the bullet does not fully penetrate the aluminum backing and the two lowest velocity tests at which the bullet fully penetrates the backing.

RESULTS AND DISCUSSION

Materials Characterization

Table II gives grain size, density, aspect ratio, and polytypes present for the six materials subjected to DOP testing. From estimates of theoretical density for each composition, percent closed porosity was calculated. Closed porosity in the parts ranged from a low of 0.3 % for the SiC-B sample to a high of 1.9 % for the solid-state SiC material. Early work on pressureless sintered SiC showed the strong dependency of Young's modulus on porosity.[23] The higher modulus of the SiC-B relative to the other materials is more related to low porosity than to secondary phases present. All of the materials were predominantly 6H after densification, but significant differences in polytypes were evident. Mean grain sizes varied by a factor of 8 between the finest (low-toughness YAG) and the coarsest (solid-state) materials. The large aspect ratio of the high-toughness ABC is apparent in the microstructures shown in Figure 1.

Table II
Microstructural Characterization and Young's Modulus Data

Designation	Density (g/cc)	Polytypes Present 3C	4H	6H	15R	Grain Size (µm)	Aspect Ratio	E (GPa)
Solid State	3.15±0.01	0.0	0.0	95.9	4.1	6.6±0.6	6.2±1.6	401
Low-Toughness ABC	3.15±0.01	0.0	18.8	68.3	12.9	4.0±0.2	4.3±1.2	401
High-Toughness ABC	3.14±0.01	0.0	21.4	62.4	16.2	2.7±0.3	8.5±1.9	395
SiC-B	3.21±0.01	0.0	6.2	85.3	8.5	4.1±0.2	2.5±0.4	448
Low-Toughness YAG	3.25±0.01	8.9	8.9	74.9	7.3	0.8±0.1	2.3±0.4	418
High-Toughness YAG	3.22±0.01	0.0	41.7	54.7	3.6	3.7±0.2	4.0±1.5	418

Strength, % intergranular fracture, fracture toughness, and hardness are listed in Table III. The fracture mode is reflected by the photomicrographs of fracture surfaces (see Figure 2) and Vickers indents (see Figure 3).

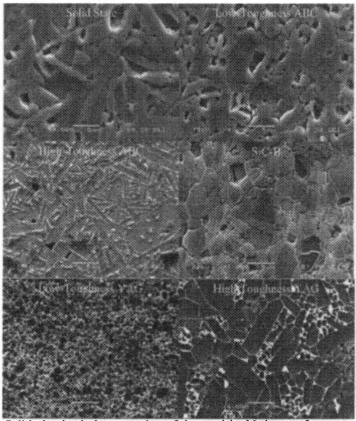

Figure 1. Polished and etched cross-sections of six materials. Markers are 5 µm.

Table III
Mechanical Characterization

Designation	Strength (MPa)	% Intergranular Fracture	Toughness (MPa-m$^{1/2}$)	Hardness (GPa) HK1	HV1
Solid State	352±81	11	2.5±0.1	20.1±0.3	24.1±1.0
Low-Toughness ABC	397±53	30	2.8±0.1	18.5±0.3	21.5±0.7
High-Toughness ABC	649±50	84	8.1±0.4	17.1±0.4	19.7±0.4
SiC-B	615±45	89	5.0±0.5	20.0±0.3	20.6±1.2
Low-Toughness YAG	507±39	93	3.9±0.1	18.1±0.3	21.4±0.4
High-Toughness YAG	466±31	74	5.6±0.5	17.5±0.5	21.9±1.2

Figure 2. Fracture surfaces of six materials. Markers are 5 μm.

135

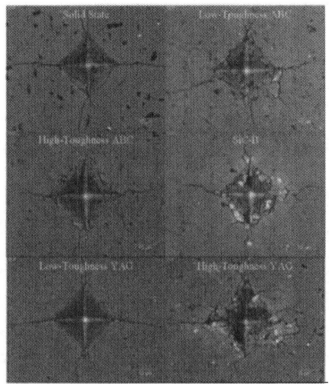

Figure 3. Vickers indents at one-kilogram load showing tendency for intergranular fracture.

The Weibull moduli varied from a low of 5 for the solid state SiC to a high of 18 for the high-toughness YAG composition. The strength and Weibull modulus of the high-toughness ABC composition was comparable to SiC-B despite having more porosity. Since the porosity in the solid-state, ABC materials and high-toughness YAG material were all similar, the effect of higher fracture toughness contributing to higher strength is evident for these materials. If the flaw size were influenced by the grain size, d, then one would expect that strength to be proportional to $K_{Ic}/d^{1/2}$. The trend is shown in Figure 4, but the slope (5.6×10^{-2}) is shallower than would be expected with the crack geometry used. High toughness combined with small flaw size increased strength, as expected.

The Knoop hardness values at one-kilogram loads were similar for the SiC-B and solid-state SiC despite a factor of two difference in toughness, as shown in Figure 5. Vicker's hardness data gave slightly different hardness values. The high-toughness ABC composition had the lowest hardness regardless of the measurement technique.

Ballistic Testing
    To establish a baseline for the M993 simulants, unprotected Al blocks were penetrated 60.4±0.4 mm when shot at 907 m/s. Moynihan, et al.[17] shot APM2 steel-cored bullets into the

136

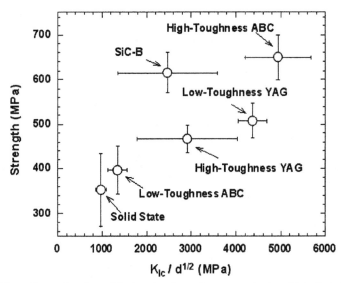

Figure 4. Strength as a function of fracture toughness/inverse grain size. Note that strength generally increases as toughness/inverse grain size increases.

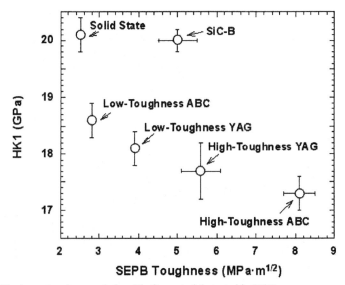

Figure 5. Hardness-toughness relationship for materials tested in DOP.

same type of aluminum at velocities ranging between 450 and 900 m/s and gave an empirical fit to their penetration data as a function of strike velocity. Using their fit at 907 m/s, one would have expected the steel bullets to have penetrated 53.4 mm into the Al. The WC-Co bullets penetrated 13 % further than their steel counterparts.

Since all of the SiC materials have similar densities and thicknesses, their areal densities all ranged between 19.9 to 20.6 kg/m^2. SiC-B when tested with the APM2 round at 841 m/s showed 6.7 mm penetration at an areal density of 12.6 kg/m^2 and no penetration at 16.8 kg/m^2.[17] As shown in Figure 6, SiC-B had penetration of 4.4±1.2 mm at an areal density of 20.4 kg/m^2 when using WC-Co bullets of similar size at 907 m/s. Higher penetration of the WC-Co bullets was desired in order to better resolve differences between materials.

There was considerable scatter in the DOP results for the two ABC-SiC materials. This scatter was due to one of the results being very different from the other two tests. The reason for the inconsistent behavior of these samples is unknown.

Depth of penetration for the six materials was plotted as a function of porosity, grain size, aspect ratio, strength, Weibull modulus, and Young's modulus and showed no correlation with any of these variables. Depth of penetration, however, was a strong function of hardness, with Knoop hardness showing a better correlation than Vicker's hardness when both were compared at a one-kilogram load.

Fracture toughness was poorly correlated with depth of penetration as seen by the identical performance of the solid-state SiC and SiC-B, which vary in toughness by a factor of two. Krell and Straussburger[24] suggested a linear relationship between ballistic efficiency and hardness of their alumina samples. While it can be debated whether a linear or non-linear fit is

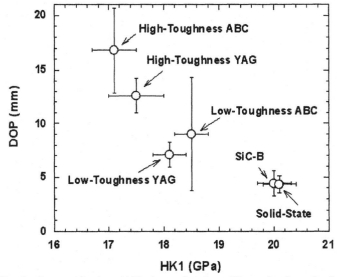

Figure 6. Depth of penetration into Al block as a function of Knoop hardness for the SiC armor. The WC-Co projectile (7.62 mm diameter by 51 mm long M993 shot at 907 m/s) had a hardness of 14.1 GPa. Note that depth of penetration in this test is controlled by the hardness of the SiC.

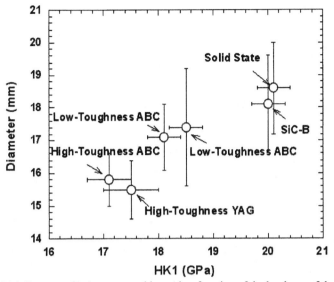

Figure 7. Initial diameter of hole penetrated into Al as function of the hardness of the SiC target. Compare with Figure 6.

better, it is very clear that depth of penetration in this test is controlled by the hardness of the SiC in spite of the fact that all of the materials were at least 3 GPa harder than the WC-Co bullets.

Another way to compare this ballistic data is to measure the initial diameter of the hole in Al backing plate (see Figure 7). By comparing the data in Figures 6 and 7 it is easy to see that when the depth of penetration is shallow, the harder ceramic has spread the bullet fragments and ceramic debris that penetrates into the Al compared to the smaller initial diameter when a softer ceramic is used. The fact that solid-state and SiC-B materials give similar diameter holes is interesting in light of their fracture mode. The solid state material fractures transgranularly and has low toughness since it has no contribution from crack bridging. The SiC-B material, on the other hand, has predominantly intergranular fracture and could potentially generate very different shaped rubble at the tip of a high-velocity penetrator.[9]

Depth of penetration testing is used as a screening test for determining which materials are worthy of more extensive evaluation. The mass efficiency, E_m, is defined as:

$$E_m = \frac{\rho_{Al} P_{Al}}{\rho_{SiC} t_{SiC} + \rho_{Al} DOP} \qquad (1)$$

where ρ_{Al} and ρ_{SiC} are the material densities, P_{Al} is the penetration depth into the unprotected Al target, t_{SiC} is thickness of the SiC target, and DOP is depth of penetration into Al after striking the SiC target. The mass efficiency increases with decreasing DOP for a given set of materials and SiC thickness, until the ceramic completely defeats the projectile. V_{50} test data were believed to be more representative of actual armor due to the thinner backing plate and the ability of the armor package to flex upon impact. Table IV compares the results of ballistic testing for DOP and V_{50} performance. Surprisingly, the high-toughness ABC and high-toughness YAG

139

Table IV
Ballistic Test Data

Designation	DOP at 907 m/s with M993		V_{50} with BS-41
	Depth (mm)	Mass Efficiency	(m/s)
Solid State	4.3±0.8	5.2±0.3	728[19]
Low-Toughness ABC	9.0±5.3	3.9±1.1	Not measured
High-Toughness ABC	16.8±3.9	2.5±0.4	722
SiC-B	4.4±1.2	5.1±0.5	750[19]
Low-Toughness YAG	7.1±1.2	4.1±0.3	Not measured
High-Toughness YAG	12.6±1.6	3.0±0.3	740

compositions, which performed very poorly in depth-of-penetration testing, had V_{50} values between those of the solid-state SiC and SiC-B materials. While it should be noted that the solid-state material tested by Lillo, et al.[19] is not the same material that was used in this study, measured hardness values (HK1=20.5±0.3 GPa and HV1=24.9±0.4 GPa) suggest that similar, or slightly lower, depth of penetration would be expected. These results would indicate that DOP testing was a very poor screening test since it only was dependent on the material's hardness. The mass efficiency of the SiC-B was 70 % higher than the high-toughness YAG composition in DOP testing, but V_{50} values for the two materials varied by only 1 %. The V_{50} results would be highly correlated with Young's modulus for these SiC materials if the solid state material tested by Lillo, et al.[19] had the same value of Young's modulus as this study's solid state material (see Table II).

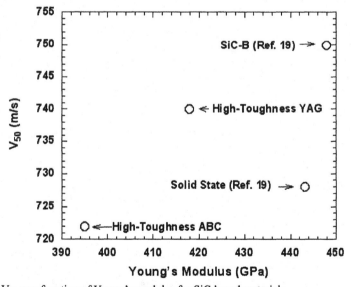

Figure 8. V_{50} as a function of Young's modulus for SiC-based materials.

Measurement of E for their material, however, showed that low porosity (i.e., high modulus) is not the only criterion that is important for obtaining a high V_{50} as shown by the data in Figure 8. The V_{50} values are lower than the muzzle velocity for the BS-41 bullets, suggesting that testing higher areal density targets would give additional insight to the performance of SiC materials. It is likely that such information would require testing at higher threats and lead to the classification of test results. The lack of readily available ballistic information makes the development of improved SiC materials difficult. It is not apparent from these tests that higher toughness materials, at the same hardness, would give substantial improvement in V_{50} performance. It is very important to determine the armor applications and corresponding ballistic tests that need to be performed to guide effective materials development work.

CONCLUSIONS

Knoop hardness, at a one-kilogram load, was a good predictor of the depth of penetration of WC-Co cored bullets (7.62 mm diameter x 51 mm long M993 simulants) into Al block after impacting 6.35 mm thick SiC. Depths of penetration ranged between 4 mm for materials with HK1=20 GPa to 17 mm when HK1=17 GPa. The depth-of-penetration screening test, however, was a poor indicator of the V_{50} performance of the same SiC materials at twice the areal density when exposed to larger WC-Co cored bullets (14.5 mm BS-41 simulants). The materials with the highest and lowest DOP varied by less than 4 % in their V_{50} values. SiC materials densified with either an Al-B-C liquid or a YAG liquid performed similarly in V_{50} testing compared to solid state (SiC sintered with B and C without a liquid phase) or SiC-B, a material densified using a small amount of AlN additive.[8] Room-temperature hardness and toughness had little effect on their V_{50} performance, with all materials ranging between 722 and 750 m/s. While high modulus appears to be beneficial for obtaining a high V_{50}, there are certainly other compounding factors that influence the performance of the SiC.

Key insight into what is required to make improved SiC materials is still lacking. Testing at higher threat levels is necessary to differentiate between SiC materials. Increasing the toughness of SiC does not appear to make improved armor, particularly when hardness is compromised in these materials.

ACKNOWLEDGEMENT

This SBIR work was performed for the U.S. Army under contract DAAD17-02-C-0052. Appreciation is expressed to Lyle Miller of Ceramatec for help with x-ray diffraction work. Technical discussions with Jane Adams (ARL), Ernie Chin (ARL), Jerry LaSalvia (ARL), Jim McCauley (ARL), and Svante Prochazka (retired GE), are gratefully acknowledged.

REFERENCES

[1]S. R. Skaggs, "A Brief History of Ceramic Armor Development," *Ceram. Sci. and Eng. Proc.*, **24**[3] 337-49 (2003).

[2]M. L. Wilkins, R. L. Landingham, and C. A. Honodel, "Fifth Progress Report of Light Armor Program," Lawrence Livermore Report UCRL-50980 (1971).

[3]M. L. Wilkins, "Use of Boron Compounds in Lightweight Armor," pp 633-48 in *Boron Refract. Borides*, ed. V. I. Matkovich (Sprinder, Berlin, 1977).

[4]D. M. Norris, V. H. McMaster and M. L. Wilkins, "Long-Rod Projectiles Against Oblique Targets: Analysis and Design Recommendations," Lawrence Livermore Report UCRL-52057 (1976).

[5]W. H. Gust, A. C. Holt and E. B. Royce, "Dynamic Yield, Compressional, and Elastic Parameters for Several Lightweight Intermetallic Compounds," *J. Appl. Phys.* **44**[2] 550-60 (1973).

[6]N. K. Bourne, "Shock Induced Brittle Failure of Boron Carbide," *Proc.. Royal Soc. Lon. A*, **458** 1999-2006 (2002).

[7]M. Chen, J. W. McCauley, and K. J. Hemker, "Shock-Induced Localized Amorphization in Boron Carbide," *Science*, **299** 1563-66 (2003).

[8]A. Ezis, "Monolithic, Fully Dense Silicon Carbide Material, Method of Manufacturing, and End Uses," U. S. Patent 5,372,978 (Dec. 13, 1994).

[9]J. Lankford, Jr., "The Role of Dynamic Material Properties in the Performance of Ceramic Armor," *Int. J. Appl. Ceram. Technol.*, **1**[3] 205-10 (2004).

[10]D. Ray, M. Flinders, A. Anderson, and R. A. Cutler, "Hardness/Toughness Relationship for SiC Armor," *Ceram. Sci. and Eng. Proc.*, **24** 401-10 (2003).

[11]M. Flinders, D. Ray and R. A. Cutler, "Toughness-Hardness Trade-Off in Advanced SiC Armor," *Ceram. Trans.*, **151** 37-48 (2003).

[12]J. C. LaSalvia, "A Physically-Based Model for the Effect of Microstructure and Mechanical Properties on Ballistic Performance," *Ceram. Sci. and Eng. Proc.*, **23**[3] 213-20 (2002).

[13]R. A. Cutler and T. B. Jackson, 'Liquid Phase Sintered Silicon Carbide," pp. 309-318A in *Third International Symposium on Ceramic Materials and Components for Engines*," ed. by V. J. Tennery (Am. Ceram. Soc., Westerville, OH. 1989).

[14]K. Y Chia and S. K. Lau, "High-Toughness Silicon Carbide," *Ceram. Eng. Sci. Proc.*, **12**, 1845-61 (1991).

[15]J. J. Cao, W. J. Moberly Chan, L. C. DeJonghe, C. J. Gilbert, and R. O. Ritchie, "In Situ Toughened Silicon Carbide with Al-B-C Additions," *J. Am. Ceram. Soc.*, **79**[2] 461-69 (1996).

[16]ASTM C 1421-99, Standard Test Methods for Determination of Fracture Toughness of Advanced Ceramics at Ambient Temperature, pp. 641-672 in *1999 Annual Book of Standards* (ASTM, Philadelphia, PA 1999).

[17]T. J. Moynihan, S. C. Chou and A. L. Mihalcin, "Application of the Depth-of-Penetration Test Methodology to Characterize Ceramics for Personnel Protection," ARL-TR-2219 (April 2000).

[18]M. J. Normandia and W. A. Gooch, "An Overview of Ballistic Testing Methods of Ceramics Materials," *Ceram. Trans.*, **134**, 113-38 (2002).

[19]T. M. Lillo, H. S. Chu, D. W. Bailey, W. M. Harrison, and D. A. Laughton, "Development of a Pressureless Sintered Silicon Carbide Monolith and Special-Shaped Silicon Carbide Whisker-Reinforced Silicon Carbide Matrix Composite for Lightweight Armor Applications," *Ceram. Eng. Sci. Proc.*, **24**[3] 359-64 (2003).

[20]H. M. Rietveld, "A Profile Refinement Method in Neutron and Magnetic Structures," *J. Appl. Crystallogr.*, **2**, 65-71 (1969).

[21]D. L. Bish and S. A. Howard, "Quantitative Phase Analysis Using the Rietveld Method," *J. Appl. Crystallogr.*, **21**, 86-91 (1988).

[22]E. E. Underwood, *Quantitative Stereology*, (Addison-Wesley, Reading, MA. 1970).

[23]R. A. Giddings, C. A. Johnson, S. Prochazka, and R. J. Charles, "Fabrication and Properties of Sintered Silicon Carbide," GE Report 75CRD060 (April 1975).

[24]A. Krell and E. Strassburger, "High Purity Submicron α-Al_2O_3 Armor Ceramics: Design, Manufacture, and Ballistic Performance," *Ceram. Trans.*, **134**, 463-71 (2002).

THE PENETRATION OF ARMOUR PIERCING PROJECTILES THROUGH REACTION BONDED CERAMICS

P J Hazell and S E Donoghue
Cranfield University
RMCS, Shrivenham
Oxfordshire, SN6 8LA, UK

C J Roberson
ADML
Sir Frank Whittle Business Centre
Butlers Leap, Rugby
Warwickshire, CV21 3XH, UK

P L Gotts
DC R&PS
DLO Caversfield
Bicester
Oxfordshire, OX27 8TS, UK

ABSTRACT

Reaction bonded ceramics can provide a viable option in protecting against steel cored ammunition such as the 7.62 x 54R mm B32 API as defined by STANAG 4569 Level 3. However, against the other type of round defined by the standardized agreement, namely the WC-Co cored 7.62 x 51 mm AP round, these ceramics are relatively ineffective for thicknesses of less than 9mm. This paper compares the performance of two reaction bonded silicon carbides and a boron carbide at increasing thickness values within the range 4-10mm to evaluate their performance and to examine the failure mechanisms of the ceramics and the projectile. This paper will discuss the relative performance of these materials against a WC-Co penetrator in light of their microstructural features.

INTRODUCTION

The reaction bonded process for the manufacture of ceramics has been around since the 1950's. The main advantages of the process is that relatively inexpensive forms of the ceramic can be manufactured to controlled tolerances and relatively low processing temperatures and no pressure are required; this reduces the cost of the final product. The main disadvantages of these ceramics for armour applications are the presence of porosity and the presence of a relatively large amount (~10-20%) of un-reacted silicon that can be a source of failure during impact and penetration. Nevertheless, these types of materials have been successfully developed for armour applications to protect against ball type ammunition and some types of harder cored amour piercing ammunition (for example, see reference [1]).

The 7.62mm AP round is now a well established munition widely fielded by several Western armies. It is also the armour defining direct fire threat for Level 3 of STANAG 4569, the standard that defines the protection levels for logistic and light armoured vehicles[2]. This paper investigates the ballistic performance of three reaction bonded ceramics: two silicon carbides and one boron carbide, to assess their ability to resist penetration against the 7.62mm threats required for Level 3 protection.

EXPERIMENTAL
The Depth of Penetration (Dop) technique as described by Rozenberg and Yeshurun[3] was used to measure the ballistic performance of the ceramic tiles. For the witness material backing, a common engineering aluminium alloy Al 6082 T651 was used (YS=250MPa). The test backing plates were 50.8 × 50.8mm pieces cut from a single 25 mm thick plate. For each ceramic tile of specific thickness (t_c), a single bullet was fired at the target and the residual penetration (P_r) into the aluminium alloy was measured (see Figure 1); at least three experiments were done for each tile thickness.

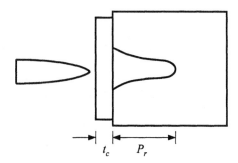

Figure 1: DOP technique for assessing each ceramic's ballistic performance.

Three ceramic materials were compared. Two reaction bonded silicon carbides: Schunk Ingenierkeramik's CarSIK NT manufactured by the uniaxial pressing method and W. Haldenwanger Technische Keramik's Halsic-I, and a reaction bonded boron carbide manufactured by M CUBED Technologies Inc.: RBBC-751. Some properties of the ceramics are provided below in Table I. Each ceramic tile was glued to the aluminium alloy backing block using Araldite 2015. This was applied to the mating surfaces and then the ceramic and aluminium block were pushed together and twisted / oscillated until an even thin adhesive line had been achieved with no gaps or obvious air inclusions.

Table I.

	Schunk CarSIK NT	Haldenwanger Halsic-I	M CUBED RBBC-751
Si content (vol%)	9.5**	10.0**	16.0*
Density** (Kg/m³)	3085	3056	2578
Hardness** (VHN)	2530	2298	2818
E (GPa)*	360	370	380

* Manufacturer's data.
**Measured data; hardness measured using an Indentec HWDM7 Digital Micro Hardness Machine; Si content acquired through X-Ray diffraction.

144

The range set up was one of a fixed test barrel mounted ten metres from the target. Bullet velocity was measured using the normal sight-screen arrangement just before the target. The test ammunition was 7.62 × 51 mm NATO FFV and the 7.62 × 54 R mm B32 was used as factory loaded and generated a mean velocity of 973m/s and 830m/s respectively. The FFV bullet core consists of tungsten carbide core with a composition by percentage weight C 5.2, W 82.6, Co 10.5, Fe 0.41[4] and of hardness 1550Hv (300gf); ρ = 14.8g/cc. The core is mounted in a low carbon steel jacket with gilding metal, on an aluminium cup. The B32 round consists of a hard steel core of hardness 920Hv; ρ = 7.8g/cc, gilding metal and incendiary composition in the nose (see Figure 2 below).

Figure 2: Steel core from the B32 bullet (left) and WC-Co core from the FFV bullet (right).

The test jig was firmly clamped to a test fixture adjustable for height and lateral position and axially aligned with the direction of shot. The jig position was accurately adjusted to ensure that the centre of the target block corresponded with the centre of the shot-line; the jig used engineering vee-blocks as clamping elements. Each of the samples was clamped in place in turn with the ceramic sample protruding out of the front of the clamps. Behind the sample, in the vee-blocks, were three more of the 25 mm blocks of aluminium giving a possible total DoP of 100 mm – effectively semi infinite for the purposes of the test ammunition. The depth of penetration into this aluminium alloy without any ceramic on the front was 75±1.5mm for the FFV round and 62±2.0mm for the B32 round.

After testing the aluminium alloy blocks were X-rayed which allowed the residual penetration to be accurately measured. Furthermore the level of fragmentation of the core and the overall shape of the penetration crater was assessed from the X-rays.

The DOP measurements taken from the X-Rays are presented below in Figure 3 (B32 results) and Figure 4 (FFV results). The mean DOP for each tile thickness is presented ± one standard deviation. A polynomial trend line is fitted through the M CUBED data in Figure 4 to separate them from the RB SiC data.

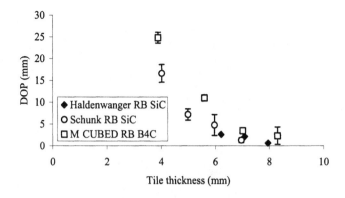

Figure 3: DOP measurements after complete penetration by the B32 round.

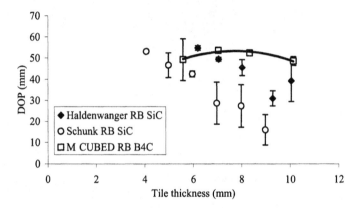

Figure 4: DOP measurements after complete penetration by the FFV round

DISCUSSION

It is clear that the M CUBED reaction bonded boron carbide is not effective against the WC-Co cored round and only just out performs the aluminium plate alone. However, against the B32 bullet the DOP drops off considerably between 4 and 8mm. These observations are consistent with work that has been done on PAD and HP B4C where it has been observed that the ceramic does not perform as well as would be expected with the WC-Co bullet considering the material's hardness value[5,6]. This is likely to be as a result of the significantly higher shock stress that the WC-Co round imparts in the ceramic on impact and the resulting reduction in ceramic strength that ensues. Little effect on the tip of the WC-Co projectile is seen.

Unfortunately, Haldenwanger samples thinner than 6mm were not available at the time of the firing trials nevertheless it appears from the results that the performance of this ceramic against the B32 round is comparable to the Schunk ceramic for thicknesses 6-8mm. Both the reaction bonded silicon carbides provided an increasing level of resistance with increasing tile thickness when penetrated by the FFV core with the Schunk silicon carbide out performing the Haldenwanger SiC. The evolution of core damage for the variety of thicknesses of Schunk ceramic is shown below in Figure 5.

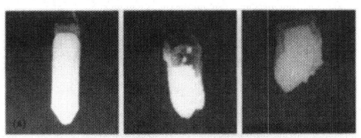

Figure 5: FFV core penetration into aluminium alloy after completely penetrating (a) 6mm, (b) 8mm and (c) 9mm of Schunk reaction bonded silicon carbide (CarSIK NT).

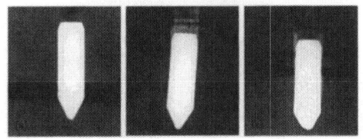

Figure 6: FFV core penetration into aluminium alloy after completely penetrating (a) 6mm, (b) 8mm and (c) 9mm of Haldenwanger reaction bonded silicon carbide (Halsic-I).

Unlike to the Schunk ceramic, the Haldenwanger silicon carbide was unable to cause significant damage to the WC-Co core of the bullet and even with relatively large thicknesses of ceramic the core was left in-tact (tip erosion is observed when the round penetrates 9mm of the ceramic – see Figure 6c; some cracking was also found when this core was retrieved from the aluminium plate). The difference in the results for the two reaction bonded silicon carbides can be explained as follows: The Schunk ceramic material exhibits a relatively dense granular packing structure when compared to the Haldenwanger ceramic that has a variable dispersion of relatively large grains intermingled with small grains of SiC (see Figure 7). An increased packing density reduces the amount of voids and silicon puddles and therefore provides and increased resistance to penetration and a larger magnitude stress wave propagation into the penetrator leading to a crushing type failure of the nose and spall failure in the shank of the core (see Figure 5b and c).

147

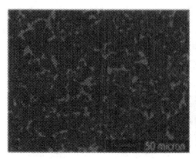

Figure 7: Micrographs from a through thickness section of a (a) Schunk tile and (b) Haldenwanger tile viewed with polarized light.

Using a calculation of a Mass Efficiency Factor it is possible to compare the performance of each ceramic taking into account the ceramic's areal density thus:

$$MEF = \frac{\rho_b P_b}{\rho_c t_c + \rho_b P_r}$$ (1)

where MEF is a mass efficiency factor, t is the thickness, ρ is the bulk density, and P is the penetration. Subscripts b and c represent the block and the ceramic respectively; P_r is the residual penetration into the witness block when there is a ceramic plate present (see Figure 1). For a nominal 6mm tile it can be seen from Figure 4 that the Halsic-I and the M CUBED RBBC-751 did not offer much more resistance to the penetrator than the same thickness of aluminium 6082 T651 alloy. The corresponding MEF's were 1.2 and 1.3 respectively. The Schunk ceramic did no fair much better with an MEF = 1.52 as opposed to a MEF = 3.02 for a 9mm tile. This is unlike the results for all ceramics that were penetrated by the B32 round (for example, Halsic-I's MEF for a 6mm tile = 6.36). Indeed, all three ceramics performed relatively poorly against the FFV core when considering the apparent hardness values with the softest ceramic (Halsic-I) possessing a measured hardness value of ~1.5 times that of the core. Nevertheless, the MEF for both RB SiCs improved as the thickness was increased through the range 6 – 9mm. This may suggest that the time of contact loading between the core and the intact ceramic is important in initiating failure in the core (for the Schunk case) and decelerating the core (for the Halsic-I) case.

Scanning electron micrographs for the fracture surfaces of both reaction bonded silicon carbides revealed large clumps of silicon particles visible on the fracture surfaces. In this instance, these are a likely source for crack growth and coalescence during the penetration of the round. Furthermore, silicon has been shown to undergo phase changes with increasing pressure. At atmospheric pressure, silicon has a cubic diamond structure up to the melting temperature (Si-I) and undergoes Si-I → Si-II transformation in the pressure range of 9-16 GPa [7]. These pressures are easily achievable in the contact zone directly in front of the penetrator tip with the rounds that are used; these pressures are not easily calculable in a 3D stress state but from 1D uni-axial strain analysis the calculated pressures are ~15 GPa with the B32 core and ~23GPa with the FFV core at muzzle velocity. The Si-I → Si-II transformation leads to a densification of the material of up to 20% [7]. Volumetric contraction of the material will inevitably lead to void growth and a source of failure. It is therefore clear that reducing the silicon content in these reaction bonded ceramics will ultimately lead to an improvement of the ballistic performance of these ceramics.

Figure 8: Fracture surface from a Haldenwanger sample; the large smooth particles are silicon.

CONCLUSIONS

Against the B32 bullet, all three ceramics behaved favourably however against the FFV there performance was relatively poor – especially as the measured hardness values were ≥ 1.5 times that of the WC-Co core of the FFV round. Nevertheless the MEF improved as the thickness increased and a 9mm Schunk tile achieved a MEF of 3.02 in these tests. This may suggest that the time of contact loading between the core and the intact ceramic is important in initiating failure in the core (for the CarSIK NT case) and decelerating the core (for the Halsic-I case). Furthermore, the different behaviour observed when the two reaction bonded silicon carbides were penetrated by the FFV round can be explained by examining the microstructure. The Schunk ceramic material exhibits a relatively dense granular packing structure when compared to the Haldenwanger ceramic that has a variable dispersion of relatively large grains intermingled with small grains of SiC. An increased packing density reduces the amount of voids and silicon puddles and therefore provides and increased resistance to penetration and a larger magnitude stress wave propagation into the penetrator leading to a crushing type failure of the nose and spall failure in the shank of the core.

Although effective at resisting penetration by the B32 round the M CUBED RBBC-751 samples did not demonstrate much ability to resist the WC-Co core and is indicative that it is behaving similarly to other boron carbides as reported by other authors.

ACKNOWLDEGEMENTS

The authors particularly wish to thank Schunk Ingenierkeramik GmbH, W. Haldenwanger Technische Keramik GmbH & Co. and M CUBED Technologies, Inc. for providing samples for this project work. We would also like to acknowledge Dave Milller at Shrivenham and Scott Newman, Dave Stretton-Smith and Ian Vickers at Caversfield for conducting the firings. We would also like to thank Paul Moth of Cranfield University for

149

helping with the hardness measurements and optical microscopy and Dr Jon Painter also of Cranfield University for the SEM work.

REFERENCES

1 M. K. Aghajanian, B. N. Morgan, J. R. Singh, J. Mears and R. A. Wolffe, "A new family of reaction bonded ceramics for armor applications," In Ceramic Armor Materials by Design, *Ceramic Transactions* Vol 134, James W. McCauley *et al*, Ed, (2002).

2 STANAG 4569 Edition 1 – Protection Levels for Occupants of Logistic and Light Armoured Vehicles North Atlantic Treaty organisation (NATO) Standardisation Agreement, AC/225-D/1463, 8 March (1999).

3 Z. Rozenberg and Y. Yeshurun, "The Relationship between Ballistic Efficiency and Compressive Strength of Ceramic Tiles," *Int. J. Impact Engng*, 7 [3], 357-62 (1988).

4 M. R. Edwards and A. Mathewson "The Ballistic Properties of Tool Steel as a Potential Improvised Armour Plate," *Int. J. Impact Engng*, 19 [4], 297-309 (1997).

5 T. J. Moynihan, J. C. LaSalvia and M. S. Burkins "Analysis of Shatter Gap Phenomenon in a Boron Carbide / Composite Laminate Armor System,". In the *Proceedings of the 20th International Symposium on Ballistics*, Orlando, FL, 23-27 September (2002).

6 C. Roberson and P. J. Hazell, "Resistance of four different ceramic materials to penetration by a tungsten carbide cored projectile," In Ceramic Armor and Armor Systems, *Ceramic Transactions* Vol 151, Eugene Medvedovski, Ed, (2003).

7 V. Domnich and Y. Gogotsi, "Phase Transformation in Silicon under Contact Loading," *Rev. Adv. Mater. Sci.* 3, 1-36 (2002).

THE EFFECTIVE HARDNESS OF HOT PRESSED BORON CARBIDE WITH INCREASING SHOCK STRESS

C J Roberson
Advanced Defence Materials Ltd.
Sir Frank Whittle Business Centre
Butlers Leap, Rugby
Warwickshire, CV21 3XH, UK

P J Hazell
Cranfield University
RMCS, Shrivenham
Oxfordshire, SN6 8LA, UK

P L Gotts
DLO Defence Clothing R&PS
Caversfield
Bicester
Oxfordshire, OX27 8TS, UK

IM Pickup
DSTL
Porton Down,
Salisbury, Wiltshire SP4 OJQ, UK

R Morrell
National Physical Laboratory
Hampton Road
Teddington
Middlesex, TW11 0LW,UK

ABSTRACT
When a sufficient shock stress has been imparted to boron carbide the resistance to penetration can significantly decrease. Using a DOP technique, an increasing shock stress is delivered to 6 mm tiles of boron carbide and their penetration resistance is evaluated by comparing the penetration into a semi-infinite aluminium backing plate. Above a threshold shock stress, the performance of the boron carbide drops off considerably and its performance decreases from that which would be similar to silicon carbide down to something similar to that of a considerably less hard alumina (~1400 HV5). This paper presents a performance map for hot pressed boron carbide comparing the effective hardness of this ceramic with increasing levels of shock (and hence threat). Conclusions are drawn on the nature of the penetration mechanics.

INTRODUCTION
Hot pressed boron carbide is the preferred high performance ceramic for use in high performance lightweight armour systems to defeat armour piercing (AP) ammunition based on the combination of properties of low density and extremely high hardness. In the past few years

the use of tungsten carbide cores for AP has become commonplace and now this form of ammunition is rapidly replacing the previous generation of steel cored AP across all of the NATO armies. US forces have adopted the M993 and M995 tungsten carbide cored AP rounds into general use in the 7.62 mm and 5.56 mm calibres respectively. 0.5″ calibre rounds with tungsten carbide cores are rapidly being adopted by European armies as a way of upgrading the AP effect of existing machine gun systems.

With respect to the interaction of these natures of ammunition with hot pressed boron carbide Moynihan et al.[1] identified a shatter gap phenomenon above a certain threshold velocity. They suggested, and subsequent workers have shown[2], that boron carbide undergoes a change in behaviour in the range of shock stress of 19 to 24 GPa associated with a phase transition. Gooch and Burkins[3] showed in reverse ballistic work that above the phase transition the boron carbide ceramic was ineffective in shattering the AP core and penetration proceeded then by a rigid body mechanism. In tests with the European equivalent projectile to the M993 our previous work[4] suggested an anomalously low level of hardness for boron carbide at these higher levels of shock stress relative to other common armour ceramics.

Using the same DOP technique as our previous work and varying the impact velocity of the bullet between 650 and 1050 m/sec it is possible to compare the penetration resistance of boron carbide relative to other ceramics in the range of shock stress between <16 GPa and > 26 GPa. Two common types of ceramic were chosen as comparators for the boron carbide; a direct sintered silicon carbide with boron and carbon as sintering aids (Morgan AM&T PS5000) and a 99.5% + grade of fine grain alumina ETEC Alotec® 99 SB. A nominal ceramic thickness of 6 mm was chosen as optimum from the standpoint that this was sufficient silicon carbide ceramic to always ensure that the core of the round was fragmented over the total velocity range. Nevertheless a sufficient DOP cavity was formed to allow accuracy of measurement. On the converse the penetration into the 6 mm alumina was normally characterised essentially by rigid body penetration. The boron carbide on a ceramic thickness basis would be expected to mirror or exceed the performance of the silicon carbide at shock stresses below the point of transition. Above this critical level the residual penetration into the aluminium block should rapidly increase and the alumina performance line will provide an indicator of relative performance of the shocked boron carbide against a well-characterised reference. From such an approach the rate and degree of change of the boron carbide in terms of reduction of effective hardness can be mapped.

EXPERIMENTAL

The Depth of Penetration (DOP) technique as described by Rozenberg and Yeshurun[5] was used to measure the ballistic performance of the ceramic tiles. As a backing for the DOP experiments, a common engineering aluminium alloy Al 6082 T651 was used (YS = 250 MPa). The test backing plates were 50.8 × 50.8 mm pieces cut from a single 25 mm thick plate. The test material was nominally 6 mm thick hot pressed boron carbide (Ceradyne Ceralloy® 546-3E). Approximately 60 samples were prepared in all.

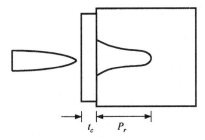

Figure 1: DOP technique for assessing each ceramic's ballistic performance.

Each ceramic tile was glued to the aluminium alloy backing block using Araldite 2015. This was applied to the mating surfaces and then the ceramic and aluminium block were pushed together and twisted / oscillated until an even thin adhesive line had been achieved with no gaps or obvious air inclusions.

The range set up was one of a fixed test barrel mounted ten metres from the target. Bullet velocity was measured using the normal sight-screen arrangement just before the target. The test ammunition was 7.62 mm × 51 mm NATO FFV: the powder load was adjusted successively to give the desired range of velocities between 600 and 1070 m/s. The FFV bullet core consists of tungsten carbide core with a composition by percentage weight C 5.2, W 82.6, Co 10.5, Fe 0.41[6] and of hardness 1550 HV0.3 (300 gf), and density ρ = 14800 kg/m^3. The core is mounted in a low carbon steel jacket with gilding metal, on an aluminium cup.

Figure 2: 7.62 × 51mm FFV bullet and WC-Co core.

The test jig was firmly clamped to a test fixture adjustable for height and lateral position and axially aligned with the direction of shot. The jig position was accurately adjusted to ensure that the centre of the target block corresponded with the centre of the shot-line; the jig used engineering vee-blocks as clamping elements. Each of the samples was clamped in place in turn with the ceramic sample protruding out of the front of the clamps. Behind the sample, in the vee-blocks, were three more of the 25 mm blocks of aluminium giving a possible total DOP of 100 mm – effectively semi-infinite for the purposes of the test ammunition. The mechanism for absorption of the kinetic energy of the bullet by this aluminium alloy is one of pure plastic strain. For the projectile penetrating a semi-infinite plate of Al 6082 T651 (without ceramic), over the velocity range of interest the DOP varies linearly with kinetic energy density

After testing the aluminium alloy blocks were X-rayed which allowed the residual penetration to be accurately measured. Furthermore the level of fragmentation of the core and the overall shape of the penetration crater was assessed from the X-rays.

The DOP measurements taken from the X-rays are presented below in Figure 3. The shots falling within a 10 m/s range have been grouped and the mean DOP for velocity range are presented ± one standard deviation. The results for boron carbide are presented against the past data[7] for sintered silicon carbide (Morgan AM&T PS5000) and 99% alumina (ETEC 99 SB). An estimate of the shock stress is also presented at velocity intervals of 100 m/s. This can be calculated from Equation 1 and assumes a 1D state of strain:

$$\sigma = \left(\frac{\rho_1 c_1 \rho_2 c_2}{\rho_1 c_1 + \rho_2 c_2} \right) v_0 \qquad (1)$$

Where ρ is the bulk density, c is the elastic wave speed for the material and v_0 is the impact velocity of the projectile. Subscripts 1 and 2 refer to the projectile and target respectively. This assumes a linear Hugoniot in σ - u_p space. Furthermore AUTODYN-2D simulations of a 5.59 mm diameter bar striking a 10 mm boron carbide plate we conducted to assess the value of shock stress that occurred in the target. Figure 3 shows how the value of the longitudinal stress varies from distance from the bar / target interface along the axis of symmetry at 0.5 µs after impact.

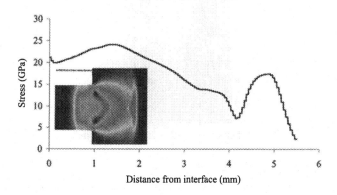

Figure 3: AUTODYN-2D simulation showing the variation of Stress TXX with distance from the bar / target interface at 0.5µs after impact.

In this case a JH-2 constitutive model was used to model the boron carbide[8] that was artificially prevented from softening / failing and therefore allowing stress relaxation. A constitutive model for the WC-Co cermet was developed from the work of Grady[9] and validated using Holmquist et al.[10]. As can be seen in Figure 3, the stress TXX increases

154

gradually from ~20 GPa to ~24 GPa before dropping off to ~7 GPa at 4.2 mm from the interface. In this case the impact velocity of the bar was 970 m/s that corresponds to a shock stress calculated from Equation 1 of ~24 GPa.

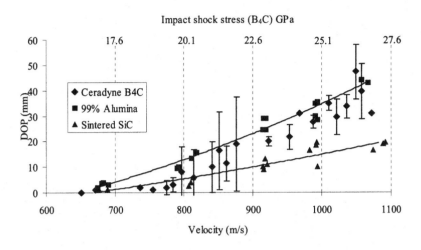

Figure 4: DOP measurements after complete penetration by the FFV round

Some physical properties for these three ceramics are shown below in Table 1.

TABLE 1 Relevant properties of ceramics

	Ceradyne Ceralloy® 546-3E HP Boron Carbide	Morgan AM&T Purbide® PS5000 Silicon Carbide	ETEC Alotec® 99 SB Alumina 99.7% content
Bulk density, kg/m³	2510	3140	3900
Elastic modulus, GPa	460	428	370
Poisson's ratio	0.14	0.15	0.22
Linear wave speed, m/s	13800	11990	10330
Vickers hardness (HV1, 1 kg load)	3412	2527	1377
Vickers hardness (HV5, 5 kg load)	Fragmented	2281	1373
Knoop hardness (HK2, 2 kg load)	2159	2105	1246
Rockwell A hardness (HRA)	67.7*	95.6	90

* not a true reading, fragmented under the indenter

Table 1: Hardness data derived from measured values, remaining data supplied by manufacturers.

DISCUSSION

Figure 4 confirms the expected reduction in the performance of hot pressed boron carbide in protecting against the FFV projectile at higher velocities. The relative DOP values for boron carbide start to increase rapidly around above 790 m/s. Initially, below this velocity, the boron carbide generally follows and has a DOP value slightly less than sintered silicon carbide, but above the 790 m/s threshold velocity the DOP moves up towards the calibration line for 99.7% alumina. Through the interval 800 to 880 m/s the boron carbide the penetration behaviour is highly erratic from shot to shot; this is can be seen from the magnitude of the error bars.

These observations are consistent with the shatter gap phenomenon observed by Moynihan et al.[1] during the testing of boron carbide faced, fibre composite backed laminate armour against the M993 round. Here they determined two values of V_{50}; a lower value at 834 m/s and a higher one at 978 m/s. These results were associated with different fragmentation behaviour of the boron carbide ceramic. The value 834 m/s is of specific interest as being coincident of where the DOP for boron carbide in our experiment has decidedly risen above the trend line for sintered silicon carbide. The degradation of the performance of boron carbide at higher impact velocities with tungsten carbide cored ammunition has been investigated and shown to be a process of crystal transformation to an amorphous state under shock stresses exceeding 20 GPa[1]

The correlation between shock stress and bullet velocity is apparent from Figure 4, and it is observed that the transformation behaviour is completed through interval between 790 m/s and 860 m/s. This corresponds to shock stress values between 19.8 GPa and 21.6 GPa. Whilst the results are somewhat dispersed, it would appear that the relative performance of boron carbide vs. alumina on a thickness for thickness basis continues to degrade with increasing shock stress. There is some suggestion that at velocities above 1020 m/s the behaviour again becomes stochastic and the penetration resistance drops below that of alumina, perhaps suggesting a further degradation of the integrity of the crystal structure of boron carbide.

Our previous work[4] has inferred a reasonable correlation at a specific thickness of ceramic facing between DOP for this round and the hardness of the ceramic. Re-evaluating the ballistic data from our previous work[11] on several types of silicon carbide and more recent work on alumina[12] has allowed a set of coherent ballistic data to be assembled. All the data has been normalised to 6mm thick ceramic impacted at a velocity of approximately 960 m/s with a 7.62 mm FFV round. The same grades of ceramic have been subjected to several differing hardness tests, Vickers HV1 and HV5 (1kg & 5kg force), Knoop HK2 (2 kg force) and Rockwell A (60 kg total force) and the results have been correlated. A good fit has been found, shown in Figure 4 for HV5, but there is a relatively poor correlation with Knoop hardness. Similarly the correlation for Rockwell A and HV1 are inferior with R^2 values of 0.7359 and 0.8256 in both cases the best curve fits are second order polynomials rather than a linear correlation.

It is likely that the good correlation between ceramic hardness and penetration resistance is specific for this projectile where the penetrator is a hard cermet material prone to brittle failure and erosion. Experience teaches that similar degrees of correlation are not seen for semi ductile penetrators like steel or tungsten alloy, and therefore the response shown in Figure 5 cannot be generalised.

156

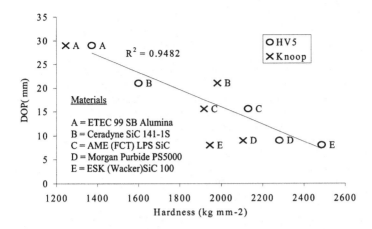

Figure 5: Correlation plot for DOP against 6mm tiles vs ceramic hardness - Vickers (HV5) and Knoop (HK2).

Cross-referencing Figures 4 and 5 it is possible to understand the penetration response of hot pressed boron carbide at high shock stress loading in terms of reduced ceramic hardness. Relative to the quasi-static hardness of measurements of the sintered silicon carbide and 99.5%+ alumina used as comparators in these experiments, the mean apparent Vickers hardness of shock transformed boron carbide can then be tabulated as follows:

TABLE 2: Apparent hardness of shock transformed boron carbide (HV5)

Impact velocity m/s	Impact Shock Stress GPa	Boron Carbide Apparent Hardness HV5
860	21.6	1870
910	22.9	1632
960	24.1	1510
1010	25.4	1420

Above 1050 m/s or 26.3 GPa shock stress it is suggested that the hot pressed boron carbide performs as if it had a hardness value lower than the ETEC alumina. Further ballistic results at elevated velocity are required to confirm this.

Whilst hardness appears to have an overwhelming influence in determining the armour performance of the ceramics in our experiment it is important not to generalise. Two factors should be taken into account when using these results; firstly one that the results were obtained by the DOP method and therefore need to be interpreted accordingly. Secondly it is difficult to attribute the response of the shock-transformed boron carbide in terms of hardness alone. Hardness and modulus of brittle materials are closely linked. It is possible, if not indeed probable, that the transformed boron carbide not only has an apparently reduced hardness, but also that simultaneously the elastic modulus and acoustic impedance of the material have also diminished. If this is the case then the stress pulse returning into the shaft of the penetrator to induce fragmentation will be modified and the damage to the penetrator may be reduced.

157

CONCLUSIONS

Hydrocode simulations of a 5.59 mm diameter WC-Co bar striking a 10 mm tile of boron carbide that is artificially not allowed to soften or fail, have shown that using Equation 1 provides a reasonable estimate of shock stress in the un-failed material.

Using a DOP method and comparing the response of boron carbide to silicon carbide and alumina it has been possible to map the response of boron carbide at high shock stress in terms of quasi-static ceramic hardness measurements. Indeed, we have shown that the effective resistance to penetration of boron carbide is equivalent or better than sintered silicon carbide below approximately 830 m/s and changes to approach that of alumina at <1000 m/s. The experiments have confirmed the shock stress whereby boron carbide transforms into a state with a reduced ballistic resistance, but also have shown that beyond the known transformation range the material continues to deteriorate in resistance to penetration by tungsten carbide cored AP projectiles relative to silicon carbide and alumina ceramics.

ACKNOWLEDGEMENTS

The authors particularly wish to thank ETEC, Morgan AM&T and Ceradyne Inc. for providing samples for this and past project work. We would also like to acknowledge the range staff at both RMCS Cranfield University and DLO DC R&PS, Caversfield for helping with the ballistic tests, and the staff at NPL for their assistance in the ceramic hardness measurements.

REFERENCES

[1] T.J. Moynihan, J.C. LaSalvia and M.S. Burkins, "Analysis of shatter gap phenomenon in a boron carbide / composite laminate armor system,". In the *Proceedings of the 20th International Symposium on Ballistics*, Orlando, FL, 23-27 September 2002.

[2] M.Chen, J.M. McCauley and K.J. Hemker, "Shock-induced localized amorphization of boron carbide, *Science* 229(5612) 1563-6 (Mar 7 2003)

[3] W. Gooch and M.S. Burkins, "Dynamic X-Ray Imaging of tungsten carbide projectiles penetrating boron carbide," Oral Presentation at American Ceramic Society, 27th Annual Cocoa Beach Conference, January 26-31, 2003.

[4] C. Roberson and P.J. Hazell, "Resistance of four different ceramic materials to penetration by a tungsten carbide cored projectile," in Ceramic Armor and Armor Systems, *Ceramic Transactions*, Vol. 151, Eugene Medvedovski, Ed, (2003).

[5] Z. Rozenberg and Y. Yeshurun, "The relationship between ballistic efficiency and compressive strength of ceramic tiles," *Int. J. Impact Engng*, **7**, [3], 357-62, (1988).

[6] M.R. Edwards and A. Matthewson, "The ballistic properties of tool steel as a potential improvised armour plate," *Int. J. Impact Engng*, **19**, [4], 297-309, (1997).

[7] K.J. Williams "Ceramic armour performance," MSc thesis, RMCS, Cranfield University, July 2003.

[8] G.R. Johnson and T.J. Holmquist, Response of boron carbide subjected to large strains, high strain rates and high pressures. *Journal of Applied Physics,* **85**, (12) pp 8061-8073 (1999).

[9] D. Grady, Impact failure and fragmentation properties of tungsten carbide, *Int. J. Impact Engng.,* **23**, 307-317, (1999).

[10] T.J. Holmquist, A.M. Rajendran, D.W. Templeton, K.D. Bishnoi, A ceramic armor material database, TARDEC Technical Report No. 13754, January (1999).

[11] C. Roberson and P.J. Hazell, "Resistance of silicon carbide to penetration by a tungsten carbide cored projectile", in Ceramic Armor and Armor Systems, *Ceramic Transactions* Vol. 151, Eugene Medvedovski, Ed, (2003).

[12] S.E. Donoghue "Defeat of tungsten carbide cored bullets," MSc thesis, RMCS, Cranfield University, July 2004.

HARDNESS AND HARDNESS DETERMINATION IN SILICON CARBIDE MATERIALS

Andreas Rendtel, Brigitte Moessner, Karl A. Schwetz
ESK Ceramics GmbH & Co. KG
Max-Schaidhauf-Strasse 25
Kempten, D-87437, Germany

ABSTRACT

The hardness of SiC ceramics plays a significant role in their performance as armor materials. Different hardness measurement techniques and loads, however, substantially influence the hardness values of ceramics, making it difficult to make material comparisons and relate these to performance. In this study hardness tests were performed with: (i) monocrystalline alpha- and beta-SiC crystals, (ii) polycrystalline, sintered (pressureless/gas-pressure sintered) alpha-SiC materials, and (iii) hot-pressed (uniaxially/isostatically) alpha-SiC materials. Hardness numbers were measured by indentation with Knoop and Vickers diamonds in the load range from 0.245 N to 4.9 N on a Leitz Miniload-2 tester. At loads over 0.98 N hardness decreased with increasing load. Highest values for Knoop and Vickers hardness at constant load were obtained with the SiC single crystals. In most cases it was found that Vickers hardness is higher than Knoop hardness. In the hot isostatically pressed SiC (encapsulated HIP-SiC) the Vickers and the Knoop numbers were found to be identical. Furthermore, it was observed that doping elements like Al and B have an effect on hardness. Al-doped SiC ceramics exhibited lower hardness numbers than pure or B-doped materials and, additionally, in these materials the Knoop hardness is higher than the Vickers hardness. Liquid phase sintered SiC showed lowest hardness numbers due to the secondary phase and solubility of Al in the SiC-lattice.

INTRODUCTION

The hardness of SiC ceramics plays a significant role in their performance as armor materials. Only diamond, cubic boron nitride, and boron carbide are harder than SiC. In the ceramic armor community the general rule "harder is better" holds.[1] Hardness testing of ceramics appears to be a simple and straightforward procedure, but there are many issues that can complicate the measurement and interpretation of the results. Such issues are taken into consideration in the corresponding standards.[2-5] Furthermore, it has been observed that hardness decreases with increasing load.[1,6,7] This phenomenon is known as the indentation size effect (ISE). Swab[1] found that a load-independent Knoop hardness can be measured at high loads (> 9.8 N) and he recommended a load of 19.6 N for determining the Knoop hardness of armor ceramics. The goal of that study was to describe the material behavior relevant to the application as armor material. In combination with that goal this recommendation is not put into question. From the scientific point of view, damage in the form of major cracking and spalling of the material during hardness testing should be avoided to obtain accurate material properties. Accordingly, the revised versions of ISO 14705 and ASTM C 1327.[3,4] recommend a reduced load of 9.8 N for the measurement of the macrohardness of ceramics. This recommendation should be followed since they are the result of extensive discussions of results obtained in round robins. A further possibility to characterize the hardness of ceramic materials that are readily damaged during hardness testing is the use of microhardness measurements.

This study is focused on microhardness testing of selected SiC ceramic grades produced by ESK Ceramics, using both Vickers and Knoop methods, with loads less than 5 N. The main objective was to rank the different SiC ceramics relative to their hardness. Additionally, using microhardness measurements allowed to compare the results obtained on SiC with earlier data measured on boron carbide[6] that showed a hardness maximum at small loads (< 4.9 N). A further objective here was to verify the existence of a similar hardness maximum at small loads for SiC.

EXPERIMENTAL PROCEDURE

Hardness testing was performed on polished surfaces (mirror-like finish, 1 µm diamond polish) using Vickers and Knoop diamonds, using the standard indentation test procedures and equations outlined in DIN EN 843-4.[2] Indentation loads ranged from 0.245 N (25 g) to 4.9 N (500 g). Vickers and Knoop hardness tests were performed using a Miniload 2 and a Durimet hardness tester, respectively, both produced by Leitz GmbH (now Leica Microsystems AG, Wetzlar, Germany). Indenters and hardness testers were regularly controlled by measurements on certified standard materials (MPA NRW 36419.22003 – HV 0.1; MPA NRW 39004.22003 – HK 0.5; MPA, Dortmund, Germany). Size of the indents was measured on a Axioplan 2 optical microscope (Carl Zeiss AG, Oberkochen, Germany) at a magnification of 1000x using an Epiplan-NEOFLUAR 100x/0.90 HD objective lens directly after the indent was produced. A minimum of ten valid indentation measurements were made at each load for each material, with mean values and standard deviation reported. All hardness measurements were performed by the same experienced person to avoid variations due to different observers.

MATERIALS

In this study a selection of different SiC-materials was investigated. In particular mono-crystalline alpha- and beta-SiC crystals, polycrystalline, sintered (pressureless/gas-pressure sintered) alpha-SiC materials, and hot-pressed (uniaxially/isostatically) alpha-SiC materials were studied. These materials can be divided into three groups, i) single crystals (SC), ii) developmental materials, and iii) commercial materials. SiC single crystal hardness refer to the (0001) crystal faces. The characterization of the different materials is listed in table I.

Table I. Dopant/impurity content [wt%], sintered density [%TD], and mean grain size [µm] of the SiC materials investigated

Material	SiC single crystals (SC)			developmental			commercial		
Element/ Property	green (α)	black (α)	yellow (β)	HP-SiC	HIP-SSiC1	HIP-SiC	S-SiC	HIP-SSiC2	LPS-SiC
B	–	–	n.d.	0.008	0.02	0.01	0.45	0.45	n.d.
Al	0.017	0.045	n.d.	2.0	0.63	0.16	0.04	0.04	2.20
Y	–	–	–	n.d.	–	–	0.001	0.001	1.80
Fe	n.d.	n.d.	n.d.	0.28	n.d.	n.d	0.03	0.03	0.025
Co	–	–	–	0.35	n.d.	n.d.	–	–	–
W	–	–	–	4.0	n.d.	n.d.	–	–	–
density [%TD]	100.0	100.0	100.0	99.3	99.6	100.0	98.7	99.6	99.9
mean grain size [µm]	–	–	–	2.1	3.5	2.0	2.4	3.0	1.2

Green and black SiC single crystals (SiC-SC) were identified by X-ray diffraction as α-SiC and the yellow SC as β-SiC. The single crystals are pure materials with minor Al-impurities.

The developmental materials are all α-SiC ceramics that differ in the doping with boron and aluminum as well as in the processing. The hot-pressed material (HP-SiC) is fine-grained and contains a high amount of aluminum as sintering aid, iron as impurity, and additionally some cemented carbide inclusions from the grinding media. The post-HIPed sintered SiC (HIP-SSiC1) contains much less sintering additive, in particular aluminum, and impurities. Its grain size is larger compared to the HP-SiC. The encapsulated HIPed SiC (HIP-SiC) is again a fine-grained material and contains the lowest amount of impurities.

The commercial materials are currently available at ESK Ceramics GmbH & Co. KG, Kempten, Germany, and are potential materials for armor applications. These materials are all fine-grained α-SiC ceramics with a mean grain size of 1 to 3 μm and with relative densities ≥ 98 %TD. The sintered (S-SiC = EKasic® F) and the post-HIPed sintered material (HIP-SSiC2 = EKasic® F+) contain the same doping (boron) and impurities. The liquid-phase sintered material (LPS-SiC = EKasic® T) contains a larger amount of oxidic sintering aids with aluminum and yttrium as the cations. Other impurities levels are comparable with that of the other two commercial materials. Compared to solid state sintered SiC, liquid phase sintered SiC (LPS-SiC) features improved fracture toughness due to an intergranular fracture mode[8].

RESULTS AND DISCUSSION

In Figures 1a) and 1b) Vickers and Knoop hardness values (mean value ± standard deviation) are plotted as a function of the applied load respectively. For loads over 1 N a decrease in hardness is observed for both techniques. This phenomenon is known as the ISE. Taking the ratio H 0.5/H 0.1 as a measure for the ISE it can be seen that the decrease in the Knoop hardness (average HK 0.5/HK 0.1 = 0.84) is more pronounced than in Vickers hardness (average HV 0.5/HV 0.1 = 0.90). The difference in the ISE is attributed to the variations in indenter geometry. At loads less than 1 N microhardness measurement is very difficult due to the small size of the indentations. The error for the measurement of the apparent diagonal lengths is about 10 %, and thus the error for the hardness is about 20 %. Due to the large error of the hardness values at very small loads these data are simply shown for information and are not used for interpretation or discussion. Hence, it was not possible to verify the existence of a hardness maximum in silicon carbide as it was described earlier for boron carbide.[6]

Highest hardness values were obtained for SiC single crystals (black symbols in figures). Hardness' of the three SC is very similar and differences are within the experimental error. For SiC-SC (black symbols) and commercial SiC ceramics (empty symbols), Vickers hardness is higher than the corresponding Knoop hardness at the same load. On the other hand, for the developmental materials (gray symbols) Vickers hardness is lower than the corresponding Knoop hardness at the same load. Only for the HIP-SiC (▲)Vickers and Knoop hardness values are identical. The developmental materials HP-SiC (◆)and HIP-SSiC1 (■) exhibit the lowest Vickers hardness values. For the Knoop hardness, lowest values were found for the LPS-SiC EKasic® T (△).

Examples of typical Vickers and Knoop indentations obtained on EKasic® F are shown in Figure 2. For Vickers indentations difficulties arose in measuring the size of the indentations due to the small size at low loads. At higher loads the formation of edge cracks was observed (indicated by arrows in Fig. 2). However, these cracks did not influence the hardness measurement since no spalling occurred. Knoop indentations were defect free over the entire load range. At low loads the Knoop indentation size was difficult to measure due to weak contrast and small size.

The significance of microhardness numbers determined at low loads is not only limited by the large error as described before, but also due to the fact that the size of the indentations is in the range of the size of larger grains of the materials.

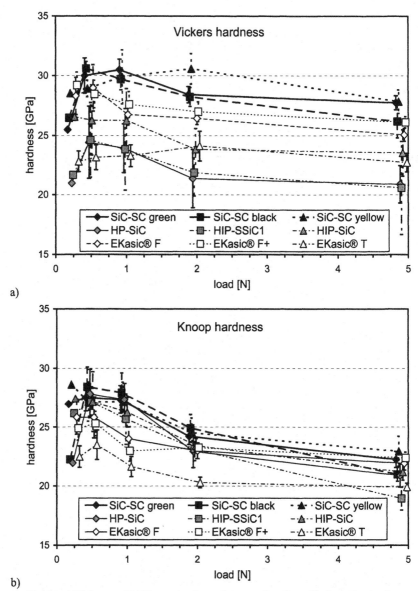

a)

b)

Fig. 1. a) Vickers and b) Knoop hardness data as a function of indentation load.
(For better identification of the data points they are plotted with constant load-offsets, different for each material. Measurements were made with identical loads of 0.245, 0.49, 0.98, 1.96; and 4.9 N. At the smallest load of 0.245 N for some materials the standard deviation is quite large in the range of ± 4.9 GPa. These standard deviations are not shown for reasons of clarity.)

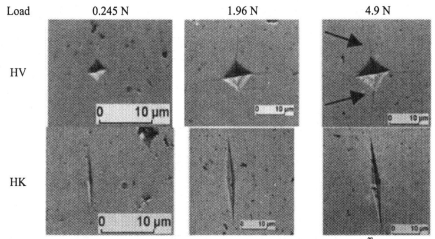

| Load | 0.245 N | 1.96 N | 4.9 N |

HV

HK

Fig. 2 Typical Vickers and Knoop indentation obtained on EKasic® F

In the following discussion only microhardness values obtained at a load of 1.96 N are used. This load was chosen in order to use hardness values with limited error as well as values representing the properties of the material and not only of single or a few grains. Taking the Knoop hardness numbers measured at a load of 1.96 N the following ranking of the materials is obtained. Highest hardness exhibit the SiC-SC (black symbols) with HK 0.2 ~ 25 GPa. The hardness of the HIPed and solid state sintered SiC ceramics is somewhat lower in the range around HK 0.2 ~ 23.5 GPa. Lowest hardness values were measured for the liquid phase sintered SiC (△) with HK 0.2 ~ 20.5 GPa.

As shown in Figure 3 for the load of 1.96 N, microhardness of SiC ceramics decreases with increasing total aluminum content of the materials. This trend is observed at most load levels and is more pronounced for Vickers hardness than for Knoop hardness. A possible reason for this feature may be a different effect of the microstructure (grain size, phase distribution, grain boundary state) on Vickers and Knoop hardness measurement due to the different indenter geometries.

It can be seen that EKasic® T exhibits lowest Knoop hardness at highest aluminum content (2.2 wt% Al-total). Recent analytical results[9] indicate approximately half of Al-total is dissolved in the SiC lattice and half is present as a softer secondary phase ($Y_3Al_5O_{12}$). This softer secondary phase is believed to be the reason for the lower hardness.

We conclude from Figures 3a) and 3b) that highest hardness numbers were obtained for the green and black single crystals, the polycrystalline SiC materials sintered with boron additives (EKasic® F and EKasic® F+) as well as with HIP-SiC, which all have very low Al impurity contents in the range of 0.02 to 0.2 wt%. HIP-SiC is a high-purity product with a SiC content of more than 99.5 %. Owing to the high isostatic pressure of 2 kbar (argon gas), compared with normal hot-pressing, no sintering aids were added[8].

It is generally accepted that in the boron-doped, solid state sintered SiC materials (EKasic® F and EKasic® F+) approximately 0.2 wt% boron is present in solid solution in the SiC lattice, the residual boron occurring as boron carbide (B_4C) inclusions. The grain boundaries are virtually free of any segregation[10]. This is the reason for the transgranular fracture mode in

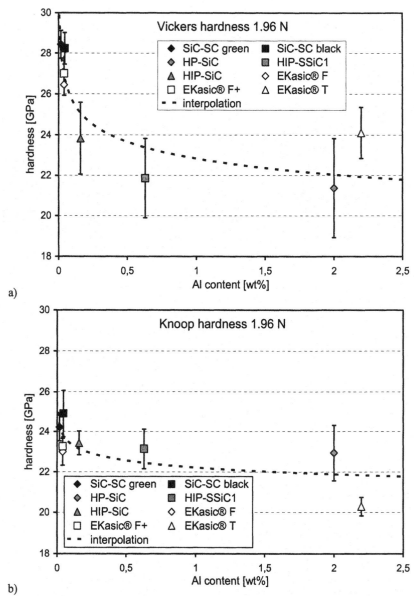

Fig. 3. Hardness versus Al content in various SiC materials at 1.96 N load for
a) Vickers hardness and b) Knoop hardness.

boron-doped SiC. The presence of boron carbide inclusions is consistent with the high hardness of the boron-doped materials.

According to the literature, during sintering of SiC with Al-containing additions, considerable amounts of aluminum, namely 0.5 to 1.4 wt% Al, can be taken into solid solution.[9,11,12] For the materials densified using metallic aluminum (HP-SiC) or Al-based compounds (HIP-SSiC1 and LPS-SiC) as sintering aids, three different occurrences of aluminum have been confirmed:

i) solution of Al in the SiC lattice,

ii) segregation of Al at the grain boundaries leading to intergranular fracture, and

iii) formation of secondary phases (viz. YAG = $Y_3Al_5O_{12}$ in LPS-SiC, Al-silicates in HP-SiC) intergranulary at triple points.[9,10,12]

For the case of the solid solution (Si,Al)C, where Al atoms substitute in the SiC lattice for Si atoms, and for the case of thin Al-containing segregation films at the grain boundaries it is likely that hardness reduction is small or limited. On the other hand, the occurrence of relatively high volume fractions of aluminum oxide compounds with significant lower hardness in HP-SiC and LPS-SiC is believed as being detrimental for hardness of SiC ceramics at total Al-contents of over 1 wt%.

CONCLUSIONS

Microhardness measurements are a useful tool to characterize the hardness of silicon carbide ceramics. The microhardness of the selected SiC materials tested depended on the applied load and the indenter shape (Vickers/Knoop), but also on the purity and phase composition of the materials. To obtain microhardness values that characterize the properties of the material with limited error a load of 1.96 N had to be used. Due to the large error of hardness values measured at low loads, the existence of a hardness maximum could not be verified.

At a relative density of above 98 % and a mean grain size of 1 to 3 μm the main factor affecting the hardness of SiC materials is the content of secondary phases. Highest hardness numbers were obtained with pure single crystals, the pure HIP-SiC, and the boron-doped solid-state sintered materials. Significant hardness reduction occurred in sintered or hot-pressed SiC using an Al-containing liquid phase (Al-silicate in HP-SiC, YAG in LPS-SiC) at total Al-contents of over 1 wt% in the sintered materials.

ACKNOWLEDGEMENT

The authors are grateful to B. Mikijelj at Ceradyne Inc. and C. Lesniak at ESK Ceramics for helpful discussions and review of the manuscript.

REFERENCES

[1]J. J. Swab, "Recommendations for determining the hardness of armor ceramics," *Int. J. Appl. Ceram. Technol.*, **1**, 219-25 (2004).

[2]DIN EN843-4 "Advanced technical ceramics – Monolithic ceramics – Mechanical properties at room temperature – Part 4: Vickers, Knoop and Rockwell superficial hardness," prEN 843-4:2002.

[3]ISO/FDIS 14705 "Fine ceramics (advanced ceramics, advanced technical ceramics) – Test method for hardness of monolithic ceramics at room temperature," 2000.

[4]ASTM C1327-96 "Standard test method for Vickers indentation hardness of advanced ceramics" 2003 Annual Book of ASTM Standards, Vol. 15.01.

[5]ASTM C1326-96 "Standard test method for Knoop indentation hardness of advanced ceramics" 2003 Annual Book of ASTM Standards, Vol. 15.01.

[6] A. Lipp, K. A. Schwetz, "Hardness and hardness determination of nonmetallic hard materials, I: Boron carbide," (in german: Härte und Härtebestimmung von nichtmetallischen Hartstoffen, I. Borcarbid") *Ber. Dt. Keram. Ges.*, **52**, 335-38 (1975).

[7] A. Kerber, S. V. Velken, "Liquid phase sintered silicon carbide: Production, properties, and possible applications," in C. Galassi (Ed.), "FOURTH EURO CERAMICS, Vol. 2," *Gruppo Editoriale Faenza Editrice S.p.A.*, 1995, pp. 177-84.

[8] K. A. Schwetz, "SiC-based hard materials," in R. Riedel (Ed.), "Handbook of ceramic hard materials, Vol. 2," *Wiley-VCH, Weinheim N.Y.*, 2000, pp. 683-748.

[9] K. A. Schwetz, H. Werheit, E. Nold, "Sintered and monocrystaline black and green SiC – Chemical compositions and optical properties," *cfi/Ber. DKG*, **80**, E37-E44 (2003).

[10] R. Hamminger, G. Grathwohl, F. Thuemmler, "Microchemistry and high temperature properties of sintered SiC," in "Proc. 2nd Conf. Science of Hard Materials (Rhodos)," *Inst. Phys. Conf. Ser. No. 75, Adam Hilger Ltd*, 1986, Chapter 4, pp. 279-92.

[11] Y. A. Vodakov, E. N. Mokhof, "Diffusion and solubility of impurities in SiC," in R. C. Marshall et al. (Eds.), "Silicon Carbide - 1973," *Univ. of South Carolina Press, Columbia*, 1974, p. 508.

[12] Y. Tajima, W. D. Kingery, "Solid solubility of Al and B in SiC," *J. Amer. Ceram. Soc.*, **65**, C27 (1982).

168

Damage Characterization: Observations, Mechanisms, and Implications

SPHERE IMPACT INDUCED DAMAGE IN CERAMICS: I. ARMOR-GRADE SiC AND TiB₂

J.C. Lasalvia, M.J. Normandia, H.T. Miller*, and D.E. MacKenzie
U.S. Army Research Laboratory - Aberdeen Proving Ground
AMSRD-ARL-WM-MD
Aberdeen, MD 21005-5069

ABSTRACT

Armor-grade SiC and TiB_2 cylinders (25.4 mm x 25.4 mm) were impacted with WC-6Co spheres (6.35 mm diameter) at velocities between 50 m/s and 500 m/s. The recovered cylinders were subsequently sectioned and metallographically-prepared to reveal the dominant sub-surface damage types and change in damage severity as a function of impact velocity. In general, both ceramics exhibited radial, ring, Hertzian cone, and lateral cracks which increased in number and length as the impact velocity increased. Furthermore, both ceramics exhibited a comminuted region directly beneath the impact center which appeared to be wholly contained between the set of Hertzian cone cracks which formed first. The comminuted region consisted of a high density of microcracks located at grain boundaries. The observed effect of impact velocity on the resulting damage and differences in damage between these two armor-grade ceramics will be presented.

INTRODUCTION

During the last decade, ceramics have emerged as an important class of armor materials for the protection of personnel and light vehicles against small arms and machine gun threats.[1-3] Ceramics are attractive as armor materials because they are more weight efficient than monolithic steel armors for a variety of threats. However, because of their inherent brittleness, ceramics must be "packaged" properly in order to take advantage of their high compressive strengths.

Unfortunately, the design of ceramic armors today is still largely a process which relies heavily upon either previous experience or a lot of testing. However, efforts have been made to utilize computational models to at least gain insight into the gross physical phenomena which occur and affect the performance of ceramic armors during the ballistic event.[4-6] Because the models are phenomenologically-based and may not incorporate the "correct" physics, the validity of the results can be subject to question. In spite of this, much can be learned and gained by intelligent use of computational models; consequently, their development and use is strongly advocated.

The ability to mitigate or contain shear and tensile driven damage generated during the ballistic event is essential to producing a high performance ceramic armor. Consequently, it is essential that ceramic models possess the ability to capture the effects of shear and tensile driven damage, including their temporal and spatial variations, for both penetrating and non-penetrating impacts. A number of experimental studies have been undertaken to determine the evolution of sub-surface damage in ceramics due to projectile impact.[7-11] These observations have proven extremely insightful into the types, spatial distribution, and governing mechanisms of sub-

*Work performed while an undergraduate student at the University of Maryland, Baltimore County with support by an appointment to the Research Participation Program at the U.S. ARL administered by the Oak Ridge Institute for Science and Education through an interagency agreement between the U.S. Department of Energy and U.S. ARL.

surface damage. The types and spatial distribution of sub-surface damage can be used to help validate computational models.

The purpose of this paper is to report on damage observations in two commercially-available armor-grade ceramics generated by non-penetrating impacts with WC-6Co (6 wt.% Co) spheres at velocities between 50 – 500 m/s. The ceramics chosen for this study are of interest because of their similarities in hardness and fracture toughness and their significant differences in grain sizes, densities, and elastic moduli. Analysis of observations will not be presented in this paper.

EXPERIMENTAL PROCEDURES

Armor-grade SiC and TiB_2 variants manufactured by Cercom, Inc. and known commercially as SiC-N and PAD[†] TiB_2 were used in this study. The densities, grain sizes, mechanical properties, and phases for these materials are listed in Table I. Densities and elastic moduli were determined using the Archimedes water immersion and pulse-echo techniques (ASTM Standard E494), respectively. Knoop hardness was determined on specimens final polished with 0.05 μm colloidal silica and in accordance with ASTM Standard C1326 using a load of 40 N. Fracture toughness was determined using the single-edge pre-cracked beam technique in accordance with ASTM Standard C1421 and specimen (bar) dimensions 3 mm x 4 mm x 50 mm. Flexural strengths were determined from bars (30) of the same dimensions as those used for fracture toughness determination. A standard 20 x 40 mm semi-articulating flexure four-point fixture was used with a cross-head speed of 0.5 mm/min in accordance with ASTM C1161. Phases were determined using X-ray diffraction (Cu Kα, 0.02° 2θ step size, 2 s dwell time) and a commercial pattern matching program.

Table I. Material Characteristics.

Material	ρ (g/cm^3)	Grain Size (μm)	E (v) (GPa)	HK4 (GPa)	K_{IC} (MPa*m$^{1/2}$)	σ_b (MPa)	Phases
SiC-N	3.22	3.3	452 (0.165)	18.6	5.1	500	6H, 15R, 3C
PAD TiB_2	4.52	15[‡]	555 (0.11)	16.8	6.9[†]	285[†]	TiB_2

These materials were machined into cylinders that were 25.4 mm in diameter and 25.4 mm in length (nominal). The impact surfaces were prepared to a 100 grit finish (10 μm down-feed rate) which resulted in a surface roughness Ra value of 0.1 μm for both SiC and TiB_2 (2 μm stylus tip diameter). The cylinders were slip-fitted (0.025 mm nominal diameter difference) into Ti-6Al-4V cups, and impacted (near center) with WC-6Co spheres (Machining Technologies Inc, Grade 25, 0.000635 mm roundness tolerance, Class C-2) 6.35 mm in diameter in the velocity range of 50 – 500 m/s. The density and hardness of the spheres are 14.93 g/cm^3 and 16.3 GPa (HV0.5)[12], respectively. According to Normandia et al.[13], these spheres do not begin to experience tensile failure until the impact velocity is above 800 m/s. Details of the ballistic test set-up and procedures are described elsewhere[14].

Following the sphere impact experiments, photomicrographs were taken of the impacted surfaces. The impact surfaces were subsequently impregnated with a cold mount epoxy in an attempt to preserve the impact surface integrity. The cylinders were extracted from their Ti-6Al-

---

[†] Pressure-Assisted Densification
[‡] Cercom Material Data Sheet

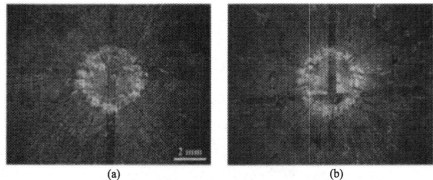

(a)	(b)

Figure 1. Impacted surfaces: (a) SiC-N (161 m/s), and (b) PAD TiB₂ (184 m/s). Radial and ring cracks are clearly evident. Magnifications for both pictures are the same.

4V cups by carefully sectioning the cups (lengthwise) into several pieces. Following extraction, the cylinders were completely encased in cold mount epoxy and then sectioned 1 – 2 mm from the center of impact. The cross-sections were remounted again and metallographically-prepared using 15 μm, 6 μm, and 1 μm diamond slurries. The final cross-sections were 0.1 – 0.3 mm from the apparent center of impact. The cross-sections were subsequently examined by optical and scanning electron microscopy.

RESULTS AND DISCUSSION
    The damage induced by the impact of the WC spheres on the top surfaces of both ceramics is shown in Figure 1. For all impacts in this study, radial and concentric ring cracks were observed. With the exception of lowest impact velocity for SiC-N, the radial cracks extended to the outer surface of the cylinders. However, the radial cracks did not extend all the way through the cylinders to their back surfaces. The depth to which the radial cracks extended was difficult to visually discern precisely and consequently was not determined in this study. It was noted that the number of radial cracks was always greater in PAD TiB₂ than SiC-N (see Figure 1). This is consistent with the lower tensile strength exhibited for the PAD TiB₂ (see Table 1).
    In addition to the radial and ring cracks noted on the impact surface, lateral cracks, Hertzian cone cracks, and a comminuted region were observed beneath the impact surface. Figure 2 shows the polished cross-section from a recovered SiC-N cylinder that was impacted at 500 m/s. The different crack types and comminuted region are clearly indicated. The crack marked "radial crack" is caused by the intersection of the polished cross-section with one of the radial cracks. The comminuted region is essentially composed of a high density of microcracks located at grain boundaries. For SiC-N, the comminuted region is more evident because of grain pull-out as a result of the metallographic preparation. Details of the localized damage near the center of impact will be examined later. In the remainder of the paper, the effect of impact velocity on the sub-surface damage will be examined for both SiC-N and PAD TiB₂.

Sub-Surface Damage: SiC-N
    The effect of impact velocity on the resulting sub-surface damage is shown in Figure 3. As expected, the severity of the damage increases with increasing velocity, and is most severe beneath the impact center. In Figures 3a and 3b, cone and ring cracks are observed.

173

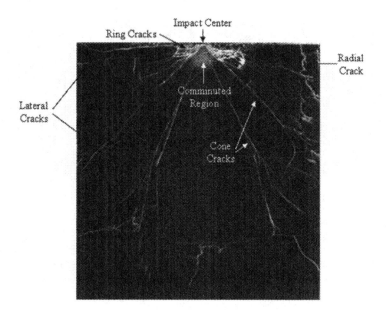

Figure 2. Polished cross-section from a recovered SiC-N cylinder that was impacted at 500 m/s. (Specimen height is equivalent to 25.4 mm).

Furthermore, the comminuted region is present though barely visible. The number of ring cracks increase with increasing impact velocity, while the comminuted region enlarges and the cone cracks extends further into the cylinder. At an impact velocity of 161 m/s, the cone cracks (curved) interact with a lateral crack from the outer surface of the cylinder, which with the aid of a radial crack, allowed the top left corner of the cylinder to separate. The comminuted region is wholly contained between the cone cracks and does not extend to the top surface. In Figure 3c (impact velocity of 322 m/s), the number of lateral and ring cracks increases significantly. Material surrounding the impact center has been lost due to erosion and spallation. This causes a truncated cone to form directly underneath the impact center. Both erosion and spallation of material is facilitated by the formation of shallow ring cracks in the region surrounding the truncated cone. It is evident that the cone cracks exhibit less curvature and clearly bifurcate into at least two distinct cracks. At an impact velocity of 500 m/s, two sets of cone cracks are clearly visible. The angle of the initial set of cone cracks does not appear to vary significantly. In addition, the amount of material removed at the top surface due to spallation has increased substantially due to the extension of some of the shallow ring cracks into lateral cracks that extend out further, as well as a general increase in the number and length of shallow ring cracks surrounding the truncated cone.

The sub-surface damage near the center of impact for the 63 m/s impact velocity is shown in more details in Figure 4. The comminuted region and cone cracks are clearly visible in Figure 4a, while the fine spacing of the ring cracks at the top surface is shown in Figure 4b. In Figure 4b, the comminuted region appears as an increase in apparent porosity due to grain pull-out and

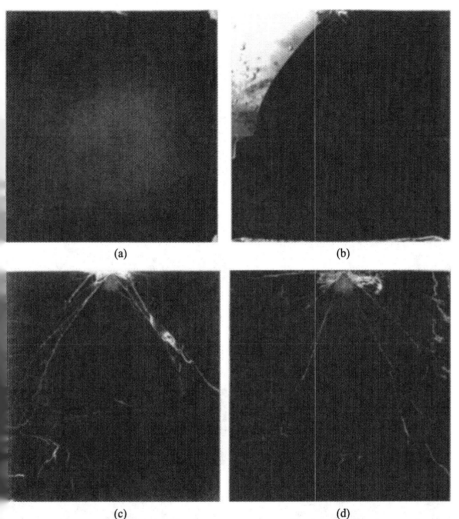

(a)	(b)
(c)	(d)

Figure 3. Polished cross-sections of recovered SiC-N cylinders showing the effect of impact velocity on damage: (a) 63 m/s; (b) 161 m/s; (c) 322 m/s; and (d) 500 m/s. (Specimen heights are equivalent to 25.4 mm).

its outline is not distinct due to the low density of microcracks. However, it is evident that the comminuted region does not extend to the top surface.

Details of the comminuted region and intensity of damage near the impact center for the cylinder impacted at 500 m/s are more clearly shown in Figure 5. The truncated cone formed by the upper cone cracks and the impact surface is more evident in Figure 5a. The comminuted region appears to be wholly contained between the upper cone cracks. Examination of the region

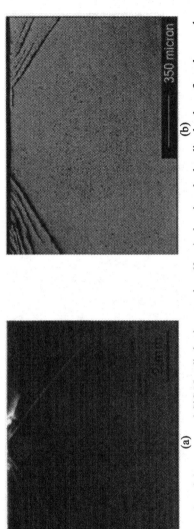

(a)

(b)

Figure 4. Micrographs of the recovered SiC-N cylinder impacted at 63 m/s showing the localized sub-surface damage beneath the impact center. (a) Lower magnification, and (b) Higher magnification.

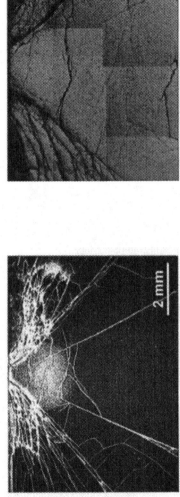

(a)

(b)

Figure 5. Micrographs of the recovered SiC-N cylinder impacted at 500 m/s showing the localized sub-surface damage beneath the impact center. (a) Lower magnification, and (b) Higher magnification.

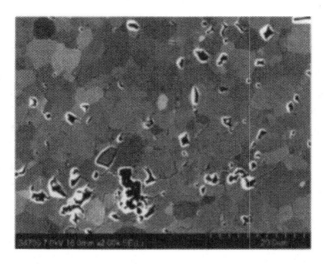

Figure 6. Comminuted region within SiC-N impacted at 500 m/s. Grain pull-outs and grain boundary microcracks are clearly evident.

adjacent to the upper cone crack does not provide conclusive evidence that the comminuted region does expand past the upper cone crack. The material that appears to be comminuted within the shallow ring cracked region that can be seen in Figure 5b may or may not be an artifact introduced during metallographic preparation. Further work is needed to determine if this is true or not. As was mentioned previously, the comminuted region is made up of grain boundary microcracks as shown in Figure 6. The density of microcracks increases with impact velocity. A number of investigators have made similar observations with respect to ceramics that were subjected to quasi-static Hertzian indentation and dynamic impact by rods.[15-17] These microcracks form as a result of shear-driven compressive failure.[18-19]

Examination of Figure 5 offers insight into the mechanisms responsible for the onset of penetration for an impacting sphere (or pointed-nose projectile). During impact, the upper cone crack forms in which the subsequent comminuted region is wholly contained. This in itself does not lead to penetration. However, as the contact area increases, numerous finely-spaced ring cracks form. Smaller cracks form normal to and in-between the ring cracks. This is most severe near the top surface than away from it. This results in the fragmentation of the material adjacent to the upper cone crack and at the top surface. This fragmented material is then displaced or lost due to erosion or spallation processes, leading to the "exposure" of the truncated cone directly beneath the impact center. However, this truncated cone is full of microcracks and therefore its strength is pressure-dependent. Removal of the surrounding material decreases the lateral confining stresses (i.e. decreases the hydrostatic pressure) while increasing the load on the truncated cone itself. Consequently, compressive failure occurs and penetration begins. In one sense, the stress-state for the material directly beneath the center of impact is changed from multi-axial compression to approximately uni-axial compression due to erosion and spallation.

177

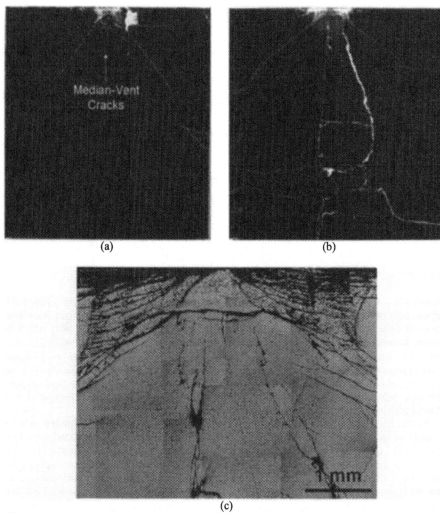

(a)　　　　　　　　　　　　　　　(b)

(c)

Figure 7. Polished cross-sections of recovered PAD TiB$_2$ cylinders: (a) 184 m/s; (b) 311 m/s; (c) Close-up view of localized damage region in 311 m/s cylinder. (Specimen heights in (a) and (b) are equivalent to 25.4 mm).

Sub-Surface Damage: PAD TiB$_2$

The effect of impact velocity on the sub-surface damage is shown in Figures 7a and 7b (184 m/s and 311 m/s, respectively). Comparison with SiC-N (Figures 3b and 3c) shows that PAD TiB$_2$ exhibits more localized damage and cracking for the same nominal impact velocities. An increase in lateral cracking is clearly visible, while the comminuted region is significantly larger. Unlike SiC-N, in which the contrast mechanism for the comminuted region is predominately

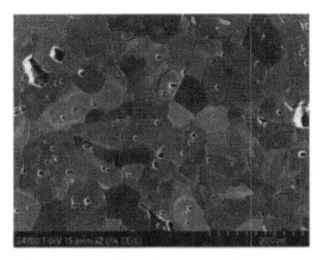

Figure 8.　Comminuted region within PAD TiB$_2$ impacted at 311 m/s.　Grain boundary microcracks are clearly evident.

grain pull-out, the main contrast mechanism for PAD TiB$_2$ is grain boundary microcracking. Two sets of cone cracks and cone crack bifurcation are observed in Figure 7b. While bifurcation was observed in SiC-N impacted at 322 m/s, two sets of cone cracks was not observed. In Figure 7a, median-vent cracks are seen (circled region), and most probably are the precursors to the inner cone cracks observed in Figure 7b.

The intensity of the damage near the center of impact is shown more clearly in Figure 7c for the cylinder impacted at 311 m/s. As in SiC-N, the comminuted region appears to be wholly confined between the upper cone cracks. The apparent comminuted damage near the top surface is most likely a metallographic artifact as discussed previously. The ring cracks are very shallow (more so than observed for SiC-N) and end abruptly at a fixed distance from the center of impact. The ring cracks also bifurcate close to the end of their tips giving the appearance that can be best described as a "zipper". The inner cone cracks propagate through the comminuted region. Based upon the apparent crack opening displacement of these cone cracks, it appears that they initiated outside the comminuted region and then propagated in both directions. This is consistent with the median-vent cracks observed in the cylinder impacted at 184 m/s (Figure 7a). Lastly, the lateral crack observed between the upper cone cracks and near the top surface in SiC-N (Figure 5b) is also seen in PAD TiB$_2$, and is believed created on unloading (i.e. rebound of the sphere).

SUMMARY AND CONCLUSIONS

Observations concerning the different types of damage in two armor-grade ceramics (SiC-N and PAD TiB$_2$) impacted by WC-6Co spheres between 50 and 500 m/s have been reported. At low velocity, the damage was localized near the center of impact. The damage included a comminuted region, and radial, ring, lateral, and cone cracks. The comminuted region consisted of grain boundary microcracks. In SiC-N, while some microcracks were evident, the primary reason for light-image contrast was grain pull-outs. In PAD TiB$_2$, microcracks were dominant.

In both ceramics, the cracking was predominately intergranular, which is reflective of their relatively high fracture toughness values. The severity of the damage increased with increasing impact velocity for both ceramics. In addition to an increase in the size of the comminuted region and density of microcracks, the number and length of ring and cone cracks increased. Furthermore, lateral cracks were also observed. Median-vent cracks were observed in PAD TiB$_2$ only, while cone crack bifurcation was observed in both ceramics. With increasing impact velocity, many of the shallow ring cracks became lateral cracks which linked back up with the top surface. The ring cracks in both SiC-N and PAD TiB$_2$ impacted at a nominal velocity of 300 m/s appeared to adopt a more shallow path and ended abruptly a fixed distance from the center of impact, yielding a damage structure that can be best described as a "zipper". The formation of a truncated cone surrounded by shallow ring cracks near the center of impact in both ceramics yielded insight into the possible mechanisms responsible for penetration onset. Keeping the material surrounding the truncated cone from being displaced should increase the penetration onset velocity.

ACKNOWLEDGEMENTS
The authors would like to thank Mr. Matthew J. Motyka (Dynamic Science International) and Mr. Steve Szewczyk (Oak Ridge Institute for Science and Education) for their technical support.

REFERENCES
1. W.A. Gooch, "An Overview of Ceramic Armor Applications," Ceramic Armor Materials By Design, eds. J.W. McCauley, A. Crowson, W.A. Gooch, A.M. Rajendran, S.J. Bless, K.V. Logan, M.J. Normandia, and S. Wax, Cer. Trans., 134, 3-21(2002).
2. B. Matchen, "Applications of Ceramics in Armor Products," Key Eng. Mat., 122-124, 333-342 (1996).
3. D.J. Viechnicki, M.J. Slavin, and M.I. Kilman, "Development and Current Status of Armor Ceramics", Am. Cer. Soc. Bull., 70 [6] 1035-1037 (1991).
4. M.L. Wilkens, "Mechanics of Penetration and Perforation," Int. J. Eng. Sci., 16 793-807 (1978).
5. C.E. Anderson, Jr. and J.D. Walker, "Ceramic Dwell and Defeat of the 0.30-Cal AP Projectile," Proceedings of the 15th Army Solid Mechanics Conference, eds. K.R. Iyer and S.C. Chou, Battelle Press, Columbus, OH, 2000, pp. 17-28.
6. T.J. Holmquist and G.R. Johnson, "Modeling Prestressed Ceramic and Its Effect on Ballistic Performance," Int. J. Impact Eng., 31 [2] 113-127 (2005).
7. A.G. Evans and T.R. Wilshaw, "Quasi-Static Solid Particle Damage in Brittle Solids – I. Observations, Analysis, and Implications," Acta Metall., 24 939-956 (1976).
8. D.A. Shockey, A.H. Marchand, S.R. Skaggs, G.E. Cort, M.W. Burkett, and R. Parker, Int. J. Impact Eng., 9 263-275 (1990).
9. H. Yoshida, M.M. Chaudhri, and Y. Hoshi, "Quasistatic Indentation and Spherical Particle Impact Studies of Turbine-Grade Silicon Nitrides," Phil. Mag. A, 82 [10] 2031-2040 (2002).
10. D. Sherman and T. Ben-Shushan, "Quasi-Static Impact Damage in Confined Ceramic Tiles," Int. J. Impact Eng., 21 [4] 245-265 (1998).
11. D.A. Shockey, D.J. Rowcliffe, K.C. Dao, and L. Seaman, "Particle Impact Damage in Silicon Nitride," J. Am. Cer. Soc., 73 [6] 1613-1619 (1990).

12. A.A. Wereszczak, "Elastic Property Determination of WC Spheres and Estimation of Compressive Loads and Impact Velocities That Initiate Their Yielding and Cracking," *Cer. Eng. Sci. Proc.*, to be published.

13. M.J. Normandia and B. Leavy, "Ceramic Armor – Ballistic Impact of Silicon Carbide with Tungsten Carbide Spheres," *Cer. Eng. Sci. Proc.*, **25** [3] 573-578 (2004).

14. M.J. Normandia, D.E. MacKenzie, B.A. Rickter, and S.R. Martin, "A Comparison of Ceramic Materials Dynamically Impacted by Tungsten Carbide Spheres," *Cer. Eng. Sci. Proc.*, to be published.

15. H.H.K. Xu, W. Lanhua, N.P. Padture, B.R. Lawn, and R.L. Yeckley, "Effect of Microstructural Coarsening on Hertzian Contact Damage in Silicon Nitride," *J. Mat. Sci.*, **30** [4] 869-878 (1995).

16. S.K. Lee, S. Wuttiphan, and B.R. Lawn, "Role of Microstructure in Hertzian Contact Damage in Silicon Nitride," *J. Am. Cer. Soc.*, **80** [9] 2367-2381 (1997).

17. J.C. LaSalvia, E.J. Horwath, E.J. Rapacki, C.J. Shih, and M.A. Meyers, "Microstructural And Micromechanical Aspects Of Ceramic/Long-Rod Projectile Interactions: Dwell/Penetration Transitions", Fundamental Issues and Applications of Shock-Wave and High-Strain-Rate Phenomena, eds. K.P. Staudhammer, L.E. Murr, and M.A. Meyers, Elsevier Science, 2001, pp. 437-446.

18. H. Horii and S. Nemat-Nasser, "Brittle Failure in Compression: Splitting, Faulting, and Brittle-Ductile Transition," *Phil. Trans. Roy. Soc. London A*, **319**, [1549] 337-374 (1986).

19. G. Vekinis, M.F. Ashby, and P.W.R. Beaumont, "The Compressive Failure of Alumina Containing Controlled Distribution of Flaws," *Acta Met. Mat.*, **39** [11] 2583-2588 (1991).

SPHERE IMPACT INDUCED DAMAGE IN CERAMICS: II. ARMOR-GRADE B₄C AND WC

J.C. LaSalvia, M.J. Normandia, H.T. Miller*, and D.E. MacKenzie
U.S. Army Research Laboratory - Aberdeen Proving Ground
AMSRD-ARL-WM-MD
Aberdeen, MD 21005-5069

ABSTRACT
Armor-grade B₄C and WC cylinders (25.4 mm x 25.4 mm) were impacted with WC-6Co (6 wt.% Co) spheres (6.35 mm diameter) at velocities between 100 m/s and 400 m/s. The recovered cylinders were subsequently sectioned and metallographically-prepared to reveal the dominant sub-surface damage types and change in damage severity as a function of impact velocity. In general, both ceramics exhibited radial, ring, Hertzian cone, and lateral cracks which increased in number and length as the impact velocity increased. The cracking was predominately transgranular for B₄C and intergranular for WC. However, unlike SiC and TiB₂ (reported in the part I[1]), no evidence of a comminuted region directly beneath the impact center was observed in either ceramic. B₄C exhibited severe spallation of material surrounding the impact center. In addition, evidence of shear localization beneath the impact center was also observed. This observation may in part explain the sharp drop in shear strength that B₄C exhibits in plate impact experiments when shocked above 20 GPa. In contrast, WC was almost unremarkable in its response to being impact with the WC spheres in that it exhibited a nice spherical crater that is more typical of the response of a metal. The effect of impact velocity on the observed damage and differences in damage between these two armor-grade ceramics will be presented.

INTRODUCTION
In the companion paper (part I)[1], the sub-surface damage generated by the impact of WC spheres on commercially-available armor-grade SiC and TiB₂ was reported. The motivation was to provide both insights into the types of damage and damage mechanisms that may govern penetration resistance, as well as provide a means of possible validation of computational models that can be used in guiding ceramic armor development. In this paper (part II), the sub-surface damage generated in two commercially-available armor-grade ceramics due to impact with WC spheres at velocities between 100 – 400 m/s is reported. The ceramics chosen for this study are of interest because of their significant differences in density, grain size, elastic moduli, fracture toughness, and four-point bend strength. Analysis of observations will not be presented in this paper.

EXPERIMENTAL PROCEDURES
PAD† B₄C and PAD WC (Cercom, Inc.) were used in this study. The densities, grain sizes, mechanical properties, and phases for these materials are listed in Table I. Densities and elastic

*Work performed while an undergraduate student at the University of Maryland, Baltimore County with support by an appointment to the Research Participation Program at the U.S. ARL administered by the Oak Ridge Institute for Science and Education through an interagency agreement between the U.S. Department of Energy and U.S. ARL.
† Pressure-Assisted Densification (i.e. Hot-Pressed)

Table I. Material Characteristics.

Material	ρ (g/cm³)	Grain Size (μm)	E (v) (GPa)	HK4 (GPa)	K_{IC} (MPa*m$^{1/2}$)	σ_b (MPa)	Phases
PAD B₄C	2.53	15	465 (0.17)	18.9	2.9	450	B₄C, C
PAD WC	15.6	0.4[2]	697 (0.20)	20.0[2]	6 - 8 [2]	1030[2]	WC, W₂C

moduli were determined using the Archimedes water immersion and pulse-echo techniques (ASTM Standard E494), respectively. Knoop hardness was determined on specimens final polished with 0.05 μm colloidal silica and in accordance with ASTM Standard C1326 using a load of 40 N. Fracture toughness was determined using the single-edge pre-cracked beam technique in accordance with ASTM Standard C1421 and specimen (bar) dimensions 3 mm x 4 mm x 50 mm. Flexural strengths were determined from bars (30) of the same dimensions as those used for fracture toughness determination. A standard 20 x 40 mm semi-articulating flexure four-point fixture was used with a cross-head speed of 0.5 mm/min in accordance with ASTM C1161. Phases were determined using X-ray diffraction (Cu Kα, 0.02° 2θ step size, 2 s dwell time) and a commercial pattern matching program.

These materials were machined into cylinders that were 25.4 mm in diameter and 25.4 mm in length (nominal). The impact surfaces were prepared to a 100 grit finish (10 μm down-feed rate) which resulted in a surface roughness Ra values of 0.18 μm and 0.12 μm for PAD B₄C and PAD WC (2 μm stylus tip diameter), respectively. The cylinders were slip-fitted (0.025 mm nominal diameter difference) into Ti-6Al-4V cups, and impacted (near center) with WC-6Co spheres (Machining Technologies Inc, Grade 25, 0.000635 mm roundness tolerance, Class C-2) 6.35 mm in diameter in the velocity range of 100 – 400 m/s. Details of the ballistic test set-up and procedures are described elsewhere[3].

Following sphere impact, photomicrographs of the impacted surfaces were taken. The impact surfaces were subsequently impregnated with a cold mount epoxy in order to preserve the impact surface integrity. The cylinders were extracted from their Ti-6Al-4V cups by carefully sectioning the cups (lengthwise) into several pieces. The recovered cylinders were then encased with cold mount epoxy and sectioned 1 – 2 mm from the impact center. The cross-sections were remounted again and metallographically-prepared using 15 μm, 6 μm, and 1 μm diamond slurries. The final cross-sections were 0.1 – 0.3 mm from the impact center. The cross-sections were examined using optical and scanning-electron microscopy.

RESULTS AND DISCUSSION

Examples of the impact surfaces for both ceramics are shown in Figure 1. Radial cracks were observed for both ceramics at all impact velocities (the radial cracks in PAD WC were not as visible as the ones in PAD B₄C). The radial cracks did not extend all the way through the cylinders to their back surfaces. The depth to which the radial cracks extended were difficult to visually discern precisely and consequently was not determined in this study. The number of radial cracks was always greater for PAD B₄C compared to PAD WC (see Figure 1). This is consistent with the lower tensile strength for PAD B₄C (see Table I). Unlike both SiC-N and PAD TiB₂ in the part I companion paper[1], concentric ring cracks were not always evident on the top surfaces of either PAD B₄C or WC. As will be seen later, the ring cracks are observed on all cross-sections. As can be seen in Figure 1a, PAD B₄C exhibited considerable spallation of material surrounding the impact center. This occurred at all impact velocities. Because of the spallation of material surrounding the impact center, the truncated cone which was exhibited by

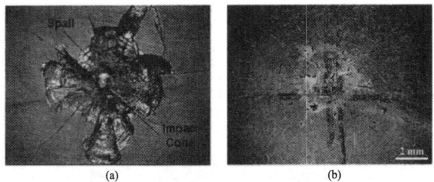

(a)   (b)

Figure 1. Impacted surfaces: (a) PAD B_4C (103 m/s), and (b) PAD WC (174 m/s). Radial cracks are evident in both ceramics. PAD B_4C exhibits severe spallation of material surrounding the impact center. Magnifications for both pictures are the same.

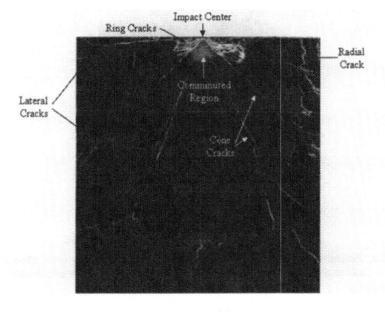

Figure 2. Sub-surface damage for a SiC-N cylinder (polished cross-section) that was impacted at 500 m/s. (Specimen height is equivalent to 25.4 mm).[1]

both SiC-N and PAD TiB_2[1], is clearly seen for PAD B_4C. As was shown previously[1], this truncated cone is the result of the formation of the Hertzian cone crack. In contrast to the brittleness exhibited by PAD B_4C, PAD WC exhibited a much more ductile response. A spherical crater that is more typical of metals was formed.

185

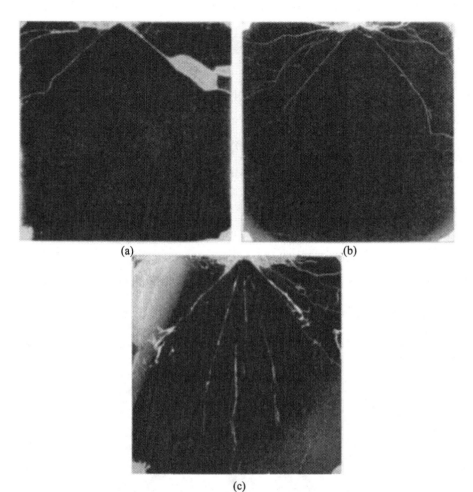

(a)   (b)

(c)

Figure 3. Polished cross-sections of recovered PAD B$_4$C cylinders showing the effect of impact velocity on damage: (a) 103 m/s; (b) 209 m/s; and (c) 312 m/s. (Specimen heights are equivalent to 25.4 mm).

Figure 2 shows the types of damage that may be observed on the PAD B$_4$C and WC cross-sections. Shown in Figure 2 is the polished cross-section from a recovered SiC-N cylinder that was impacted at 500 m/s. Radial, ring, lateral, and Hertzian cone cracks are clearly seen. In addition, a comminuted region is also indicated. The radial crack is due to its intersection with the plane of the cross-section. As was reported in the companion paper (part 1)[1], the comminuted region is composed of a high density of microcracks located at grain boundaries. However, as will be shown later, no comminuted region was observed in either PAD B$_4$C or

186

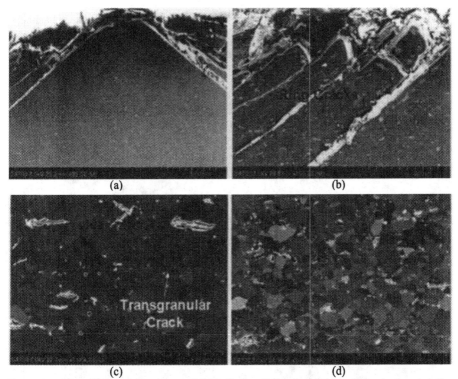

Figure 4. Micrographs of the recovered PAD B$_4$C cylinder impacted at 103 m/s showing the localized sub-surface damage. (a) Truncated cone beneath impact center; (b) Ring cracks adjacent to truncated cone; (c) Transgranular crack-tip from one of the ring cracks; and (d) No comminuted region!

PAD WC. In the remainder of this paper, the effect of impact velocity on the sub-surface damage will be examined for both PAD B$_4$C and PAD WC.

Sub-Surface Damage: PAD B$_4$C

The effect of impact velocity on the resulting sub-surface damage is shown in Figure 3. As expected, the severity of the damage increases with increasing velocity, and is most severe in the region surrounding the impact center. As can be seen, ring, cone, and lateral cracks are visible. In addition, a significant amount of material has been lost due to spallation at all impact velocities. This can be understood from the large number of shallow lateral cracks that can been seen on the polished cross-sections. The truncated cone that was observed in both SiC-N and PAD TiB$_2$[1] has been eroded (note that the top of the truncated cone is below the top surface). No comminuted region is evident in any of the images.

Details of the sub-surface damage beneath the impact center for the 103 m/s impact velocity is shown in Figure 4. The ring and cone cracks are clearly visible in Figures 4a and 4b. The

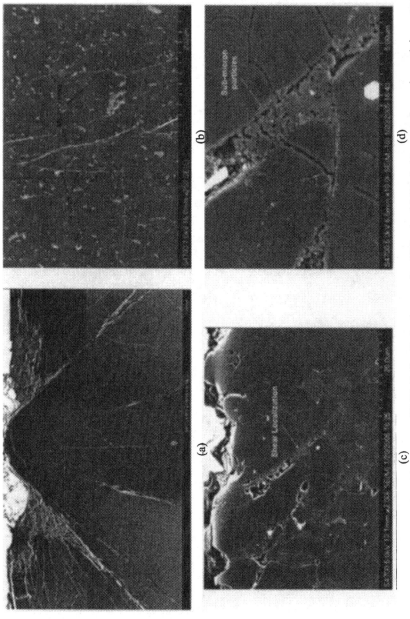

Figure 5. Micrographs from the recovered PAD B$_4$C cylinder impacted at 312 m/s. (a) Eroded truncated cone beneath impact center; (b) "Axial-splitting" microcracks (transgranular) where comminuted region would be expected; (c) Shear localization regions at the top surface of the eroded truncated cone; and (d) Close-up of shear localization regions showing sub-micron particles.

crack-tip from one of the ring cracks is shown in Figure 4c. In general, cracks in PAD B₄C were predominately transgranular, which is reflective of its low long-crack fracture toughness value (Table I). The long lenticular particles are graphite. As will be shown later, these did not appear to serve as crack initiators. Figure 4d was taken from the area of the cross-section where a comminuted region was expected. However, none was apparent.

Details of the sub-surface damage beneath the impact center for the 312 m/s impact velocity is shown in Figure 5. The eroded truncated cone, inner cone cracks, and intensity of damage adjacent to the truncated cone is clearly seen in Figure 5a. In the area of the cross-section where comminuted regions were observed for both SiC-N and PAD TiB₂[1], microcracks aligned predominately parallel to the impact loading direction were observed (Figure 5b). These microcracks appeared to follow "sharp" pores, perhaps initiated by these pores and then coalescing together. Examination of the region near the top surface of the eroded truncated cone revealed the existence of what appears to be shear "bands" (or regions of shear localization) as shown in Figure 5c. Shear localization in ceramics under dynamic loading conditions was first observed by Nesterenko et al.[4] and Shih et al.[5-6] for granular Al₂O₃ and both granular and fully-dense SiC, respectively. Using a technique known as the "thick-walled cylinder collapse" technique, the shear bands formed as a result of the inward collapse of ceramic tubes whose external surfaces were indirectly loaded explosively. If this initial assessment proves correct, then it would be the first time shear localization of a ceramic has been observed in a ballistic experiment and may be the main reason why B₄C exhibits a sharp drop in its shear strength in plate impact experiments when shocked above its Hugoniot Elastic Limit[7-8]. The fineness of the particles within the shear localization regions is shown in Figure 5d. The sub-micron size of the particles relative to the average grain size (15 μm) would seem to indicate a large build-up of stored elastic energy followed by its sudden release due to the formation of localized deformation regions.

Sub-Surface Damage: PAD WC

The effect of impact velocity on the sub-surface damage is shown in Figure 6. The apparent "cloudiness" of the images is due to a combination of grain pull-out and regions of anomalous grain growth. Hertzian cone, ring, lateral, and median-vent cracks can be seen. No comminuted region can be seen. The Hertzian cone cracks (including bifurcation) are well-defined in Figures 6a and 6b (impact velocities 174 and 263 m/s, respectively). However, in Figure 6c (impact velocity 383 m/s), they are not as one would expect. The reason for this is not known, but it is speculated that at the higher velocity, the impact event is significantly more dynamic and that outwardly propagating Rayleigh waves (surface shear waves) initiate flaws away from the region of maximum tensile stress (contact radius) as predicted from quasi-static contact theory.[9-10]. On the other hand, it may be the result of surface flaw population statistics for this particular cylinder coupled with the ceramic's high fracture toughness.

A close-up view of the impact center region for the cylinder impacted at 383 m/s is shown in Figure 7. The permanent indent of the impact crater is clearly visible. It was interesting to note that many of the cracks (including cone cracks) observed exhibited "kinks" as indicated in Figure 7. The reason for the kinks is not understood, but may involve mixed-mode crack propagation. Another reason may be due to the interaction of reflected stress waves with the propagating cracks. However, if this were true, one would expect to see these kinks in the other ceramics of this study (including part I[1]), which they apparently were not.

189

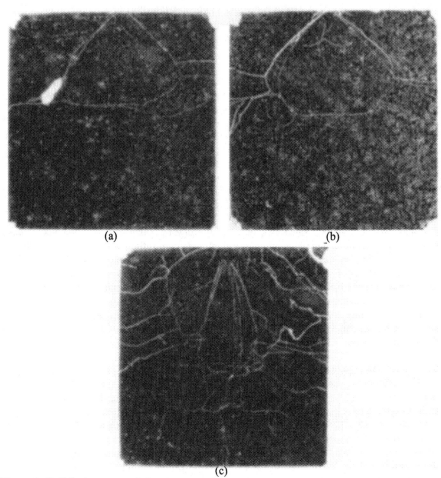

(a) (b)

(c)

Figure 6. Polished cross-sections of recovered PAD WC cylinders showing the effect of impact velocity on damage: (a) 174 m/s; (b) 263 m/s; and (c) 383 m/s. (Specimen heights are equivalent to 25.4 mm).

Lastly, Figure 7b is a close-up of one of the cracks seen in Figure 7a. The crack is predominately intergranular, reflective of the high fracture toughness value exhibited by this ceramic. Grain pull-out is also evident.

SUMMARY AND CONCLUSIONS

Observations concerning the different types of damage in two armor-grade ceramics (PAD B_4C and PAD WC) impacted by WC-6Co spheres between 100 and 400 m/s have been reported. As expected, the severity of damage increased with increasing impact velocity. Damage

190

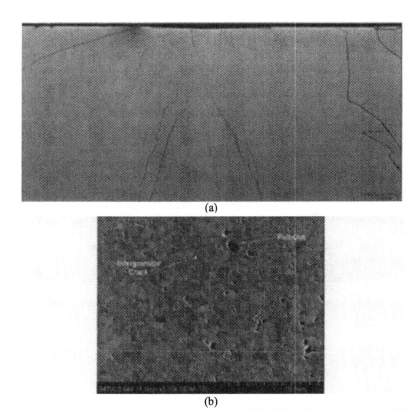

(a)

(b)

Figure 7. (a) Detailed view of sub-surface damage in PAD WC cylinder impacted at 383 m/s. (b) Higher magnification image showing grain pull-out and intergranular crack. Crack deflection is responsible for this ceramic's high fracture toughness.

generally included radial, ring, Hertzian cone, and lateral cracking. Unlike the ceramics in part 1[1], comminuted regions were not observed. In addition, PAD B_4C exhibited severe spallation (brittle response) on the impact surface, while PAD WC formed a well-defined spherical crater (ductile response). Examination of the eroded truncated cone in PAD B_4C revealed evidence of shear localization. While this must still be confirmed (additional testing and characterization), if it is true, then it may be one physical mechanism by which B_4C loses its shear strength when impacted dynamically (or even quasi-statically). For the most part, the damage observed for PAD WC was almost unremarkable and with the exception of shear localization in PAD B_4C and the presence of comminuted regions in SiC-N and PAD TiB_2[1], not much different than the other ceramics in the study even though its tensile strength was significantly higher.

ACKNOWLEDGEMENTS
The authors would like to thank Mr. Matthew J. Motyka (Dynamic Science International) and Mr. Steve Szewczyk (Oak Ridge Institute for Science and Education) for their technical support.

191

REFERENCES

1. J.C. LaSalvia, M.J. Normandia, H.T. Miller, and D.E. MacKenzie, "Sphere Impact Induced Damage in Ceramics: I. Armor-Grade SiC and TiB₂," *Cer. Eng. Sci. Proc.*, to be published.

2. J.J. Swab, <u>Hardness and Damage Associated with Pointed Indentation of Armor Ceramics</u>, Ph.D Dissertation, State University of New York at Stony Brook, 2004. 168 pp.

3. M.J. Normandia, D.E. MacKenzie, B.A. Rickter, and S.R. Martin, "A Comparison of Ceramic Materials Dynamically Impacted by Tungsten Carbide Spheres," *Cer. Eng. Sci. Proc.*, to be published.

4. V.F. Nesterenko, M.A. Meyers, and H.C. Chen, "Shear Localization in High-Strain-Rate Deformation of Granular Alumina," *Acta Mater.*, **44** [5] 2017-2026 (1996).

5. C.J. Shih, M.A. Meyers, and V.F. Nesterenko, "High-Strain-Rate Deformation in Granular Silicon Carbide," *Acta Mater.*, **44** [11] 4037-4065 (1996).

6. C.J. Shih, V.F. Nesterenko, and M.A. Meyers, "High-Strain-Rate Deformation and Comminution of Silicon Carbide," *J. Appl. Phys.*, **83** [9] 4660-4671 (1998).

7. D.E. Grady, "Shock-Wave Compression of Brittle Solids," *Mech. Mat.*, **29** [3] 181-205 (1998).

8. N.K. Bourne, "Shock-Induced Brittle Failure of Boron Carbide," *Proc. R. Soc. London A*, **458** 1999-2006 (2002).

9. K.F. Graff, <u>Wave Motion in Elastic Solids</u>, Dover Publications, Inc., New York, 1975, 649 pp.

10. K.L. Johnson, Contact Mechanics, Cambridge University Press, New York,1985, 452 pp.

SPHERE IMPACT INDUCED DAMAGE IN CERAMICS: III. ANALYSIS

J.C. LaSalvia, M.J. Normandia, D.E. MacKenzie, and H.T. Miller*
U. S. Army Research Laboratory – Aberdeen Proving Ground
AMSRD-ARL-WM-MD
Aberdeen, MD 21005-5069

ABSTRACT

In parts I and II of this study, the impact surface and sub-surface damage generated in four commercially-available hot-pressed armor-grade ceramics as the result of high-velocity impact with Tungsten Carbide spheres was presented. In this paper, a preliminary mechanical analysis is presented. The ceramic response is assumed to be elastic, perfectly-plastic; hence, the dependence of compressive strength on hydrostatic pressure is largely ignored. Based upon this assumption, the impact response of the ceramic can be divided into three regimes: (I) Elastic; (II) Elastic-Plastic; and (III) Fully-Plastic. These three regimes are represented by the following ranges for the mean impact stress p_m: (I) $p_m/Y \leq 1.1$; (II) $1.1 \leq p_m/Y \leq 3$; and (III) $p_m/Y = 3$, where Y is the compressive "yield" strength. The analysis considers regimes (I) and (II) only, and examines the following issues: (1) Quasi-Static Impact; (2) Critical Velocities for the Onset of Plasticity and Full-Plasticity; and (3) Impact Stress and Contact Diameter. The analysis for impact stress and contact diameter is made with respect to Silicon Carbide.

INTRODUCTION

It had been shown previously that "thick" ceramics when impacted by dense Tungsten-Alloy long-rod penetrators at over a kilometer a second remain impenetrable.[1-2] Examples of the sub-surface damage that was generated in several recovered ceramics by long-rod Tungsten Alloy penetrators impacting at approximately 1500 m/s are shown in Fig. 1 (polished cross-sections).[3-4] Clearly, the long-rod penetrators did *not penetrate*; instead, they dwelled on the impact face of the ceramic, becoming shorter and decelerating in the process. Both Silicon Carbide and Titanium Diboride exhibit a region of severe compressive damage (microcracking) beneath the impact center, while Tungsten Carbide did not. This region is often referred to as the comminuted region or Mescall zone, and has recently received attention because of its potential connection with the onset of penetration.[4-5]

In parts I and II of this study[6-7], our interest was in determining the type and evolution of sub-surface damage in several armor-grade ceramics subject to high-velocity impact with Tungsten Carbide spheres. This information is valuable for the development and validation of both physically-based and phenomenologically-based ceramic damage models. In addition, knowledge may be gained as to the dominant damage mechanisms which govern both inelastic response and penetration onset. This may provide insight into how to improve the performance of ceramics. Consequently, we chose four commercially-available hot-pressed variants of Boron Carbide, Silicon Carbide, Titanium Diboride, and Tungsten Carbide because of their significant differences in density, grain size, elastic moduli, fracture toughness, and bend strength (their Knoop hardness values at 40 N are similar). It has often been advocated that hardness is a

Work performed while an undergraduate student at the University of Maryland, Baltimore County with support by an appointment to the Research Participation Program at the U.S. ARL administered by the Oak Ridge Institute for Science and Education through an interagency agreement between the U.S. Department of Energy and U.S. ARL.

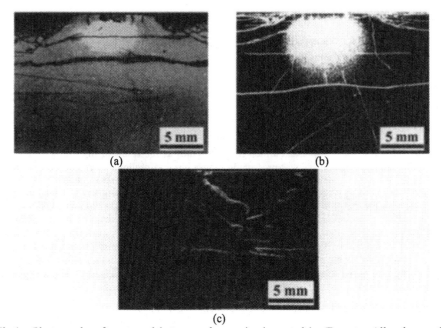

(a)   (b)

(c)

Fig.1. Photographs of recovered hot-pressed ceramics impacted by Tungsten-Alloy long-rod penetrators at 1500 m/s (nominal). In each case, the penetrator completely dwelled on the impact face of the ceramic and did not penetrate. Both Silicon Carbide (a) and Titanium Diboride (b) shows highly damaged, comminuted region, while Tungsten Carbide (c) shows no evidence of this region.[3-4]

primary determinant of material behavior in ballistic impact environments if the fracture toughness is sufficiently high.[5,8] However, the observed modes of damage for the materials tested, suggested completely different behaviors under similar ballistic loading conditions. In this paper, preliminary analysis with respect to several issues associated with observations made in parts I and II is attempted. These issues include the velocity below which the impact event can be considered quasi-static, the critical velocities for the onset of plasticity (or inelasticity) and full-plasticity, and estimations for the impact stress and contact diameter as a function of impact velocity.

EXPERIMENTAL PROCEDURES
    Armor-grade Boron Carbide, Silicon Carbide, Titanium Diboride, and Tungsten Carbide manufactured by Cercom, Inc. were used in this study. Material properties and characteristics are reported parts I and II.[6-7] These ceramics were machined into cylinders, nominally 25.4 mm in diameter and 25.4 mm in length. The cylinders were slip-fitted (0.025 mm nominal diameter difference) into Titanium Alloy (6 wt. % Aluminum, 4 wt. % Vanadium) cups, and centrally impacted with Tungsten Carbide (6 wt. % Cobalt) spheres 6.35 mm in diameter between 50 – 500 m/s (see Fig. 2a). The spheres were obtained from Machining Technologies

194

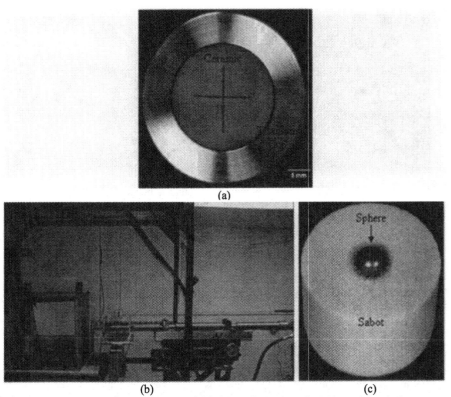

Fig. 2. (a) Ceramic cylinder sitting in Titanium Alloy cup. (b) Photograph of air-gun test facility. (c) Tungsten Carbide sphere (6.35 mm diameter) attached to foam sabot.

Inc, Grade 25 (0.635 μm roundness tolerance), Class C-2 material (Rockwell A 92) with a 14.93 g/cm³ measured density, Young's modulus of 618 GPa, Poisson's ratio of 0.21, and hardness of 16.3 GPa (HV0.5)[9]. The spheres remained intact and were recovered after each experiment. The quasi-static yield stress of the spheres is between 3 and 4 GPa, while the uni-axial compressive failure stress is approximately 5 GPa.[10]

The test facility shown in Fig. 2b consists of a compressed gas-gun launch tube, laser velocimeters, and dual x-rays. The test procedure consists of firing the spheres glued onto a foam sabot (see Fig. 2c) from the launch tube, measuring the time between two lasers at the end of the gun barrel to determine velocity, and observing (when desired) the penetrator in flight, prior to impact, with two 150 keV x-radiographs, which also served as a velocity measurement backup. In these experiments, the foam sabot also impacted the target. In retrospect, a discarding sabot prior to impact would have been preferred, as the added mass affected the load. However, this added mass was constant in all tests conducted. The launch package weight was measured as 7 g including the Tungsten Carbide sphere.

195

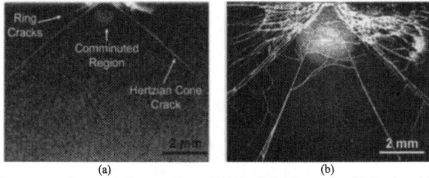

(a)	(b)

Fig. 3. Images from polished cross-sections of SiC-N cylinders impacted at: (a) 63 m/s, and (b) 500 m/s.[6]

Following sphere impact, photomicrographs were taken of the impacted surfaces. The impacted cylinders were extracted from their Titanium Alloy cups, sectioned, and metallographically prepared. Details of the extraction, sectioning, and polishing procedures are given part I.[6] Final cross-sections were approximately 0.1 mm from the impact center. The cross-sections were examined using both optical and electron-optical microscopy, and the results are presented in parts I and II.[6-7]

RESULTS & DISCUSSION
The resulting sub-surface damage for Silicon Carbide impacted at 63 m/s and 500 m/s are shown in Fig. 3a and 3b, respectively.[6] Hertzian cone crack, ring cracks, and the comminuted region are clearly visible in both images. As was reported previously[4,6], the predominate damage mechanism in the comminuted region is grain boundary microcracking. These microcracks form as a result of a shear-driven compressive failure.[11] As can be seen, the size of the comminuted region is significantly larger in the cylinder impacted at 500 m/s. However, even at this velocity, penetration is not observed. Based upon previous work[12], the penetration onset velocity for this particular Silicon Carbide was reported to be 397 m/s. The discrepancy between this value and the current result is due to the fact that the penetration depths used to determine the previous value were not measured at the impact center, but rather where the crater was deepest. As can be seen in Fig. 3b, this occurs in the region adjacent to the impact center, where material was lost due to spallation and erosion. Clearly, this methodology was incorrect and the reason for the discrepancy. Recent re-evaluation of these tiles indicate penetration onset to occur between 500 – 550 m/s. In the following sections, several issues that have arisen from the observations reported in parts I and II[6-7] and that are naturally associated with sphere impact, will be examined.

Quasi-Static Impact
The similarity of the sub-surface damage shown in Fig. 3a for the cylinder impacted at 63 m/s with that observed for ceramics that have been indented quasi-statically[13] brings into question the velocity below which the impact event can be considered quasi-static. The impact event is essentially quasi-static if the total contact time $\tau_{contact}$ is significantly greater than the

196

Table I. Critical Velocities for Quasi-Static Impact

Ceramic	E (GPa)	v	C_L (m/s)	$V_{1\%}$ (m/s)	$V_{2\%}$ (m/s)
Boron Carbide	465	0.17	14,080	101	320
Silicon Carbide	452	0.165	12,250	80	256
Titanium Diboride	555	0.11	11,000	65	206
Tungsten Carbide	697	0.20	7,060	30	94

relaxation time $\tau_{relaxation}$ of the larger body. It has been shown that the dynamic response of an elastic half-space can be modeled as an elastic spring in parallel with a dashpot, where the energy dissipated by the dashpot is associated with that lost by stress waves.[14] Based upon this model, the impact event can be considered quasi-static if the impact velocity V is less than:[14]

$$\frac{V}{C_L} \approx 7 \left(\frac{\rho_{sphere} C_L^2}{E^*} \right)^{\frac{1}{3}} \left(\frac{\tau_{relaxation}}{\tau_{contact}} \right)^{\frac{5}{3}} \tag{1}$$

$$\frac{1}{E^*} = \frac{1-v_{sphere}^2}{E_{sphere}} + \frac{1-v_{ceramic}^2}{E_{ceramic}} \tag{2}$$

C_L is the longitudinal wave velocity of the ceramic, ρ_{sphere} is density of the sphere, E is Young's modulus, and v is Poisson's ratio. The critical velocities for the ceramics used in this study are listed in Table I. $V_{1\%}$ and $V_{2\%}$ correspond to the velocities for which relaxation time is 1% and 2% of the total contact time, respectively. As can be seen for Silicon Carbide, impacts between 80 m/s and 256 m/s may be considered nominally quasi-static. The slight concave curvature of the Hertzian cone crack in Fig. 3a is a feature indicative of quasi-static loading because of its similarity to trajectories that are normal to principal tensile stress profiles predicted under Hertz's theory for elastic contact.[15] This curvature was also observed for the Hertzian cone crack in the Silicon Carbide cylinder impacted at 161 m/s.[6]

Onset of Plasticity
Observation of a comminuted region in the Silicon Carbide cylinder impacted at 63 m/s was initially surprising since its formation in Silicon Nitride, also impacted with WC spheres, was previously reported at 231 m/s.[16] This brings into question as to the impact velocity above which the comminuted region would be expected to form. Assuming that the formation of the comminuted region corresponds approximately with the distribution of maximum shear stress associated with Hertzian contact[4], the impact velocity V_Y corresponding to the onset of plasticity is given by the following expression:[14]

$$V_Y \approx 5.284 \sqrt{ \left(\frac{Y_{ceramic}}{\rho_{sphere}} \right) \left(\frac{Y_{ceramic}}{E^*} \right)^4 } \tag{3}$$

197

Table II. Impact Velocities for the Onset of Plasticity.

Ceramic	$HK4$[6] (Gpa)	$Y_{ceramic}$ (Gpa)	V_Y (m/s)
Boron Carbide	18.9	7.3	2.6
Silicon Carbide	18.6	7.1	2.5
Titanium Diboride	16.8	6.5	1.6
Tungsten Carbide	20	7.7	1.9

$Y_{ceramic}$ is the compressive "yield" strength for the ceramic. Values for V_Y are listed in Table II assuming that $Y_{ceramic}$ is approximately $H_{ceramic}/2.6$[15], where $H_{ceramic}$ is the Knoop hardness of the ceramic at 40 N. Based upon the values listed in Table II, the impact velocities used in this study are at least one-order of magnitude higher; therefore, the occurrence of plasticity should be expected. As a side note, it was indicated previously that the compressive strength of ceramics is pressure-dependent. As will be shown later, the mechanical pressure at the point below the surface where the onset of plasticity is expected to occur is quite large and $Y_{ceramic}$ should be larger than the values used in Table II. However, even with more reasonable estimates for $Y_{ceramic}$, values for V_Y are still significantly smaller than the impact velocities used in this study.

Elastic-Plastic Transition: Impact Stress and Contact Diameter

One of the goals of a mechanical analysis of the problem of a sphere indenting or impacting another body is to determine the relationship between load and indent depth or mean stress and contact radius (or diameter). The methodology for this type of analysis is relatively well established for ductile materials. In the conventional view, the indentation response of a ductile material can be divided into three regimes: (I) Elastic; (II) Elastic-Plastic; and (III) Fully-Plastic.[15] These three regimes are represented by the following ranges for the mean impact stress p_m: (I) $p_m/Y \leq 1.1$; (II) $1.1 \leq p_m/Y \leq 3$; and (III) $p_m/Y = 3$.[15] The analysis for response regimes (I) and (III) are based upon Hertz's theory of elastic impact and cavity expansion models, respectively.[14-15,17] Analysis of response regime (II) is somewhat more complicated because both mean impact stress and contact area change as the elastically-constrained plastic region grows. However, it is also amenable to the cavity expansion approach as well. For ceramics, complications to this conventional view of the indentation process include tensile cracking (e.g. Hertzian cone cracking), compression-driven microcracking (e.g. comminuted region), and pressure-dependent strength for both intact and compressively-damaged ceramic. However, for simplicity, it is instructive to view the ceramic's response as that of a ductile material with at least partial consideration for the pressure-dependent strength of the intact ceramic.

Utilizing expressions derived for the normal impact of an elastic sphere on an elastic half-space and the experimental observation that p_m/Y varies linearly with $\ln(E^* a/YR)$ within response regime (II)[18], the following relationship between the impact velocity V and the mean impact stress p_m can be derived:

$$\frac{V}{V_{pl}} \approx 0.4 \left(\frac{p_m}{Y_{ceramic}} \right)^{\frac{5}{6}} \left(\frac{a_Y}{a_{pl}} \right)^{\frac{5}{3} \left[\frac{3 - \left(p_m / Y_{ceramic} \right)}{1.9} \right]} \tag{4}$$

198

Table III. Estimated Contact Radius at the Onset of Full-Plasticity.

V_{pl} (m/s)	$a_{pl(7.1)}$[a] (mm)	$a_{pl(9.5)}$[b] (mm)	$a_{pl(11.8)}$[c] (mm)	a_{Hertz} (mm)
400	2.74	2.37	2.12	1.64
450	2.94	2.54	2.28	1.72
500	3.13	2.71	2.43	1.79
550	-	2.87	2.57	1.86
600	-	3.02	2.71	1.93

[a]Based upon Knoop hardness at 40 N load.
[b]Uni-axial compressive yield strength derived from HEL.
[c]HEL.[21]

R is the sphere radius, a_Y is the contact radius at the onset of plasticity, a_{pl} is the contact radius at the onset of full-plasticity, and V_{pl} is the impact velocity that corresponds to a_{pl}. a_{pl} and a_Y are given by:

$$\frac{a_{pl}}{R} \approx 0.578 \sqrt{\left(\frac{E^*}{Y_{ceramic}}\right)^{2/5} \left(\frac{\rho_{sphere} V_{pl}^2}{Y_{ceramic}}\right)^{3/5}} \qquad (5)$$

$$\frac{a_Y}{R} \approx 2.535 \left(\frac{Y_{ceramic}}{E^*}\right) \qquad (6)$$

Values of a_{pl} for Silicon Carbide are listed in Table III as a function of V_{pl} and several assumed values of $Y_{ceramic}$ (i.e. 7.1, 9.5, and 11.8 GPa).

In Equations (3), (4), and (5), $Y_{ceramic}$ and V_{pl} are unknowns and must be estimated. In calculation of V_Y (Table II), $Y_{ceramic}$ was estimated from Knoop hardness measurements at 40 N assuming elastic, perfectly-plastic behavior. However, as was indicated earlier, $Y_{ceramic}$ is pressure-dependent, and the validity of estimating $Y_{ceramic}$ from Knoop hardness values is not known. The complication of using hardness to estimate $Y_{ceramic}$ is due to the non-uniform stress state beneath the indenter, as well as the possible influence from a number of different types of tensile cracking that can occur.[19] A better estimate for $Y_{ceramic}$ might be given by its Hugoniot Elastic Limit (HEL).[20] The HEL of a material represents its compressive yield strength under uni-axial strain conditions. Under these conditions, the mechanical pressure is $((1+\nu)/(1-\nu))\sigma_{applied}/3$ instead of $\sigma_{applied}/3$ under uni-axial stress conditions (for $\nu = 0.2$, $0.5\sigma_{applied}$ versus $0.33\sigma_{applied}$). Based upon Hertz's theory for elastic contact, the mechanical pressure is approximately $0.5\sigma_{applied}$ at the point where the onset of plasticity is expected to occur.[14] Consequently, the HEL represents a better estimate for $Y_{ceramic}$ rather than that used in Table II. The HEL for the Silicon Carbide used in this study is reported to be 11.8 GPa.[21-22]

V_{pl} may be determined experimentally (not easy), estimated based upon cavity expansion models, or estimated from Equation (5) by setting $a_{pl} = R$. However, in this investigation, V_{pl} was estimated based upon the observation that penetration did not occur at an impact velocity of 500 m/s for Silicon Carbide. Based upon this value for V_{pl} and several values for $Y_{ceramic}$, the variation of the mean impact stress p_m with impact velocity is shown in Fig. 4. The predicted

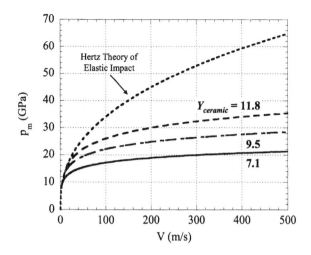

Fig. 4. Predicted mean impact stress for Silicon Carbide as a function of impact velocity and compressive yield strength (V_{pl} = 500 m/s).

variation based upon Hertz theory of elastic impact is also included for comparison.[14,23] Hertz's theory of elastic impact predicts a very high mean impact stress because the contact area is small. However, because of plasticity (or quasi-plasticity), the material is able to decrease the concentration of stress by spreading the load over a larger contact area. It can be shown that the contact radius as a function of mean impact stress is given by:

$$\frac{a}{R} \approx \left(\frac{a_{pl}}{R}\right)\left(\frac{a_Y}{a_{pl}}\right)^{\left(\frac{3-p_m/Y_{ceramic}}{1.9}\right)} \tag{7}$$

where a_{pl} and a_Y are given by Equations (5) and (6), respectively. The dependence on impact velocity through p_m is given by Equation (4). The predicted contact diameters as a function of impact velocity and several values for $Y_{ceramic}$ are shown in Fig. 5 assuming V_{pl} = 500 m/s. Based upon image analysis of the impact surface of the Silicon Carbide cylinder impacted at 161 m/s, the contact diameter is estimated to be between 2.6 and 3.3 mm.[6] As can be seen in Fig. 5, for an impact velocity of 161 m/s, the contact diameters predicted from the elastic-plastic analysis (2.7 – 3.4 mm) agree much better with observations than that predicted from Hertz's theory of elastic impact (2.2 mm). This lends some credence for the validity of the elastic-plastic analysis, at least for Silicon Carbide impacted at low velocities.

SUMMARY AND CONCLUSIONS

In part III of this study, preliminary analysis with respect to several questions which arose based upon observations made in parts I and II was made. These questions included estimating the impact velocities for which the impact event could still be considered nominally quasi-static, as well as those for the onset of plasticity and full-plasticity. In addition, estimations for the

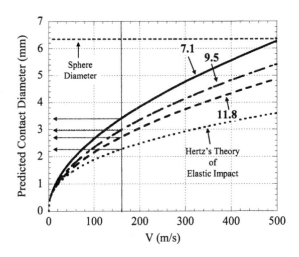

Fig. 5. Predicted contact diameter for Silicon Carbide as a function of impact velocity and compressive yield strength (V_{pl} = 500 m/s).

mean impact stress and contact diameter as a function of impact velocity were made. These estimations were based upon analytical results for the impact of an elastic sphere on an elastic half-space and those derived assuming the impact response of the ceramic to be elastic, perfectly-plastic. Critical to these estimates are the values for $Y_{ceramic}$ and V_{pl} (or a_{pl}) which are unknowns. An argument for using the HEL for $Y_{ceramic}$ was put forth, and may be validated by experimentally determining V_Y or through instrumented indentation. The value for V_{pl} (or a_{pl}) may be estimated by measuring the contact radius (by surface profilometry) for several values of impact velocity, calculating the mean impact stress by using Equations (4), (5), and (7), and then fitting the resulting mean impact stress – impact velocity curve using $Y_{ceramic}$ and V_{pl} as fitting parameters. Unfortunately, surface profilometry was not performed on the impacted cylinders before they were destructively characterized. Consequently, the analysis for mean impact stress and contact diameter as a function of impact velocity was performed using an assumed value for V_{pl}. It is noted that the value of V_{pl} may be greater than that assumed in this paper. It is expected that the analysis applied to Silicon Carbide for the mean impact stress and contact diameter is also applicable to Titanium Diboride and Tungsten Carbide since they also exhibit deformation mechanisms that can be considered as quasi-plastic (e.g. formation of a comminuted region in Titanium Diboride). However, it probably has only limited applicability to Boron Carbide because of its apparent lack of a quasi-plastic behavior.

REFERENCES

1. G.E. Hauver, P.H. Netherwood, R.F. Benck, and L.J. Kecskes, "Ballistic Performance of Ceramic Targets"; pp. 23-34 in *Proceedings of the 13th Army Symposium on Solid Mechanics*, Plymouth, MA, August 17-19, 1993.
2. P. Lundberg, R. Renstrom, and B. Lundberg, "Impact of Metallic Projectiles on Ceramic Targets: Transition Between Interface Defeat and Penetration," *Int. J. Impact Eng.*, 24 259-75 (2000).

3. C.J. Shih "Dynamic Deformation of Silicon Carbide", Ph.D Dissertation, 1998, University of California, San Diego.

4. J.C. LaSalvia, E.J. Horwath, E.J. Rapacki, C.J. Shih, and M.A. Meyers, "Microstructural and micromechanical aspects of ceramic/long-rod projectile interactions: dwell/penetration transitions", in Fundamental Issues and Applications of Shock-Wave and High-Strain-Rate Phenomena, eds. K.P. Staudhammer, L.E. Murr, and M.A. Meyers, Elsevier Science Ltd., New York, 2001, 437-446.

5. J.C. LaSalvia, "Recent Progress on the Influence of Microstructure and Mechanical Properties on Ballistic Performance," Ceramic Armor Materials By Design, eds. J.W. McCauley, A. Crowson, W.A. Gooch, A.M. Rajendran, S.J. Bless, K.V. Logan, M.J. Normandia, and S. Wax, Ceram. Trans., 134, 2002, 557-570.

6. J.C. LaSalvia, M.J. Normandia, H.T. Miller, and D.E. MacKenzie, "Sphere Impact Induced Damage in Ceramics: I. Armor-Grade SiC and TiB$_2$," Ceram. Eng. Sci. Proc., to be published.

7. J.C. LaSalvia, M.J. Normandia, H.T. Miller, and D.E. MacKenzie, "Sphere Impact Induced Damage in Ceramics II. Armor-Grade B$_4$C and WC," Ceram. Eng. Sci. Proc., to be published.

8. J. Sternberg, "Material Properties Determining the Resistance of Ceramics to High Velocity Penetration," J. Appl. Phys., 65 [9] 3417-424 (1989).

9. A.A. Wereszczak, "Elastic Property Determination of WC Spheres and Estimation of Compressive Loads and Impact Velocities That Initiate Their Yielding and Cracking," Ceram. Eng. Sci. Proc., to be published.

10. T. Weerasooriya, Private Communication.

11. H. Horii and S. Nemat-Nasser, "Brittle Failure in Compression: Splitting, Faulting, and Brittle-Ductile Transition," Phil. Trans. R. Soc. London A, 319 337-374 (1986).

12. M. J. Normandia, "Impact Response and Analysis of Several Silicon Carbides," Int. J. Appl. Ceram. Tech., 1 [4] 226-234 (2004).

13. B.R. Lawn, "Indentation of Ceramics with Spheres: A Century after Hertz," J. Am. Ceram. Soc., 81 [8] 1977-94 (1998).

14. K.L. Johnson, Contact Mechanics, Cambridge University Press, New York, 1985, 452 pp.

15. A.C. Fischer-Cripps, Introduction to Contact Mechanics, Springer Verlag, New York, 2000, 243 pp.

16. D.A. Shockey, D.J. Rowcliffe, K.C. Dao, and L. Seaman, "Particle Impact Damage in Silicon Nitride," J. Am. Ceram. Soc., 73 [6] 1613-1619, (1990).

17. W.J. Stronge, Impact Mechanics, Cambridge University Press, New York, 2004, 300 pp.

18. H.A. Francis, "Phenomenological Analysis of Plastic Spherical Indentation," Trans. ASME, Series H, J. Eng. Mats. Tech., 98 [3] 272-281 (1976).

19. R.F. Cook and G.M. Pharr, "Direct Observation and Analysis of Indentation Cracking in Glasses and Ceramics, "J. Am. Ceram. Soc., 73 [4] 787-817 (1990).

20. D.E. Grady, "Shock-Wave Compression of Brittle Solids," Mech. Mats., 29 [3-4] 181-203 (1998).

21. G. Yuan, R. Feng, and Y.M. Gupta, "Compression and Shear Wave Measurements to Characterize the Shocked State in Silicon Carbide," J. Appl. Phys., 89 [10] 5372-80 (2001).

22. D. Dandekar, private communication.

23. Y.M. Tsai, "Dynamic Contact Stresses Produced by the Impact of an Axisymmetrical Projectile on an Elastic Half-Space," Int. J. Solids Struct., 7 [6] 543-558 (1971).

A COMPARISON OF CERAMIC MATERIALS DYNAMICALLY IMPACTED BY TUNGSTEN CARBIDE SPHERES

M. J. Normandia, S. R. Martin, D. E. Mackenzie, and B. A. Rickter
U. S. Army Research Laboratory
Armor Mechanics Branch
Aberdeen Proving Ground, MD 21005

ABSTRACT

Ballistic comparisons of several armor-grade, hot-pressed ceramics are made using data obtained from dynamic impact experiments with tungsten carbide (WC-6%Co) spheres. Titanium Diboride, Silicon Carbide, Tungsten Carbide and Boron Carbide were tested and all ceramics show the same three distinct groupings, characterized by impact velocity. A lower velocity regime, to as high as 500 m/s, characterized by little or no visual surface damage (other than ring cracks), while significant sub-surface damage is generated (see companion papers). A higher velocity regime is characterized by inelastic impactor response, yield or shattering, which typically occurs above 1 km/s. The damage is evident by craters that form and the impactor penetrates the damaged material. This regime is representative of typical armor applications. An intermediate velocity range exists, characterized as an onset to penetration where the damage results in craters that are not necessarily fully developed, but where the impactor remains elastic. The penetration response curves generated for each of these materials are compared on an areal density basis. Less areal density penetrated at a given impact velocity implies greater target resistance to penetration.

The rate of crater depth with increasing velocity is governed by target properties related to ballistic penetration resistance, which is typically used within analytic penetration formulations. Penetration resistance may be obtained from correlation with the data. This is a lumped parameter that is related to the constitutive behavior of comminuted ceramic material, which is typically assumed to behave as a granular material with a pressure-dependent Mohr-Coulomb behavior. The slope of the pressure-shear dependence can be approximated from the testing technique.

INTRODUCTION

Dynamic impact experiments using tungsten-carbide spheres generate sub-surface damage up until a critical velocity, above which the damage is visible on the surface in the form of displaced ceramic. At high velocities, a crater is formed and the crater depth increases with increasing velocity. These depths are used to determine target resistance by using analytic models for the penetration process, and often include cavity expansion formulations for brittle materials. However, at these high velocities, the tungsten-carbide spherical penetrator typically deforms or shatters, making the analysis of target resistance dependent on the accuracy of the models used to describe the penetrator deformation. Computational numerical simulation tools have been used to correlate material constitutive properties using this data. It is preferable to utilize data obtained below the velocity at which the penetrator deforms inelastically. This velocity depends upon the material, as the impact stress generated depends upon the material strength. Targets with greater strength plastically deform the penetrator at lower impact velocities. Knowledge of the dynamic yield strength of the various tungsten-carbide penetrators used can then be used to deduce the target strength by determining the velocities at which the penetrator receives permanent deformation. The dynamic yield strength is estimated to be 4.9 or 5.4 GPa for tungsten carbide materials with 6% weight of Cobalt binder, as determined by Weerasooriya[1-2].

Unfortunately, for these intermediate velocities, the crater is not fully formed making the measurement of crater depth somewhat subjective. The deepest depth is typically near the impact point, but not directly under it. At high enough velocities, the maximum depth is under the impact. For impact velocities where the sphere remains elastic and above which visible surface damage occurs, a comparison of four hot-pressed ceramics purchased from Cercom, Inc. of Vista, CA are compared.

The materials from these experiments are recovered and can be examined for damage. This has been done for impact velocities below which surface "penetration" is observed and is the subject of companion papers by LaSalvia[3-5] et al. in this symposium. The damage progression observed with increasing impact velocity is similar to that discussed by Shockey[6-7] for silicon nitride and Kim[8-9] for other ceramics by generating indentation stress-strain material response curves. However, the appearance of a comminuted region of damaged material beneath the impact location is very evident in both silicon carbide and titanium diboride at impact velocities below 100 m/s, but is not evident in tungsten carbide and boron carbide at impact velocities exceeding 300 m/s.

BALLISTIC IMPACT BACKGOUND

The ballistic performance of a ceramic is system dependent; no one design is optimal for all ceramics or ceramic variations. Therefore, it is important to minimize the effects of target design when trying to assess ceramic potential for armor applications or to determine properties required for numerical and analytic models. Normandia and Gooch[10] reviewed earlier attempts to screen ceramics, including the use of tungsten-carbide spherical penetrators by Donaldson[11-12]. Critical review by Sternberg[13] of the particular flow models derived from ductile metal flow theories to determine target resistance of brittle materials were discussed, but the experiments themselves were not questioned.

Ceramics ballistically impacted with tungsten-carbide spheres generate response curves of penetration depth as a function of impact velocity. Recent experimental results for several silicon carbide variants were discussed by Normandia and Leavy[14] and Normandia[15] which revealed common features with ballistic response curves of metal (steel) impacts generated by Martineau[16]. Significant information is obtained when comparing response curves of different ceramics based on areal density penetrated, with or without any particular theoretical analysis of the data. A lower areal density penetrated at a given impact velocity implies a more efficient ceramic and a greater target resistance. The target resistance may be quantified in terms of a ballistics penetration model, of which there are many.

Ballistic limit experiments that determine a v_{50} using ceramics on composites or metal backings are common methods of ranking candidate materials. The attempts at using data generated from sphere impacts to rank candidate armor materials have had mixed results. In very light systems that rely on penetrator fracture or shatter, the results do not correlate very well. Several reasons for this are that the velocity is varied in v_{50} tests, and the material allocation is not optimized for a given material system, making the results dependent on the system properties and not necessarily strictly with the properties of the ceramic. However, for confined ceramics systems (similar to these tests) the results are more encouraging.

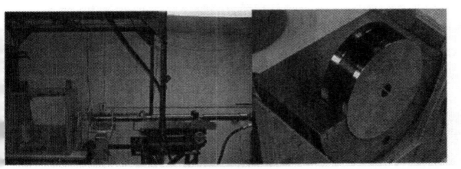

Fig.1. Photographs of air-gun test facility (left) and target within a cut-away of the fixture (right).

EXPERIMENTAL PROCEDURES

The test facility depicted in Figure 1 consists of a compressed gas gun launch tube, laser velocimeters, and dual x-rays. Also shown is a cutaway of the Ti6Al4V fixture confining the ceramic. The target components are depicted in Figure 2, which show an impacted Cercom hot-pressed SiC-N at 800 m/s. The crater is well defined as is the radial extent of the damage region. The rear surface of the front plate captures the damage features in the ceramic showing radial and circumferential cracks. All metal components are Ti6Al4V circular stock. The 3.125 mm thick front plate has an entry hole to accommodate the 2 g, 6.35mm diameter, WC spherical penetrator purchased from one of two vendors, Machining Technologies Inc., Grade 25 (.000025" roundness tolerance), Class C-2 material (Rockwell A 92) and contained 6% Co binder with a 14.93 g/cc measured density, or New Lenox Machining Company with similar properties with a measured density of 14.89 g/cc. The 9.525 mm thick back plate confines a 19.05 mm thick, 76.2 mm diameter ceramic, held to tight tolerances and enclosed in a 101.6 mm diameter ring. Four hot-pressed ceramic materials were purchased from Cercom Inc., and their measured elastic properties are provided in Table I. Microstructures of these materials are shown in LaSalvia[3-4] et al.

The test procedure consists of launching the spheres embedded in a flared sabot from the launch tube, measuring the time between two lasers at the end of the launch tube to determine velocity, discarding the sabot prior to impact, and observing the penetrator in flight, prior to impact, with two 150keV x-radiographs, which also served as a velocity measurement backup. A significant benefit of this technique is material recovery for microscopic analyses. This has been done for similar targets using these same materials, 25.4 mm thick and 25.4 mm in diameter and is reported in LaSalvia[3-4] et al. Depths were measured directly under the impact point and at the maximum crater depth. A Faro gage arm was used to measure depths within an accuracy of 0.02 mm.

Fig. 2. Target components: 101.6 mm diameter Titanium 6Al4V cover with hole (inside shown), backing plate and confining ring containing 76.2 mm ceramic (typically 19.05 mm thick). This SiC-N tile was impacted at 800 m/s and exhibits well defined crater and radial damage extent.

205

Material (Cercom)	Density (g/cm^3)	Grain Size (μm)	E (GPa)	ν	HK4 (GPa)	K_{IC} $(MPa\text{-}m^{1/2})$	σ_b (MPa)	Phases
SiC-N	3.22	3.3	452	0.165	18.6	5.1	500	6H, 15R, 3C
PAD TiB_2	4.52	15*	555	0.11	16.8	6.9*	285*	TiB_2
PAD B_4C	2.53	15	465	0.17	18.9	2.9	450	B_4C, C
PAD WC	15.6	0.4**	697	0.20	20**	6 - 8**	1030**	WC, W_2C

*Cercom Material Data Sheet
**J.J. Swab, "Hardness and Damage Associated with Pointed Indenters in Armor Ceramics", PhD Dissertation, 2004

Table I. Measured and Reported Properties For Ceramic Materials Tested.

RESULTS & DISCUSSION

Shown in Figure 3 is a sliced and polished cross-section of a recovered boron carbide specimen impacted at 312 m/s, LaSalvia et al[4]. There was no penetration under the impact point, but material was removed (or spall formed) at the surface at the edges of the load. When penetration under the impact point begins, the deepest measurable depth is often away from the impact point. At higher velocities, the maximum depth is under the impact point as can be seen in Figure 2.

Figure 3. SEM of sliced and polished cross section of Boron Carbide impacted at 312 m/s.

Typically test firings conducted below ~350 m/s produced little or no visible "penetration" damage to the ceramics impacted, other than radial cracks. As reported in Normandia[15] (indicated by framed pictures in Figure 6), the first visible signs of damage for SiC-N was observed at 393 m/s using Machining Technologies WC-6% Co spheres. The current data was generated using similar WC-6% Co spheres manufactured by New Lenox (indicated by unframed pictures). For all of the hot-pressed ceramics tested, the damage extent (diameter) can be identified as a function of impact velocity. This was measured, but not reported here. Imprints on the backing plate surface revealed radial cracks, until the highest velocities, when circumferential cracks clearly appeared. In all tests, the rear surface of the ceramic remained smooth and the backing plates had no permanent deformation.

After careful removal from the fixture, the front plate was removed and photographs were taken. Careful excavation of the surface revealed the deepest point of penetration and the penetration directly under the impactor, which were not always identical at the low velocities. After excavation, well defined craters were observed and indentation/penetration depths and crater diameters were measured, as were the radial extent of the circumferential damage. Crater diameter correlated to the diameter of the opening in the cover plate, which implies the confiment provided by the cover plate influenced the surface damage. All targets have preserved for microscopic analysis prior to, and after, sectioning. All pictures shown are after excavation, but photos are available pre-excavation as well. Data generated are shown in Figures 4-7, respectively for Titanium Diboride, Tungsten Carbide, Boron Carbide, and Silicon Carbide. Well defined craters are evident, particularly at higher velocities. See for example, Figure 5 for a close up of a crater impacted at 750 m/s.

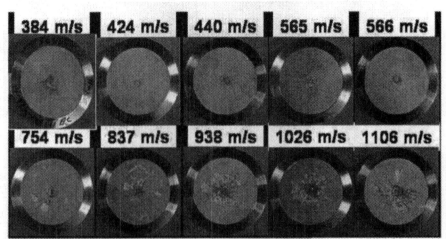

Figure 4. Photographs of damage progression in confined Cercom TiB$_2$ ceramic disk impacted by 6.35 mm WC-6% Co New Lenox spheres at impact velocities shown.

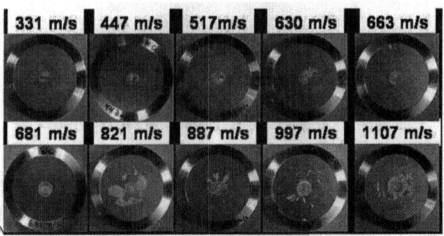

Figure 5. Photographs of damage progression in confined Cercom WC ceramic disk impacted by 6.35 mm WC-6% Co New Lenox spheres at impact velocities shown.

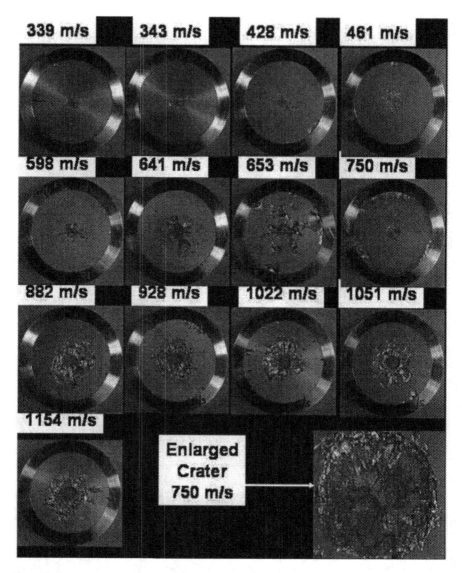

Figure 6. Photographs of damage progression in confined Cercom B_4C ceramic disk impacted by 6.35 mm WC-6% Co New Lenox spheres at impact velocities shown.

208

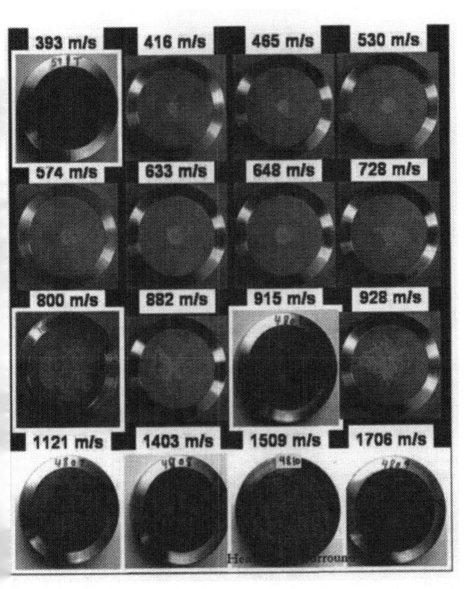

Figure 7. Photographs of damage progression in confined Cercom SiC-N ceramic disk impacted by 6.35 mm WC-6% Co Superior Graphite spheres (framed in white) and New Lenox WC-6% Co spheres (unframed) at impact velocities shown. Data for impact velocities >1200 m/s appears in Normandia[15].

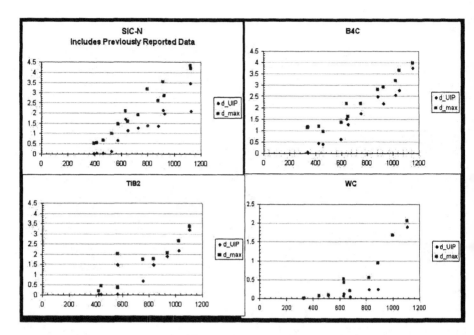

Figure 8. Measured depths for various Hot –pressed ceramics impacted by WC-6% Co New Lenox spheres at impact velocities shown.

Measurements for all of the ceramics, in mm of depth are shown in Figure 8. The depth directly under the impact point is denoted as d_UIP or diamond symbols, and the maximum crater depth is denoted as d_max, or square symbols. As the velocity increased and the crater was well formed, these gave identical measurements (slight differences are expected due to excavation process). At lower velocities, the depths varied, as discussed with Figure 3, in that the top of the "mountain-peak" was often accompanied by failed material (in tension) that was ejected as spall at the edges of the load. Both measurements are presented.

All of the maximum crater depths are shown in Figure 9 for the four materials tested. These are compared on an areal density basis. This is the product of the maximum crater depth and the target material density (see Table I) divided by the product of the sphere diameter and the sphere density. This enabled comparisons using the two different WC sphere variants used. For the materials tested here, the ranking based upon least areal density penetrated changes with impact velocity. Boron carbide penetrates the deepest on an areal density basis for velocities below 400 m/s, but penetrates the least for velocities above ~700 m/s. Tungsten carbide, on the other hand, penetrates least for velocities below ~600 m/s and penetrates the most for velocities above ~850 m/s. Both SiC-N and Titanium Diboride penetrate about the same on an areal density basis throughout the entire velocity range and fall in between the boron carbide and tungsten carbide data. The slope of the data prior to the deformation of the WC is linear. The velocity at which penetrator deformation occurs can be used to estimate the penetrator dynamic yield strength, if the target strength is known, or can be used to estimate the target strength with the aid of correlation to the analytic models. The dynamic yield strength for the WC was chosen as 5.4 GPa (slightly higher than the 4.9 +/- .25 GPa used in earlier work, Normandia[15]), based on recent Split-Hopkinson Pressure Bar experimental data by Weerrasooryria[2]. This analysis is not presented here. The slope of the penetration data changes above the velocity at which the WC sphere deforms. The slope of the penetration data below this velocity is determined solely on target material strength and elastic penetrator properties, while above the critical velocity, the slope is determined by a relative strength difference between the target and the deforming penetrator strengths, hence is lower. The deformation increases the contact area with the target and the penetration can decrease. Failure strain limits the maximal extent of penetrator deformation.

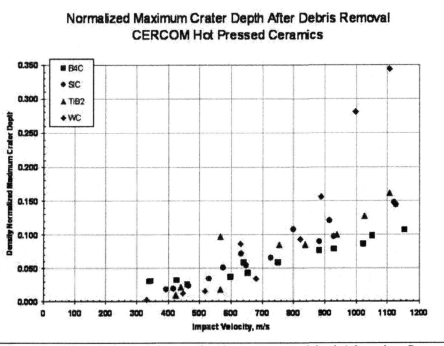

Figure 9. Areal density penetrated (normalized by penetrator areal density) for various Cercom Hot –pressed ceramics impacted by WC-6% Co New Lenox spheres at impact velocities shown.

SUMMARY

Four armor-grade, hot-pressed and pressure densified ceramic variants were impacted using tungsten-carbide spheres containing 6% by weight, Cobalt binder. Three distinct impact regimes were observed. The first occurs until ~400 m/s, where visible surface damage was limited to radial cracks and sometimes spall away from the impact point. However, significant subsurface damage is evident and is discussed elsewhere[3-5]. A second regime occurs above this velocity where craters begin to form, and a third regime where the craters are fully formed, typically at ~600-800 m/s, depending on the material impacted. Measurements of depth under the impact point, and the maximum crater depth were reported for use in subsequent analytic and numerical analyses.

REFERENCES
[1]T. Weerasooriya, P. Moy, and W. Chen, "Effect of Strain-Rate on the Deformation and Failure of WC-Co Cermets under Compression," presented at the 29th International Conference on Advanced Ceramics and Composites, *Cocoa Beach*, Jan. 2004.
[2]T. Weerasooriya, P. Moy, and W. Chen, "Effect of Strain-Rate on the Deformation and Failure of WC-Co Cermets under Compression," presented at the 29th International Conference on Advanced Ceramics and Composites, *Cocoa Beach*, Jan. 2005.
[3]J.C. LaSalvia, M.J. Normandia, H.T. Miller, and D.E. MacKenzie, "Sphere Impact Induced Damage in Ceramics: I. Armor-Grade SiC and TiB$_2$," *Ceram. Eng. Sci. Proc.*, to be published.
[4]J.C. LaSalvia, M.J. Normandia, H.T. Miller, and D.E. MacKenzie, "Sphere Impact Induced Damage in Ceramics II. Armor-Grade B$_4$C and WC," *Ceram. Eng. Sci. Proc.*, to be published.
[5]J.C. LaSalvia, M.J. Normandia, H.T. Miller, and D.E. MacKenzie, "Sphere Impact Induced Damage in Ceramics III. Analysis," *Ceram. Eng. Sci. Proc.*, to be published.
[6]D. A. Shockey, A.H. Marchand, S.R. Skaggs, G.E. Cort, M.W. Burkett, and R. Parker, "Failure Phenomenology of Confined Ceramic Targets and Impacting Rods," *Int. J. Impact Eng.*, 9 [3] 263-75, 1990.
[7]D. A. Shockey, D. J. Rowcliffe, K. C. Dao, and L. Seaman, "Particle Impact Damage in Silicon Nitride," *Journal of the American Ceramic Society*, 73 1613-19, 1990.
[8]D. K. Kim and C-S Lee, "Indentation Damage Behavior of Armor Ceramics," *Proceedings of the Symposium on Ceramic Armor Materials by Design*, PAC RIM 4, pp. 429-440, Wailea, Maui, HI, November 4-8, 2001.
[9]D. K. Kim, C-S Lee, and Y-G Kim, "Dynamic Indentation Damage of Ceramics," *Proceedings of the Symposium on Ceramic Armor Materials by Design*, PAC RIM 4, pp. 261-268, Wailea, Maui, HI, November 4-8, 2001.
[10]M. J. Normandia and W. Gooch, "An Overview of Ballistic Testing Methods of Ceramic Materials," *Proceedings of the Symposium on Ceramic Armor Materials by Design*, PAC RIM 4, pp. 113-138, Wailea, Maui, HI, November 4-8, 2001.
[11]C. duP. Donaldson and T. B. McDonough, " A Simple Integral Theory for Impact Cratering by High Speed Particles," *Aeronautical Research Associates of Princeton, Inc. A. R. A. P. Report No 201*, Dec. 1973.
[12]C. duP. Donaldson, R. M. Contiliano, and C. V. Swanson, "The Qualification of Target Materials using the Integral Theory of Impact," *A. R. A. P. Report No 295*, Nov. 1977.
[13]J. Sternberg, "Material Properties Determining the Resistance of Ceramics to High Velocity Penetration," *J. Appl. Phys.*, 65 [9] 3417-424 (1989).
[14]M. J. Normandia and B. Leavy, "Ballistic Impact of Silicon Carbide with Tungsten Carbide Spheres," *Proceedings of ACER Cocoa Beach*, Jan. 2004.
[15]M. J. Normandia, "Impact Response and Analysis of Several Silicon Carbides," *International Journal of Impact Engineering*, Vol. 1. Issue 4 Jul. 2004.

Non-Destructive Evaluation

ULTRASONIC TECHNIQUES FOR EVALUATION OF SIC ARMOR TILE

J. Scott Steckenrider
Illinois College
1101 W. College Ave.
Jacksonville, IL 62650

William A. Ellingson, Jeffery Wheeler
Argonne National Laboratory
9700 S. Cass Avenue
Argonne, IL 60439

ABSTRACT

Obtaining reliable ballistic performance is important to full utilization of ceramic armor tile. One possible way to reduce scatter in ballistic performance is through utilization of a reliable, low-cost, high-speed nondestructive evaluation (NDE) method that could act as a screening tool to select those tiles that could produce poor ballistic performance. Ultrasonic methods utilizing advanced phased array technology would seem to be a possible candidate. As a step towards evaluation of this technology, recent single transducer water-immersion tests have been conducted using both focused and unfocused transducers on a variety of SiC armor materials. These materials had grain sizes from 0.89 um to 5.26 um. These different grain sizes likely impact the signal-to-noise ratio (SNR) at different frequencies and therefore these tests provide information on which to select the frequencies to use for phased arrays. In addition to these empirical studies, an analytical ultrasonic performance prediction model has been evaluated. This paper will present the ultrasonic methods under study and review results to date.

INTRODUCTION

Poor ballistic performance of SiC armor tile for vehicles is a concern for reliable protection of military personnel. One approach to reduce the possibility of poor ballistic performance would be to develop an effective non-destructive screening method that could sort "good" tile from "bad" tile[1]. This implies that the knowledge exists for judging a "good" tile from a "bad" tile. To some extent, this is indeed known and the "good", "bad" differentiation can likely come from the presence of flaws in the form of inclusions, voids or large individual grains[2]. While a correlation between the size/position of these processing flaws and ballistic performance is still being established, such a correlation would require a method for detecting and characterizing such flaws to establish ballistic performance. Based on this information, the use of phased array ultrasonic technology would seem to offer a possible solution[3]. Such technology allows fast scanning over a large region, provides high SNR data for defect depth discrimination and would likely be cost effective[4,5].

APPROACH

In order to properly explore ultrasonic methods, the first issue is selection of the best ultrasonic frequency (i.e., the one with the greatest Signal-to-Noise Ratio, or SNR). By comparing these frequencies using defect types similar to those expected in "real" specimens and at a variety of defect depths the key components of detectability and penetration depth are respectively incorporated into such a SNR. Two approaches, one empirical and the other theoretical, are being undertaken to address this question. The empirical approach is being accomplished using Argonne's water-coupled SONIX ultrasonic system. In this work, both focused and unfocused transducers were evaluated. The unfocused transducers used were 6.4 mm diameter immersion transducers from R/D Tech (formerly Panametrics NDT) of nominally identical designs but at operational frequencies of 10, 15, 20 and 25 MHz. The focused transducer was also from R/D Tech but was 19mm in diameter and had a nominal 12.7 cm focal length in water. (Note that because of the extremely high ultrasonic velocities in these SiC materials a very long focal length was required in order to focus into the depth of the specimen.) All data have been recorded using an A/D interface operated at a double sampled 400 MHz acquisition rate. In addition to these efforts for establishing the S/N ratio, initial work was done using an 128-element phased array. The theoretical analytical modeling effort used a 2-dimensional Finite Element Modeling software package called Wave-2000.

TEST SAMPLES

In order to establish the "best" ultrasonic transducer frequency, several test specimens with different known grain sizes and geometries were used. The four materials were: Cercom SiC-N, Cercom SiC-SC1R, Coors SC300 and Kyocera SC1000. The last three of these specimens had nominal dimensions of 101.6 mm x 101.6 mm x 30 mm thick while the first was only 25.4 mm thick.

In addition to having various grain sizes, one sample, the Cercom SiC-N, was altered by drilling 3.2 mm diameter holes (nominally flat-bottomed) into the back surface (see Fig. 1.). Although it is acknowledged that the size of these holes represent defects far larger than what is considered critical in application, these provided sufficient comparison of relative signal strength for comparison of the transducers' signal-to-noise ratio (SNR). However, a comparable specimen containing more realistic defects (i.e., processing inclusions) is being fabricated and will be evaluated in future work.

Materials properties for all specimens, as measured ultrasonically, are presented in Table I. As would be expected, the materials with equiaxed grains show no statistically significant anisotropy while the elongated grain materials do show anisotropy of greater than 3%.

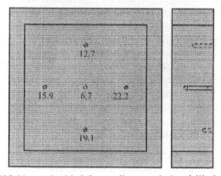

Figure 1. Diagram of SiC-N panel with 3.2 mm diameter holes drilled to various depths in the back side. Darker shaded region indicates location where ultrasonic C-scan was taken. Depths shown are mm below the top surface.

216

Table I. Properties of silicon carbide materials evaluated (measured at 10 MHz).

Mechanical Property	Cercom SiC-N	Cercom SiC-SC1R	Coors SC300	Kyocera SC1000	Units
Mfg. Method	Hot-pressed	Hot-pressed	Sintered	Sintered	---
Average Grain Size	3.18	0.89	5.26	3.37	μm
Grain Type	Equiaxed	Equiaxed	Elongated	Elongated	---
Density	3.241	3.211	3.108	3.119	g/cc
V_L Thru-Thickness	12,223	12249 ± 44	11863 ± 47	12230 ± 29	m/s
V_T Thru-Thickness	7,674	7321 ± 52	6875 ± 37	7105 ± 20	m/s
λ (Lamé constant)	102,476	137,581	94,502	86,213	MPa
μ (Lamé constant)	190,883	172,096	146,903	157,418	MPa
V_L Along Plate	N/A	12231 ± 12	11512 ± 19	11837 ± 13	m/s

RESULTS

Empirical – C-scan Evaluation

A fast way to ultrasonically interrogate a volume of material for bulk defects using conventional transducers is to perform a pulse-echo C-Scan. Figure 2 shows a typical C-Scan of the SiC panel with back-drilled holes. The scan is oriented to correspond to the diagram shown in Fig. 1. The color scale represents relative response amplitude as shown in the accompanying scale. Scans were performed with unfocused transducers of 10, 15, 20 and 25 MHz as well as with the focused 10 MHz transducer. For each transducer a relative SNR was measured by comparing the average response over each back-drilled hole to the average background response. For all transducers the C-scan window was set to begin 0.65 μs after the front surface reflection and end immediately before the back-surface reflection. Thus, the depth inspected was from 4.0 mm below the front surface to the back surface. Note, however, that in application this "blindness" could be overcome by repeating the scan from the backside.

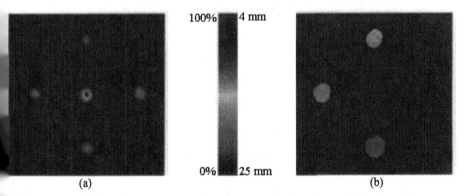

(a)   (b)

Figure 2.   10 MHz C-scan results of SiC panel with back-drilled holes using unfocused transducer. a) Peak amplitude and b) Time-Of-Flight (TOF) response. Percentage scale is for (a) and represent % of full-scale amplitude. Depth scale is for (b) and represents depth below front surface.

217

Figure 3a shows the relative SNR for the 10, 15, 20 and 25 MHz transducers at each hole depth, indicating that the 20 MHz transducer appears to provide the greatest SNR. It should be noted that this is more related to an increase in "noise" in the higher frequency transducer. The 25 MHz transducer appeared to be somewhat less damped than the 20 MHz transducer so that there was still a remnant of the front-surface reflection adding "noise" to the C-scan response. However, the similar shape of each curve indicates that, in the range of thicknesses evaluated, penetration depth does not contribute significantly to any variation in SNR between frequencies.

The effect of focusing is shown in Figure 3b, which plots the improvement in SNR for the 10 MHz focused transducer at two different focal depths (relative to the unfocussed transducer). Note that the focused 10 MHz transducer has a much greater SNR than the unfocused 10 MHz transducer for the holes closest to the focal depth. However, its response for the shallowest hole (several mm short of the focus) is significantly diminished, as would be expected. The figure shows that there is an improvement of approximately 4-5 dB in SNR for the focused transducer relative to the unfocussed transducer at the focal depth.

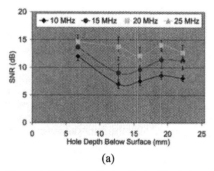
(a)

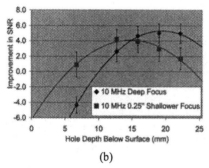
(b)

Figure 3. Signal-to-Noise Ratio (SNR) for C-scan data comparing (a) different acoustic frequencies, (b) effect of focus and focal. Error bars represent 1 σ.

Empirical – B-scan Evaluation

Unlike the C-scan, which records only the peak of the reflection within the window of analysis, a B-scan records the entire reflected waveform within that window, thereby allowing visualization of multiple defects at different depths. This essentially gives a "slice" through the specimen thickness, resulting in an x-z image (where z is depth into the specimen). Also, because the entire waveform (within the window) is recorded, a B-scan can be further filtered after the data have been acquired to enhance sensitivity. However, because it records the entire waveform, a B-scan is much more memory intensive on a per-measurement basis (two to three orders of magnitude or more), and is therefore often impractical to use for inspecting a large area. A representative B-Scan over the 0.5" deep hole is shown in Fig. 4a.

One means of additional analysis available with the B-scan is the opportunity to subtract out systematic variation that can contribute significantly to the "noise" floor. As can be seen in the top of the B-scan of Fig. 4a, there are horizontal lines at the top of the image that represent the "ringing" of the transducer from the front surface reflection. In a C-Scan these high spots are the peaks detected in the absence of any other defects. Thus, a defect must reflect a pulse with amplitude greater than these in order for the C-scan to detect it. However, in the B-scan an

218

average of this "defect-free" response can be subtracted from the image, thereby lowering the "noise" floor and increasing the effective sensitivity of the scan. This is shown in Fig. 4b where an average of 100 "defect free" scans was subtracted from the B-scan image. Although the difference appears to be subtle in Fig. 4b the impact on the effective SNR is shown in Fig. 5, where the 25 MHz probe shows the highest SNR and all frequencies show an improved SNR.

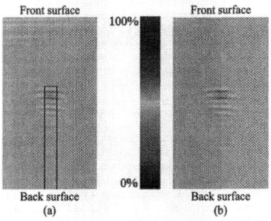

Figure 4.   10 MHz B-scan results for 0.5" deep back-drilled hole using unfocused transducer (a) as-scanned and (b) after subtracting a "background" signal. The scale again represents percent of full scale response.

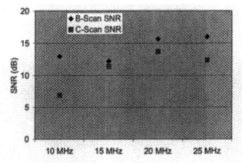

Figure 5.   SNR response for B-scan data after subtraction of background and comparison to C-scan SNR.

Theoretical
    The boundary conditions selected for the model were slightly modified from the experimental conditions in order to reduce computational time. Transducer standoff distances used in the model were 7 mm rather than the 23.5 mm used experimentally. This allows enough time-delay in the transmit/receive transducer to avoid multiple reflections from the front surface of the specimen and thus not interfere with sub-surface signals.

Correlation of the theoretical model data to the empirical data was first confirmed by comparing the amplitude of a reflection from a 3.2 mm diameter back-drilled hole. The model was then applied to a "real" defect type for simulation. Alumina media are used in the process of grinding the SiC powders before sintering or hot-pressing and small chips of this alumina can become embedded in the SiC component. A nominal defect size of 0.06 mm^2 cross section was chosen based on discussions with others in the field. Because this is a 2-dimensional model only breadth and thickness were considered. Two defect geometries were evaluated: a 100 μm thick x 600 μm wide plate and a 250 μm diameter circle (i.e., 2-D representation of a spherical defect). While it is acknowledged that these defects will be effectively larger than the real defects they represent (since this is only a 2-D model, so that the defects effectively extend infinitely into the third dimension) this approximation was accepted for initial evaluations, as a 3-D version of the model was not available. All data shown used a 10 MHz transducer even though other frequencies were explored.

Figure 6a shows an example of a predicted waveform for a thin planar alumina inclusion at the mid-point of a 25.4 mm thick SiC panel. From this waveform the amplitude of the reflection from the alumina plate relative to the amplitude of the back-surface reflection can be use to compare against the experimental SNR results to predict the experimental sensitivity to such a defect. In the case of the focused transducer the reflection from the alumina plate is approximately 3.3% of the amplitude of the back surface reflection. While this is too small to be detected by the full-window C-scan presented above it would be more than 10 σ above the mean response in the B-scan analysis (with a 4 dB SNR), and thus would likely be detectable experimentally. In addition, digital narrow-band filtering of the detected signal would further enhance the SNR (which is inversely proportional to the square root of bandwidth), thereby increasing the ability to detect such an alumina inclusion.

A similar analysis modeling the unfocused transducer (Fig. 6b) shows that the reflected amplitude from the alumina plate inclusion would be only 0.85% of the back-surface reflection amplitude. This would be less than 2 σ above the mean B-scan response, and thus would not likely be detectable.

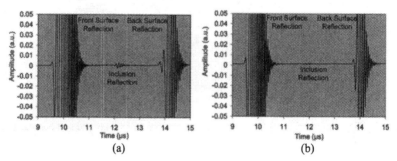

Figure 6. Theoretically predicted waveforms for (a) 10 MHz focused (at panel mid-plane) and (b) 10 MHz unfocused transducers when inspecting a SiC panel with a 100 μm x 600 μm alumina inclusion at its center.

Phased Array

The initial tests of phased array ultrasound incorporated a 128-element array nominally 76.2 mm long operating at 10 MHz (as a higher frequency transducer was not commercially available at the time) in which 32 elements were active at any given time (giving an effective transducer diameter equal to that of the focused transducer used in the C-scan results presented above). Fig. 7 shows these initial results for a C-scan (upper left quadrant), B-scans (upper right and lower left quadrants) and an A-scan (a single waveform from a "background" region in the lower right quadrant) of the SiC tile with back-drilled holes. Table II shows the SNR for each hole using this system.

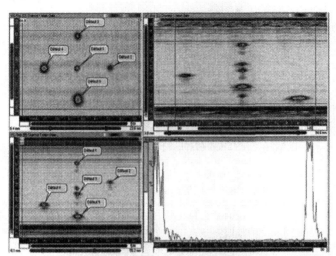

Figure 7.   Representative phased array C-scan and B-scan data of SiC panel with back-drilled holes.

Table II. SNR data from phased array data for each of the holes (in order of increasing depth below the front surface).

Defect Depth (mm)	SNR (dB)
3.2	27.9
6.4	21.9
9.6	24.2
12.7	28.7
18.7	27.7

DISCUSSION

Initial C-scan data clearly indicates that the optimum frequency for maximum SNR would be 20 or 25 MHz for the SiC materials tested. Further, as would be expected, data also suggested that a focused transducer enhanced the SNR by approximately 5 dB by concentrating the acoustic energy at the defect. However, because that enhancement is observed only at the

221

depth of the transducer's focus it has essentially narrowed the inspection window, so that implementation of a focused transducer would require an additional axis of scan, thereby dramatically increasing the total scan time required.

Another means of improving SNR by an average of more than 3 dB can be demonstrated by recording the entire acoustic waveform using a B-scan. By averaging and subtracting systematic background "noise" from each waveform the process becomes less sensitive to response variations from one transducer to the next (as well as any additional geometrically induced waveform variation). However, as mentioned above, the requirement to save an entire waveform for each point scanned makes this an extremely memory-intensive approach which would again increase the scan time required.

The potential of phased array of ultrasonic probes was demonstrated using a 76.2mm long array. Two of the three scan dimensions (lateral and depth) required for a 3-D C-scan can be accomplished electronically, leaving only one dimension of mechanical scanning. Thus the enhancement of using a focused transducer can be achieved without sacrificing scan time. In addition, because the phased array system only records signals at the arrival time corresponding to its focal point, the remainder of the waveform can be excluded from falsely triggering the system, thereby accomplishing the same "background" subtraction achieved in the B-scan.

CONCLUSIONS AND FUTURE WORK

Ultrasonic methods have been evaluated for their potential for detecting defects which compromise ballistic performance in ceramic armor. Preliminary results from both empirical tests and analytical modeling indicate that 20-25 MHz would be the best frequencies to select. In addition, phased array inspection has been demonstrated to improve upon the SNR of conventional ultrasonic inspection by as much as 10 dB even at non-optimum frequencies.

Future efforts will be focused in two parallel efforts. First, application of the modeling work will be extended to higher frequencies and will incorporate different SiC materials. These results will then guide the selection of conditions for future empirical work. Second, test specimens with more "realistic" defects are being produced so that we can more directly correlate empirical and theoretical analyses and so that we can determine the degree to which phased array methods can be extended to these defect types.

REFERENCES

[1] J.M. Wells, W.H. Green and N. L. Rupert—""On the Visualization of Impact Damage in Armor Ceramics", Eng. Sci. and Eng. Proc. Vol, 22, Issue 3, H.T. Lin and M. Singh, eds, pgs 221-230, 2002

[2] Y. Tanabe, T. Saitoh, O. Wada, H. Tamura, and A. B. Sawaoka, "' An Overview of Impact Damages in Ceramic materials---For Impact below 2km/s" in Review of the Research Laboratory of Engineering materials, Tokyo Institute of Technology 19, 1994

[3] G.P. Singh and J. W. Davies, "Multiple Transducer Ultrasonic Techniques: Phased Arrays" In Nondestructive Testing Handbook, 2nd Ed. , Volume 7, pgs284-297. 1991

[4] D. Lines, J. Skramstad, and R. Smith, " Rapid, Low-Cost, Full-Wave Form Mapping and Analysis with Ultrasonic Arrays", in Proc, 16th World Conference on Nondestructive Testing, September , 2004.

[5] J. Poguet and P. Ciorau, " Reproducibility and Reliability of NDT Phased Array Probes", in Proc, 16th World Conference on Nondestructive Testing, September , 2004.

NON-DESTRUCTIVE EVALUATION (NDE) OF CERAMIC ARMOR: FUNDAMENTALS

Raymond Brennan
Rutgers University
607 Taylor Road
Piscataway, NJ 08854-8065

Richard Haber
Rutgers University
607 Taylor Road
Piscataway, NJ 08854-8065

Dale Niesz
Rutgers University
607 Taylor Road
Piscataway, NJ 08854-8065

James McCauley
US Army Research Laboratory
Aberdeen Proving Ground, MD 21005-5066

Mahesh Bhardwaj
The Ultran Group
1020 Boal Avenue
Boalsburg, PA 16827, USA

ABSTRACT

Anomalous defects are known to initiate fracture in ceramic armor during quasi-static testing. Results will show the feasibility of using non-destructive evaluation (NDE) techniques as a means of locating anomalous defects within dense, ceramic materials used for armor applications. Implementation of a cost-effective NDE technique can reduce ballistic performance scatter by establishing some "flaw-population/flaw-size" threshold level above which the part will be rejected and below which it will be accepted. High frequency ultrasound techniques will be used to locate and identify defects within ceramic armor materials.

High frequency ultrasound NDE (generally >10 MHz) is based on the relationship between acoustic measurements and material properties. Ultrasound wave velocities are directly dependent on physical characteristics of the medium through which they are transferred. Reflection, transmission, and scattering of ultrasonic waves can be studied to detect discontinuities and calculate elastic properties. This technique will provide crucial information about critical flaw sizes and populations in ceramic armor, serving as a major step for nondestructively gauging the success and failure of commercial armor plates.

DESTRUCTIVE TESTING

Critical-sized defects and flaws such as pores and inclusions can lead to failure in ceramic armor. This can mean the difference between life and death for the soldiers who depend on armor plates in their vests to protect them during combat. Currently, the most common

223

method for evaluating the material integrity of ceramic armor materials is destructive testing. A combination of static and ballistic testing is often conducted to optimize a ceramic composition, process technology, or component design. The results from testing are used to form assumptions about similar plates from the same batch. By this method, there is no direct testing of the plates that are used by the soldiers in combat. Assumptions about destructively tested armor plates are not reliable enough to ensure that the untested manufactured parts meet predetermined specifications. Anomalous or critical defects that could prove detrimental to the performance of the armor plates could be present, but these flaws cannot be detected without proper testing. Destructive testing also incurs additional cost to the manufacturer. Armor plates that undergo ballistic or static testing are destroyed and rendered useless, leading to the eradication of potentially usable products. The development of a method for 100% manufactured part inspection is crucial for every armor plate before it is used for protection on the battlefield.

NON-DESTRUCTIVE TESTING (HIGH FREQUENCY ULTRASOUND)

Keeping in mind the goals of nondestructive location and identification of micron-size defects in ceramic armor, one technique that meets these needs is high frequency ultrasound. Like other techniques, such as x-ray characterization, which operate on the principle of wave-material interaction, ultrasound is used to propagate an acoustic wave into a given medium, and the transmitted and reflected signals are collected and analyzed. Ultrasonic waves travel by exerting oscillating pressure on particles in a medium, corresponding to the frequency of the incident wave.[1] In a solid medium where particles are tightly bound, oscillation of one particle generates corresponding vibrations in adjacent particles, causing ultrasound propagation.[1] Ultrasound wave velocities are directly dependent on the composition and physical characteristics of the medium through which they are transferred.[1] Reflection, transmission, refraction, diffraction, interference, and scattering of the ultrasonic waves can be studied to detect defects, evaluate elastic properties, and produce surface and internal images.[1]

The advantages of ultrasound evaluation are that it is nondestructive, does not require special sample preparation, and can be adapted to on-line manufacturing environment, making it useful for ceramic armor production inspection.[1] It is non-hazardous, unlike x-ray characterization methods, and can be used to test transparent or opaque materials.[1] By investigating characteristics of ultrasonic waves transmitted through a material, important properties can be obtained. Variations in material density cause corresponding variations in longitudinal, shear, and surface wave velocities, from which elastic properties can be determined.[1] Variations in particle size introduce scattering as a function of frequency-dependence of ultrasound attenuation, from which microstructure can be evaluated.[1] Anisotropic characteristics can be established by measuring direction-dependent velocities.[1] Manipulation of reflected or transmitted signals can be used for surface and internal imaging.[1]

BASIC HIGH FREQUENCY ULTRASOUND PRINCIPLES

There are many important principles that are vital to understanding the concepts of ultrasound NDE. Critical ultrasonic parameters such as incident frequency, wave characteristics such as velocity and wavelength, and material characteristics such as acoustic impedance and attenuation, are evaluated for the location of various defects and features within a specimen. A system for generating an ultrasonic pulse, transmitting and receiving acoustic waves, and collecting and analyzing received data, is necessary for calculating elastic properties and producing internal images of highly dense armor samples.

Acoustic Wave Principles

Waves are the basis of ultrasound, and some of their important characteristics include wavelength and frequency. The frequency of a wave (f) is defined as the number of oscillations of a given particle per second. The wavelength (λ) is the distance between two planes in which the particles are in the same state of motion. The frequency and wavelength are inversely proportional, and related by the equation:

$$c = f \cdot \lambda \tag{1}$$

where c represents the speed of sound, or velocity of ultrasonic wave propagation, through a material. According to the equation, a high frequency will correspond to a short wavelength and a low frequency will correspond to a long wavelength. Shorter wavelengths enable the detection of smaller features within the sample. The sound velocity is a material property, and is constant for a given material at a specific wavelength and frequency. It can also be calculated by the equation:

$$c = 1/(\rho\kappa)^{1/2} \tag{2}$$

where ρ is the density and κ is the volume compressibility of the material. To propagate, the wave must displace a volume of material against the elastic constraints of its bonds with other volume elements. The ease of propagation is a function of the density, which represents how much material must be moved, and the elastic modulus, which represents how difficult it is to move the material.

There are two main types of acoustic waves that are analyzed for ultrasound NDE, the longitudinal wave and the shear wave. In a longitudinal wave, the oscillations occur in the direction of wave propagation, and particle motion also occurs in the same direction. Longitudinal waves are similar to audible sound waves in that they are compressional in nature.[2] Compressional sound waves transmit the oscillations as a source of acoustic energy through the air into our ears. In a shear wave, the particles no longer oscillate in the direction of propagation, but at right angles to it in a transverse direction. In solid bodies, the shear force can be transmitted to the particles in adjacent planes, but their transverse oscillations will show a lag in time, so longitudinal waves will propagate faster than shear waves in the same medium.[3] Ultrasonic longitudinal and shear wave velocities can be calculated by the equations:

$$c_l = t/TOF_l \text{ and } c_s = t/TOF_s \tag{3}$$

where c_l is the longitudinal wave velocity, c_s is the shear wave velocity, t is the thickness of the sample, and TOF is the time-of-flight, or time it takes for the wave to be transmitted through the sample. Longitudinal velocity values of common ceramic armor materials are shown in Table I.

Acoustic Impedance

Acoustic impedance (Z) is a material property defined as the product of density of material (ρ) and speed of sound (c) in that material (Z=ρc). This property is important in ultrasound NDE for identifying the nature of defects, since variations in acoustic impedance between two materials cause reflection of ultrasound waves, and large mismatches cause strong scattering of ultrasound and refraction of the beam. As the impedance ratio of two dissimilar materials increases, the amount of sound coupled through the interface decreases.[2] The reflections caused by acoustic impedance mismatch occur at material boundaries and can aid in the detection of flaws such as pores and inclusions. Air has an extremely low acoustic impedance value (Table I), so if there are pores within a ceramic, reflections and scattering will readily occur at the air-material boundary. The same trend occurs for inclusions within samples of various materials, as even a slight difference in acoustic impedance can be detected. The sound pressures of transmitted and reflected ultrasound waves are related to the pressure of the incident wave.[4] Reflection (R) and transmission (T) coefficients of the sound pressure can be determined by using the Z values on each side of the boundary. If the ultrasound beam is going from a material with Z_1 to a material with Z_2:

$$R = (Z_2-Z_1)/(Z_2+Z_1) \qquad (4)$$

The portion coefficient of transmitted sound pressure through the boundary is given by:

$$T = 2Z_2 / (Z_2+Z_1) \qquad (5)$$

This relationship is shown in Figure 1. This material property must also be considered for a transducer face in contact with a test material. When ultrasound travels from a medium with low Z to one of high Z, only a fraction of energy is transmitted into the test material.[5] Transmission losses are reduced when the Z of the coupling medium is very similar to the test material. One technique to reduce loss is to add a thin coupling agent of intermediate Z between the transducer and the sample to enable optimum transmission of ultrasonic energy. Acoustic impedance is a critical factor for detecting defects, and must be considered for optimizing ultrasound performance by proper matching. Z values for typical armor materials are shown in Table I.

Table I. Acoustic impedance and longitudinal velocity values for typical armor ceramics.[6,7]

Materials	Density (g/cm³)	Longitudinal Velocity (m/s)	Acoustic Impedance ($\times 10^5$ g/cm²s)
Air	-	330	0.0004
Water	1.00	1,480	1.48
Al_2O_3 (sintered)	3.98	10,600	43.0
SiC (sintered)	3.18	12,000	37.5
B_4C (sintered)	2.51	14,500	26.4

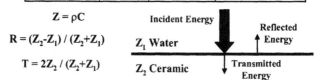

$$Z = \rho C$$
$$R = (Z_2-Z_1) / (Z_2+Z_1)$$
$$T = 2Z_2 / (Z_2+Z_1)$$

Figure 1. Reflection and transmission of acoustic energy due to Z mismatch.

Attenuation

Attenuation is a general term that refers to any reduction in strength of a signal. In ultrasound NDE, it can be described as the loss in acoustic energy that occurs between two points of travel. There are many factors that lead to attenuation, including beam spreading (initial pulse distribution as a wave front advances), field effects, wave reflections from coupling agent mismatch, beam diffraction, ultrasonic energy absorption (conversion of mechanical energy to heat as wave front passes), and wave scattering (reflections at grain boundaries and material inhomogeneities).[3] Since the sound pressure of an ultrasound wave decreases as a result of attenuation, the attenuation coefficient, α, can be calculated by the equation:

$$p = p_0\, e^{-\alpha d} \tag{6}$$

where p_0 is the initial sound pressure, p is the final sound pressure, and d is the distance of travel. In general, since higher attenuation results in a loss of signal strength and a poor signal-to-noise ratio, minimization of attenuation is desirable.

Attenuation of sound is also proportional to frequency. As higher frequency ultrasound pulses are utilized, attenuation increases. The high loss from attenuation reduces the maximum penetration depth into the material under analysis. This trend leads to a tradeoff between detectability and penetration depth.[3] Higher frequencies can be used to detect smaller features at the expense of lower penetration depth (high attenuation). Lower frequencies can be used to achieve deeper penetration (low attenuation), but the sensitivity required to detect smaller features will be lost. For this reason, it is important to choose parameters such as transducer frequency based on the desired result. Attenuation plays a key role in this relationship.

HIGH FREQUENCY ULTRASOUND SYSTEM

There are several important components that are necessary for the implementation of a complete ultrasound system with imaging capabilities, including a pulser-receiver unit, one or more ultrasonic transducers, an oscilloscope, an ultrasonic digitizer, and a computer. The pulser drives the system by sending short and repetitive electrical pulses to the ultrasonic transducer. The high voltage electrical pulses are used to drive the piezoelectric crystal in the transducer, which converts the electrical pulses into mechanical energy. The piezoelectric crystal rings at its resonant frequency, producing short bursts of high frequency vibrations which are propagated into the sample in the form of an ultrasonic beam. The beam is transmitted through the sample, and if there are any acoustic impedance mismatches in the form of flaws or defects, part of the incident ultrasonic energy will be reflected or scattered as additional mechanical waves. The waves are picked up by either the same transducer that transmitted the beam or by a second receiving transducer, where they are transformed from scattered vibrations back into electrical pulses. These signals are amplified by the receiver and displayed as a voltage versus time trace on an oscilloscope, which is synchronized with the pulse repetition frequency of the pulser-receiver. The received signals require further processing for evaluation, analysis, and imaging, so the analog signal must be converted into a digital signal by an ultrasonic digitizer, which is sometimes a component of the oscilloscope itself. After digital conversion, the signal can be transferred to a computer for further evaluation and imaging with the appropriate software.

There are three main categories of ultrasonic testing configurations that are commonly used.[8] The first configuration involves a single transducer that is used as both the transmitting and receiving transducer, and this is referred to as the pulse-echo technique. The second configuration, known as through-transmission, requires two transducers directly facing each other. The sample is placed between the transducers, and flaws and defects are detected by the presence or absence of a signal picked up by the receiving transducer. The third configuration,

227

known as the pitch-catch method, also involves two transducers, but they are not directly across from each other. This technique is used specifically for samples from which either the top or bottom surface is inaccessible. In this case, the receiving transducer is placed in an accessible location to collect the signal.

There are three different ways to display the ultrasound signal information.[3] The first and most common is the A-scan, which is another name for the voltage verse time reading produced on the oscilloscope. This type of scan can be performed through point analysis of a sample at any position. In a B-scan, or line scan, the A-scan results are collected over either the x or the y dimension of the sample, and ultrasonic data representative of the cross-section of that area is displayed. In the final type of ultrasound display, the transducer moves in raster-like fashion so both x and y coordinates drive a recording device. A time gate is set to consider only signals within the chosen time range, and this produces a two-dimensional downward view of flaw responses, known as a C-scan. The C-scan is often used as an image map to locate flaws and defects within a sample based on acoustic differences. Changes in amplitude of the acoustic signal or time-of-flight across the sample are set to various color scales and mapped as the transducer is rastered across the sample. This can be used as a valuable tool for visual analysis of ceramic armor to nondestructively detect porosity and defects.

HIGH FREQUENCY ULTRASOUND MEASUREMENTS

With an optimized ultrasound NDE system in place, ceramic armor samples can be effectively characterized. If a point measurement is the desired outcome, a high frequency longitudinal transducer can be prepared for contact with the sample. A thin layer of coupling agent can be applied to the face of the transducer in order to provide additional Z matching between the transducer and the sample. If a C-scan image of the sample is the desired outcome, a high frequency focused immersion transducer can be set up. In this case, the sample is immersed in water, commonly in a tank or container, and the water serves as the coupling agent. Since the acoustic impedance for water ($Z=1.48\text{x}10^5$ g/cm^{2}s) is much higher than the acoustic impedance of air ($Z=0.0004\text{x}10^5$ g/cm^{2}s) and much closer to the acoustic impedance value of an armor grade ceramic such as SiC ($Z=37.5\text{x}10^5$ g/cm^{2}s), a stronger signal can be obtained in water (Table I). An x-y motor-controlled stage can be set up for the transducer to scan and collect data over the entire sample area. In both cases of water immersion scanning and point analysis, the signals are collected and displayed on the oscilloscope as a plot of time verse voltage. For the contact transducer setup, the transducer trigger signal is the first high amplitude signal displayed on the plot, followed by the top surface signal from the sample. If there are no resolvable defects resulting in reflections due to Z differences, the next signal is the reduced (due to attenuation) bottom surface signal from the sample. The distance between the top and bottom surface reflections is measured as the TOF value in units of time. Any resolvable defects result in signals between the top and bottom surface reflections (Figure 2), and their positions in the sample can be determined by their location on the plot. For the water immersion setup, the position and tilt of the immersion transducer must be optimized so that the proper focal length is established and the transducer head is perpendicular to the sample surface to give the maximum acoustic signal. After taking all of the necessary measurements with the high frequency longitudinal transducer, the same measurements are taken with a high frequency shear wave transducer. The TOF values gathered from the longitudinal and shear wave transducer measurements are then used for making further calculations.

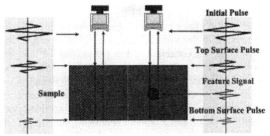

Figure 2. Ultrasound signal reflections from top surface, bottom surface, and defect in sample.

Elastic Property Calculations

The longitudinal and shear wave velocities are first calculated by dividing sample thickness by TOF (c=t/TOF). If the pulse-echo configuration is used, in which the same transducer is transmitting and receiving, the value is multiplied by a factor of two to account for the round-trip. If the through transmission configuration is used, in which one transducer transmits and the other receives, it is multiplied by a factor of one. The other variables that should be determined before measuring the elastic properties are the gravitational acceleration, g, which is typically 9.81 m/s^2 and the density, ρ, of the sample.

As mentioned earlier, variations in material density cause corresponding variations in longitudinal and shear wave velocities from which elastic properties can be determined. The velocities can be used directly to calculate Poisson's ratio, elastic modulus, shear modulus, and bulk modulus.[9] Poisson's ratio, v, which is the negative ratio of lateral and axial strains resulting from applied axial stress, can be calculated from c_s and c_l, using the equation:

$$v = [1-2(c_s/c_l)]^2 / [2-2(c_s/c_l)]^2 \qquad (7)$$

Shear, or Young's modulus, E, which is the ratio of applied stress to the change in shape of a material, can be calculated by using the equation:

$$E = [(c_l)^2(\rho)(1-2v)(1+v)] / [g(1-v)] \qquad (8)$$

The shear modulus, G, which is the ratio of shear stress to shear strain, can be calculated by using the equation:

$$G = (c_s)^2(\rho) \qquad (9)$$

The bulk modulus, K, which is the ratio of the change in pressure acting on a volume to the fractional volume change, can be calculated by using the equation:

$$K = E / [3(1-2 v)] \qquad (10)$$

Good elastic properties are critical in ceramic armor materials that deal with the impact of high stress projectiles, so thickness and time-of-flight measurements obtained from ultrasound NDE are crucial, since they can be used to calculate a variety of elastic properties.

229

High Frequency Ultrasound Scans

In addition to measuring elastic properties from point analysis, an internal image of the sample can also be generated to analyze differences in TOF or pulse amplitude. Instead of collecting data at a single point, the collection of data points over the sample area can be mapped as a C-scan. A color scale can be set according to TOF or amplitude ranges, so that visual differences in either of these variables can be distinguished. Differences in pulse amplitude or TOF can be correlated to changes in attenuation or acoustic impedance, which represent density changes or defects in the sample. The defects contained within an armor sample can be visually or quantitatively compared in this manner without destroying the sample. Full evaluations can be conducted on armor plates using ultrasound NDE before they are utilized by the soldiers who will use them as their last line of defense.

CONCLUSIONS

Historically, time-of-flight and ultrasound attenuation techniques have been useful for imaging defects that are acoustically different from the matrix material. Both porosity and inclusions with different densities than the matrix can be detected within ceramics such as Al_2O_3, SiC, or B_4C. Until recently, signal processing and transducer technologies have limited the ability of ultrasonic imaging to accurately map small defect populations. Recent advances in signal processing as well as novel high power transducers and transducer configurations have led to focused transducers and transducer arrays that permit real-time imaging. By optimizing key parameters, ultrasound NDE can be adapted to the study of highly dense ceramic armor. Ultrasound NDE has great potential, and the technology exists to turn the idea into reality. It is of vital importance in a field that produces components for the preservation of human life.

REFERENCES

[1]M.C. Bhardwaj; "Evolution, Practical Concepts and Examples of Ultrasonic NDC", *Ceramic Monographs – Handbook of Ceramics*, 41, 1-7, (1992).

[2]P.E. Mix, *Introduction to Nondestructive Testing*, John Wiley & Sons, 104-153, (1987).

[3]D.E. Bray, R.K. Stanley, *Nondestructive Evaluation: A Tool in Design, Manufacturing, and Service*, CRD Press, Inc., 53-178 (1997).

[4]J. Krautkramer, H. Krautkramer, *Ultrasonic Testing of Materials*, Springer-Verlag, (1990).

[5]M.G. Silk, *Ultrasonic Transducers for Nondestructive Testing*, A. Hilger Ltd, (1984).

[6]NDT Systems, Inc. Velocity Table (http://www.ndtsystems.com/Reference/Velocity_Table/velocity_table.html)

[7]P.T.B. Shaffer, "Engineering Properties of Carbides", Engineered Materials Handbook: Ceramics and Glasses, ASM, 4, 806-808 (1991).

[8]L.W. Schmerr, *Fundamentals of Ultrasonic Nondestructive Evaluation*, Plenum Press, 1-14, (1998).

[9]A. Brown, "Rationale and Summary of Methods for Determining Ultrasonic Properties of Materials at Lawrence Livermore National Laboratory" (http://www.llnl.gov/tid/lof/documents/pdf/225771.pdf)

[10]J. Duke, *Acousto-Ultrasonics: Theory and Application*, Plenum Press, (1988).

NON-DESTRUCTIVE EVALUATION (NDE) OF CERAMIC ARMOR: TESTING

Raymond Brennan
Rutgers University
607 Taylor Road
Piscataway, NJ 08854-8065

Richard Haber
Rutgers University
607 Taylor Road
Piscataway, NJ 08854-8065

Dale Niesz
Rutgers University
607 Taylor Road
Piscataway, NJ 08854-8065

James McCauley
US Army Research Laboratory
Aberdeen Proving Ground, MD 21005-5066

ABSTRACT
    Nondestructive evaluation (NDE) techniques have been used to analyze ceramic armor materials such as silicon carbide (SiC). Research has been conducted using contact and water immersion 25-100 MHz longitudinal and shear wave ultrasonic transducers in pulse-echo configurations to analyze defects in the form of acoustic differences in these materials. Time-of-flight (TOF) measurements were collected from oscilloscope traces of voltage verse time, and the thickness of the samples (t) at each point was recorded. Longitudinal (c_l) and shear wave velocities (c_s) of the SiC samples were calculated using the equation, $c = t/TOF$. The velocities were used to calculate elastic properties of the material, including Young's modulus, shear modulus, and bulk modulus, which compared favorably to results reported in the literature. C-scan imaging was also utilized to map acoustic differences and identify locations of flaws and defects in the armor materials. This data, combined with ballistic testing, will provide crucial information about critical flaw sizes and populations in ceramic armor, serving as a major step for nondestructively gauging the success and failure of commercial armor plates.

DESTRUCTIVE TESTING
    Destructive testing is currently the most common method for evaluating the material integrity of ceramic armor. A combination of static and ballistic testing is often conducted to optimize a ceramic composition, process technology, or component design. Assumptions about destructively tested armor plates are not reliable enough to ensure that the untested manufactured parts meet predetermined specifications. Anomalous or critical defects that could prove to be detrimental to the performance of the armor plates could be present, but these flaws cannot be detected without proper testing. The development of a method for 100% manufactured part inspection is crucial for every ceramic armor plate before it is used in the field. Work by Michael Bakas at Rutgers University (CCMC/ARL 2004) on anomalous defects in ceramic

armor has utilized high resolution field emission scanning electron microscopy (FESEM) on ballistically tested SiC samples to identify a variety of large (>75 μm) flaws that were believed to have caused failure. Nondestructive evaluation of these anomalous defects would be a huge step forward to replacing destructive testing as a means of determining the material integrity of ceramic armor. Historically, time of flight ultrasound has been a useful technique for imaging defects that are acoustically different from the matrix material. Features such as pores and inclusions with different acoustic impedance values than the matrix can be detected within ceramic armor materials such as Al_2O_3, SiC, or B_4C. High frequency ultrasound NDE can be used to resolve micron-size defects in a dense ceramic matrix, image the defects based on acoustic impedance mismatch or changes in time of flight, and determine material velocities, which can be used to calculate elastic properties for each sample.

HIGH FREQUENCY ULTRASOUND PROPERTY MEASUREMENTS

Ultrasound NDE point analysis and C-scan imaging were performed on both commercial and experimental ceramic armor samples using facilities at Ultran Laboratories in Boalsburg, Pennsylvania, and Rutgers University in Piscataway, New Jersey. Commercial and experimental SiC samples were analyzed with 50-100 MHz longitudinal transducers and 25-50 MHz shear wave transducers to obtain time-of-flight (TOF) values which were used to calculate longitudinal and shear wave velocities through the materials. The material velocities were used to calculate elastic properties including Poisson's ratio, elastic (Young's) modulus, shear modulus, and bulk modulus. A 50 MHz focused immersion transducer was used to perform C-scans on some of the SiC samples to generate image maps in terms of peak amplitude for analysis of attenuation changes in the acoustic signals. These results were collected in order to initialize the development of a database containing material data that will eventually be used to help establish a defect threshold for ceramic armor.

The first commercial sample analyzed by NDE ultrasound was a four by four-inch, one-inch thick, hot-pressed SiC ceramic armor plate. Initial evaluation of the sample was performed using a 100 MHz longitudinal wave ultrasonic contact transducer in pulse-echo configuration. In this type of transducer, the wave oscillations occurred in the direction of propagation. Before testing the sample, the bandwidth of the transducer was measured, and it was found that the actual value of input frequency recorded by this particular transducer was 78 MHz rather than 100 MHz. A thin layer of coupling agent was applied to the scanning region before placing the transducer in contact with the sample in order to enable closer acoustic impedance match between the two surfaces. This step was taken to reduce scatter and reflection of the ultrasonic beam before analysis, since these effects are enhanced by large differences in acoustic impedance. Table I shows typical acoustic impedance differences between various materials.

Table I. Acoustic impedance and longitudinal velocity values for typical armor ceramics.[1,2]

Materials	Density (g/cm³)	Longitudinal Velocity (m/s)	Acoustic Impedance (×10⁵ g/cm²·s)
Air	-	330	0.0004
Water	1.00	1,480	1.48
Al_2O_3 (sintered)	3.98	10,600	43.0
SiC (sintered)	3.18	12,000	37.5
B_4C (sintered)	2.51	14,500	26.4

The pulser feature of the pulser-receiver unit was used to send the initial electrical pulse to the transducer, and the piezoelectric crystal in the transducer converted the pulse into sound waves with a frequency of 78 MHz. The ultrasonic sound waves were transmitted through the SiC sample, and the resulting scattered and reflected waves were received by the same transducer. When one transducer is used for both transmission and reception of ultrasound waves, it is referred to as a pulse-echo configuration. After receiving the waves, the piezoelectric crystal in the transducer converted the signals back into electrical pulses, which were sent to the receiver of the pulser-receiver to amplify the signals. The amplified signals were displayed as a voltage verse time trace on an oscilloscope for further evaluation. The oscilloscope displayed a set of peaks including the trigger signal from the transducer, the top surface reflection peak from the sample, a reduced peak signal representing the bottom surface reflection, and additional reflections and echoes from the primary signals. If there were any significant defects or flaws, scattering or reflection of the initial ultrasonic energy would have occurred between the top and bottom surface reflections, resulting in additional signals from any features on the voltage versus time trace. Figure 1 shows the set of signals from the SiC sample, denoting the top and bottom surface reflection peaks.

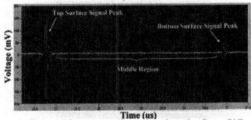

Figure 1. Top and bottom surface signal peaks from SiC sample.

The distance between the starting point of the top and bottom surface peak reflection signals was measured, and the value was recorded as time-of-flight (TOF). The TOF, measured as 4.18 μs, was representative of the transit time for the elastic waves to travel from the top to the bottom surface of the sample and back to the transducer. The pulse width of each peak, in this case 12.6 ns for the top surface pulse and 16.8 ns for the bottom surface pulse, and the pulse amplitudes, in this case 4.5 V for the top surface pulse and 274 mV for the bottom surface pulse, were also measured. These values were recorded during point analysis of the center of the SiC sample. In order to look for consistencies in the data, point analysis was also performed on eight additional points throughout the sample. The sample thickness at each point was also measured in order to calculate longitudinal velocity. The equation $c_l = 2t$ / TOF was utilized, with c_l representing the longitudinal velocity, t the thickness, and TOF the measured time-of-flight. Since the configuration of the system was pulse-echo, a multiplier of two was added to the equation to account for the time of transmission through the sample and back to the transducer. The calculated c_l for the SiC sample was 12,200 m/s, which was consistent over the nine measured points. This value compared favorably to the literature data, in which a c_l value of 12,000 m/s for hot-pressed SiC was reported.[3] Longitudinal velocities of typical armor materials are shown in Table I. Measured and calculated values of the SiC sample are shown in Table II.

The same technique was applied with a 50 MHz shear wave contact ultrasound transducer. With this type of transducer, transverse waves, which oscillate at right angles to the direction of propagation, were transmitted into the sample. Again, the bandwidth measurement revealed an initial frequency of 44 MHz, as opposed to the expected 50 MHz. Time-of-flight, top surface pulse widths and amplitudes, and bottom surface pulse widths and amplitudes were measured at each of the nine points. The TOF value at each point was 6.58 μs. Using the thickness measurements, the equation $c_s = 2t$ / TOF was used to calculate the shear wave velocity in the material. The calculated shear wave velocity for the SiC sample was approximately 7,750 m/s at each measured point. This value compared favorably to the literature data, in which a c_s value of 7,700 m/s for hot-pressed SiC was reported. [4]

Table II. Calculated SiC properties compared to values reported in the literature.[3,4,5]

Material	C_l [m/s]	C_s [m/s]	Poisson's Ratio	Young's Modulus E [GPa(psi*10⁶)]	Shear Modulus G [GPa(psi*10⁶)]	Bulk Modulus K [GPa(psi*10⁶)]
SiC (calculated)	12,200	7,800	0.16	462 (66.9)	199 (28.8)	226 (32.8)
Hot Pressed SiC (reported)	12,000	7,700	0.17	430 (62.3)	184 (26.7)	217 (31.4)

The longitudinal and shear velocities of the elastic waves were used to make elastic property calculations in the materials. First, Poisson's ratio, ν, which is the negative ratio of lateral and axial strains resulting from applied axial stress, was calculated from c_s and c_l, using the equation:

$$\nu = [1-2(c_s/c_l)]^2 / [2-2(c_s/c_l)]^2 \qquad (1)$$

The average Poisson's ratio for the SiC sample was calculated as 0.16, which was comparable to the reported value of 0.17 for hot-pressed SiC. With the ν, c_l, and c_s values, the density of the sample, $\rho = 3.18$ g/cm³, and the gravitational acceleration constant g = 9.81 m/s², equations were used to calculate elastic moduli in the sample. Young's modulus, E, which is the ratio of applied stress to the change in shape of a material, was calculated by using the equation:

$$E = [(c_l)^2(\rho)(1-2\nu)(1+\nu)] / [g(1-\nu)] \qquad (2)$$

The average Young's modulus value for the SiC sample was calculated as 462 GPa (66.9*10⁶ psi), as compared to the reported value of 430 GPa (62.3*10⁶ psi). The shear modulus, G, which is the ratio of shear stress to shear strain, was calculated by using the equation:

$$G = (c_s)^2(\rho) \qquad (3)$$

The average shear modulus value for the SiC sample was calculated as 199 GPa (28.8*10⁶ psi), compared to the reported value of 184 GPa (26.7*10⁶ psi). The bulk modulus, K, which is the ratio of the change in pressure acting on a volume to the fractional volume change, was calculated by using the equation:

$$K = E / [3(1-2\nu)] \qquad (4)$$

The average bulk modulus value for the SiC sample was calculated as 226 GPa (32.8*106 psi), compared to the reported value of 217 GPa (31.4*10⁶ psi). Measured properties of the SiC sample are compared to reported values in the literature in Table II. High elastic properties are critical in ceramic armor materials that deal with the impact of high velocity projectiles, so thickness and TOF measurements obtained from ultrasound NDE are crucial, since they can be used to calculate many of these properties.

Point analysis was repeated for the nine points of the SiC samples using a 50 MHz longitudinal velocity focused immersion transducer. The sample was laid flat in a water immersion tank and the ultrasound probe was aligned with the sample surface to ensure consistent data over the entire thickness. An x-y stage was used to manually position the transducer probe over each of the points in succession, and the voltage versus time trace on the oscilloscope was used to make measurements. For immersion transduction, the coupling agent was water, which had a higher acoustic impedance match than the coupling agent used in contact transduction, resulting in generation of a stronger overall signal. For the SiC sample, the signals were much broader, so more accurate values of TOF, pulse width, and pulse amplitude could be obtained. The TOF values measured by water immersion transduction in this case were identical to those measured by contact transduction, with TOF = 4.18 μs, and since the thickness and density of the sample were the same, c_l, E, G, and K values remained the same as well. The consistency of the data at the nine different points throughout the SiC sample showed that the sample was highly dense and homogeneous with little to no detectable defects. The signals from this sample, which were used to measure TOF, pulse width, and pulse amplitude, are also known as A-scans. A C-scan of the sample in three-dimensions provides more complete data and imaging for full characterization and location of defects, and this will be discussed in the following section.

HIGH FREQUENCY ULTRASOUND C-SCANS

Ultrasound C-scan imaging combines the principles of point analysis with the use of a scanning system, so changes in signal amplitude and/or TOF can be analyzed over the entire sample rather than just a few points. In this case, an immersion tank was once again set up in order to contain the sample in a medium where the acoustic impedance mismatch was small enough to generate a large signal. The transducer was mounted to the stage so that its position could be controlled in both the x and y directions. A manual z-positioning device was also set up so that the position of the transducer could be adjusted relative to the sample. When the transducer was positioned at its optimum focal length, completely perpendicular to the sample surface at the optimum tilt, the signal amplitude was maximized. The signals were obtained in the same manner as described for point analysis, but in this case, the position of the stage was synchronized with the signal. As the transducer was rastered over the desired sample area, the transducer collected signals which were assigned to x and y coordinates. An example of an ultrasound NDE scanning system is shown in Figure 2. There were several inputs that were decided before the scan was conducted. The step size of the scan dictated the smallest increment over which a data point was collected. The gated region or regions dictated which signals were analyzed as the scan was run. Some examples of gated signals included the top surface peak reflection, which was gated to study the amplitude difference at the surface of the sample, the bottom surface peak reflection, which was gated to study density variations and look for defects through the sample, and the region between the top and bottom surface peak reflections, which was used to closely analyze defects throughout the bulk of the sample. If TOF, rather than amplitude was varied, both the top and bottom surface peak reflections would have been gated separately and subtracted from each other to measure the change in TOF over the sample area. The collected pulse amplitude or TOF data was assigned a color scale, which represented changes in the variable of interest. Depending on the assigned scale, one color shade difference represented a difference in voltage, commonly several mV (for pulse amplitude) or a difference in time, commonly several parts of μs (for TOF). The data was mapped in relation to both the assigned color scale and the x and y positions of the points to produce a C-scan image of the sample over the gated signal or signals.

Figure 2. Ultrasound NDE scanning system and closeup of immersion transducer.

C-scan imaging was performed on the aforementioned SiC sample using a 50 MHz longitudinal focused immersion transducer. The sample was placed into an immersion tank and the z-position was manually adjusted until the maximum signal was obtained. The top and bottom surface peak reflections were identified on the oscilloscope, and the bottom surface signal was gated so that the variation of the signal could be collected as it traveled through the sample. The maximum amplitude of the signal was found to be approximately 140 mV in the center of the sample, and the position was changed randomly to get an idea of how the pulse amplitude would vary across the sample. It was determined that an amplitude range of 0 to 150 mV would be sufficient, so the color scale was set to reflect these values. An amplitude of 0 mV represented and dark blue color (lowest possible amplitude or no signal) and an amplitude of 150 mV represented a dark red color (highest possible amplitude or strong signal). Any color shades in between represented a scale of intermediate amplitudes, which could be used to determine attenuation differences in the sample. The x and y positions were selected so that the entire area of the sample could be scanned, and the step size was chosen as 0.2, which resulted in a scan composed of 20 μm pixels. The scan was initiated, and the entire area of the sample was scanned (Figure 3). Since the SiC sample under investigation was previously determined to be highly dense, there were few variations in the color of the image, as most of the C-scan was covered by red and dark orange, representative of high amplitude signals. The perimeter of the sample showed a color gradient from the dark blue of the area surrounding the sample (water only) to the red of the interior of the sample. This gradient was known as an edge effect. The edge effect occurred because the spot size of the ultrasound beam was larger than the step size. When part of the focal point was focused on the sample and the other part was focused on the water, an average of the two colors was taken, and appeared as an edge effect on the C-scan image. Since the bottom surface signal was gated for this scan, the data was representative of minor changes in attenuation, or acoustic signal loss, over the area of the sample. The image was rather uniform since this particular SiC sample was highly dense, but a lower density sample would have resulted in lighter color regions over the areas where porosity and defects were present. Portions of scanned regions from an armor grade sample and an experimental SiC sample with a low density region were compared in Figure 4. The C-scan image of the experimental sample showed evidence of yellow patterns that represented slightly lower density areas in relation to the dense armor sample.

Figure 3. C-scan images of top surface(left), middle(middle), and bottom surface gated regions.

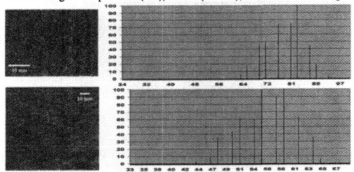

Figure 4. C-scans of armor grade SiC sample (top) and experimental SiC sample (bottom) with low density region and histogram showing occurrences of different amplitude values.

Quantitative evaluation of the raw data was also performed. Image data from the scans of different SiC samples was incorporated into histograms (Figure 4) that detailed the number of occurrences of each different color shade, each representing a specific amplitude value. This data was useful for comparing samples of varying densities. The distribution range of frequently occurring colors was broader for the low density sample, which had a wider range of amplitudes than the high density armor grade sample. The broader distribution range in the experimental sample signified the presence of more frequent occurrences of lower amplitude, while the armor grade sample stayed within a more narrow distribution of higher amplitudes. Studying the variations in these histograms for a variety of different samples could initialize the establishment of a correlation between "good" and "bad" ceramic armor samples.

Another C-scan was run on the SiC sample under the same conditions, but instead of gating the bottom surface signal, the top surface signal was gated (Figure 3). The image scan for the top surface signal showed a pattern of lighter colored lines over certain regions of the sample. Since the top surface signal provided only data relating to the top surface, the patterns seemed to represent variations in surface roughness across the SiC sample. While it was not possible to resolve surface roughness by visual study of the sample, the variations were evident in the ultrasound C-scan image. A final C-scan was run on the SiC sample under the same conditions,

237

but instead of gating the top surface signal, the middle region between the top and bottom surface signals was gated (Figure 3). Since the SiC sample was highly dense, there were very few pronounced variations in the image. If there were pores or inclusions of different acoustic impedance, they would have been evident in these C-scans as color variations. Since the scan was taken over the entire region between the top and bottom surface, B-scans were also derived from these C-scan images to determine the variations across a chosen line. The C-scan represented the x and y positions, which were averaged over a z range from the top to the bottom of the sample. A software program was used to draw a line across the x or y dimension of the C-scan to obtain B-scan data over that line. The B-scan showed amplitude variations across the drawn line (Figure 5). So far, most of the ceramic armor samples under investigation have been highly dense samples of SiC with very few detectable defects. The velocity, elastic property, and image data has been consistent with what can be expected of this type of sample. The next step is to fabricate samples with defects of known sizes and populations and to determine whether these engineered features can be resolved by ultrasound NDE.

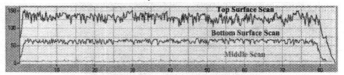

Figure 5. B-scans generated from C-scan of SiC sample.

CONCLUSIONS

High frequency ultrasound was utilized to test and evaluate SiC ceramic armor samples using 25-100 MHz longitudinal and shear wave transducers. Elastic property values of v, E, S, and K were determined by ultrasound NDE and compared favorably to values reported in the literature. C-scan images of the armor plates showed attenuation differences in armor grade and experimental samples. Image data from the C-scans was quantified into histograms, which confirmed the visual evidence, with higher density samples showing a more narrow amplitude distribution. Initial testing of SiC using high frequency ultrasound highlighted the potential of utilizing this form of NDE to establish critical flaw size and population data in ceramic armor in order to establish go/no-go criteria for commercial armor testing in the future.

REFERENCES

[1]NDT Systems, Inc. Velocity Table (http://www.ndtsystems.com/Reference/Velocity_Table/velocity_table.html)
[2]K. Petropolous, P. Peikrishvili, "Characterization of B_4C and LaB_6 by Ultrasonics and X-ray Diffraction", *International Journal of Modern Physics B*, **17**, 2781-2788 (2003).
[3]P.T.B. Shaffer, "Engineering Properties of Carbides", Engineered Materials Handbook: Ceramics and Glasses, ASM, **4**, 806-808 (1991).
[4]M.C. Bhardwaj, "Evolution, Practical Concepts and Examples of Ultrasonic NDC", *Ceramic Monographs – Handbook of Ceramics*, **41**, 1-7 (1992).
[5]A. Brown, "Rationale and Summary of Methods for Determining Ultrasonic Properties of Materials at Lawrence Livermore National Laboratory" (http://www.llnl.gov/tid/lof/documents/pdf/225771.pdf)

ON NON-DESTRUCTIVE EVALUATION TECHNIQUES FOR BALLISTIC IMPACT DAMAGE IN ARMOR CERAMICS

Joseph M. Wells
JMW Associates
102 Pine Hill Blvd
Mashpee, MA , 02649-2869

ABSTRACT
Traditional nondestructive examination, NDE, techniques are frequently satisfactory for pre-impact quality inspections of armor ceramics as well as the qualitative detection of impact created damage. However, detailed ballistic impact damage features and their volumetric distributions are generally quite difficult to characterize, spatially visualize, and quantify since real damage is frequently complex, non-uniform, and generally not conducive to accurate 3D interrogation by most common NDE modalities. Physical sectioning (destructive testing) and polishing can reveal accurate damage details only on the particular sectioned plane and risks the possibility of introducing extrinsic damage features inadvertently cause by the sectioning and polishing operations themselves. The application of x-ray computed tomography, XCT, has enabled dramatic 2-D and 3-D in situ diagnostic and visualization capabilities in both the pre- and post- impact damage assessment of ceramic and metallic targets. An overview and update of the ballistic impact damage characterization and visualization methodology and several dramatic results obtained to date utilizing the XCT NDE modality are presented and discussed. Further improvements in XCT damage analysis and assessment capability and its potential synergistic compatibility with other NDE modalities should be possible in the future.

INTRODUCTION
Ceramics are excellent materials for resisting penetration by high speed projectiles due to their high hardness and shear strength properties. However, they are also quite brittle and posses relatively little fracture toughness which is quite detrimental in terms of their damage tolerance. A single impact can shatter an unconstrained ceramic target making it ineffective for resisting subsequent hits even if full penetration is prevented on the initial hit. Techniques for encapsulating armor ceramics have been proven effective in constraining and containing the ceramic during and post impact. In fact, laboratory experiments by Hauver et al.[1] have demonstrated the feasibility of causing the projectile to be destroyed on the impact surface of encapsulated ceramic targets, a condition known as "interface defeat". Yet even in this situation, the encapsulated target ceramic sustains considerable internal cracking damage even though no penetration has occurred. The cracking damage occurs on both the micro- and the meso-scale and most probably contributes to the decay of penetration resistance in the ceramic.

Destructive sectioning, polishing and microscopy techniques have been utilized to study the ceramic impact damage in both dynamic and quasi-static experiments. However, such techniques have the limitation of being confined to a 2D perspective of the damage on the sectioned surface, often leading to unjustified assumptions of volumetric damage symmetry, as well as the possibility of introducing extraneous damage by the sectioning process itself. High resolution 3D volumetric NDE and characterization tools are thus desired in order to ultimately better understand the ballistic impact damage tolerance of armor ceramics.

BACKGROUND

Traditional NDE modalities such as immersion ultrasonics, eddy current, infrared thermography, and digital x-rays have been used for both pre-impact ceramic quality inspections as well as for post-impact damage detection. They work well for manufacturing and handling flaws and for certain types of damage detection. There are serious limitations, however, in their ability to provide high resolution volumetric details of complex and through thickness ballistic damage. Resolution and depth discrimination of 3D damage feature details are essential if one is to characterize, visualize and understand the complexities of the actual impact damage.

While both micro- and meso-scale cracking are present in impacted ceramics, it has not yet proved practical to successfully examine them with the same NDE approach. Most of the published characterization work on micro-cracked (< 10μm resolution) armor ceramics has, in fact, been conducted using destructive microscopy approaches in the absence of adequate NDE techniques with comparable resolution. For the examination of meso-scale cracking damage with minimum features ≥ 0.25 mm or 250 μm, the use of industrial x-ray computed tomography, XCT, has been successfully applied by the author and his collaborators[2-8] at ARL since 1997.

NDE CANDIDATE MODALITIES

Several commercially available NDE modalities are available which initially can be considered as candidates for the inspection and characterization of ballistic impact damage. A selection of these NDE candidate modalities is shown in figure 1 along with several highly desirable attributes for accurate impact damage characterization and visualization. While all NDE modalities share the attribute of being non-invasive (by definition), they do not all share the remaining listed attributes.

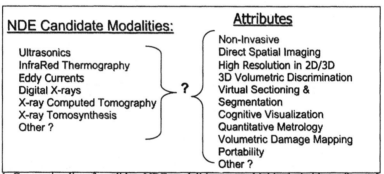

Figure 1. Contrasting list of candidate NDE modalities versus highly desirable attributes for ballistic impact damage characterization and visualization.

Unfortunately, there is neither time nor space here in this paper to present a comparative analysis of each of these NDE candidates for detailed impact damage characterization. It is apparent, however, that relatively little detailed nondestructive armor ceramic impact damage characterization has been conducted and published in the open literature. The prominent exceptions to this have been the above mentioned publications of the author and his collaborators at ARL using the NDE modality of x-ray computed tomography, XCT. These XCT impact damage analyses and assessments were limited to the meso-scale (> 250 μm) cracking damage and were conducted on encapsulated TiB_2 and TiC armor ceramic targets. Targets were impacted with laboratory sub-scale kinetic energy tungsten alloy projectiles. Further details of the ballistic testing and the XCT evaluation procedures can be found in references 2-5 respectively.

UNIQUE CAPABILITIES OF XCT

Industrial XCT has benefited greatly from the medical CT field in both the hardware and diagnostic software innovations of the past few decades. The major differences today between the industrial and the medical XCT/CT technologies include the rotation of the industrial subject and the considerable higher voltage available with some industrial machines to penetrate the larger size and higher density of many engineering structures. Also, an important distinction needs to be made between the more conventional x-ray digital radiography, DR, and the XCT. This practical distinction relates to the 2D capture of the DR information which is entirely lacking in discrimination of details in the third (z-axis) dimension. Consequently, volumetric feature discrimination, visualization and resolution is severely limited by this lack of 3D depth capability with DR. Interestingly, while high speed flash x-rays have seen considerable use in imaging ballistic events in real time (microseconds), they are actually DR images and also lack a true 3D damage interrogation capability.

XCT, on the other hand, involves a digital triangulation technique to measure the material density dominated x-ray absorption and attenuation properties through out the volume of the interrogated object. The result of the XCT scanning process is a volumetric density map with accurate and high resolution 3D data captured for the entire object. This data can be most simply reconstructed into 2D sequential and contiguous slices (XCT scan files) representing the complete interrogated object volume. Damage features such as voids and cracks are easily detected since their zero density characteristic provides substantial contrast to the density of the surrounding object structure. Moreover, the further advantage of the XCT approach for damage analysis and assessment lies in the supplemental software capability of image processing and reconstructions to create: 3D solid object images, virtual planar sectioning, variable transparency and 3D damage visualizations, virtual metrology, and quantification and 3D mapping of the volumetric damage fraction. Examples of these remarkable capabilities are provided in the following section for ballistic impact damage TiB_2 and TiC armor ceramics.

DAMAGE ASSESSMENT OF ARMOR CERAMICS WITH XCT

The experimental approach and the initial damage analysis of the TiB_2 and the TiC armor ceramic samples were reported earlier in detail [3,4,5]. The main objective here in this section is to provide a brief overview and update of the more interesting impact damage results found in these ceramic target materials to demonstrate the broad capabilities of these XCT techniques.

TiB_2 Armor Ceramic Results

Three circular TiB_2 sample ceramic disks, measuring 72 mm in diameter by 25 mm in thickness, were each encapsulated in a Ti-6Al-4V alloy welded case. The first target sample, S1wo, did not have a 17-4 PH steel ring shrunk fit on its outer diameter before encapsulation, while the other two did. The purpose of the shrunk fit ring was to provide a pre-stress on the second, S1w, and third, S2w, encapsulated TiB_2 samples. Each TiB_2 ceramic target was removed from the encapsulation package prior to XCT examination due to the limited x-ray penetration capability of the XCT system utilized. Figures 2A through 2C show a macro-photograph of the impacted surface of these three target disks. Figures 2D through 2F show a 2D XCT axial scan at

241

the approximate mid-thickness of each of these three target disks respectively. While there are obvious penetration features and residual tungsten alloy projectile fragments visible in figures 2D and 2F, only mild cracking damage features are observed in the center of figure 2E. Thus the additional compression of the 17-4 PH steel ring significantly reduced the penetration and the damage level in sample S1w over that of sample S1wo. Furthermore, the target S2w, also with a compressive ring, was able to sustain two separate and sequential impacts with their distinct mid-thickness agglomerated tungsten fragments shown in figure 2F.

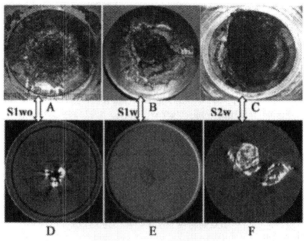

Figure 2. Macrophotos (A,B,C) and XCT mid-thickness axial scans (D,E,F0 of TiB2 armor ceramic samples S1wo, S1w and S2w respectively.

The 3D solid object reconstruction for sample S1wo is shown in figure 3. Among the topological features observed on the top impact surface are 3 distinct raised circular steps surrounding the central impact cavity. Several radial cracks transverse to these steps are also observed on the top impact surface, several of which do not extend to the same point in the central cavity.

Figure 3. Reconstructed 3D solid object images showing topological features of three circular stepped rings surrounding the penetration cavity and the radial surface cracking damage on the impact surface of TiB$_2$ S1wo sample.

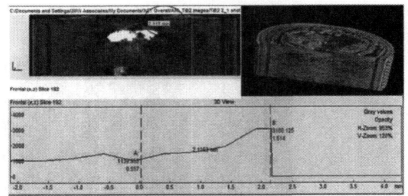

Figure 4. Analysis of inner step height on virtually sectioned impact surface of TiB$_2$ S1wo sample on through thickness plane.

Height and comparative density analysis of the surface rings was conducted as shown in figure 4. The graph in the lower section of this figure shows the increasing density (gray value) from the impact surface (A) up to the ring surface (B). The much higher density of these surface steps relative to the lesser ceramic density signifies a radial flow with substantial tungsten alloy projectile content on the impact surface. The bulk of the residual projectile fragments reside in the interior of the penetrated ceramic target, however, as indicated in figure 5. This figure includes images of a 2D axial XCT scan as well as virtual sectioned and reconstructed 3D half disk and quarter disk images. The high density tungsten alloy projectile fragments appear in white localized at the center of each image and extending through the target ceramic thickness. Also visible are various modes of cracking damage on the observable 2D planar surfaces.

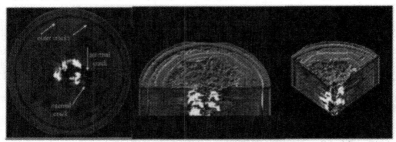

Figure 5. Shown is a 2D XCT scan image (left) and images of the 3D solid object TiB2 S1wo disk sample as half (center) and quarter (right) section reconstructions with each revealing localized internal residual projectile fragments (white) and cracking (dark).

A further capability of visualizing the 3D morphology of the internal residual projectile fragments is shown in figure 6, whereby the virtual opacity of the ceramic target is gradually reduced until only the higher density XCT data is observed. This image processing technique is known as virtual transparency. In figure 6 a., the agglomerated single shot fragment is visible in the center of target S1 as are two circular ring boundaries on the impact surface of this sample. In

243

the case of the double projectile impact on target S2w, the fully opaque 3D solid object reconstructed image is shown in figure 6b., while the fully transparent view of the dual agglomerated projectile fragments are isolated and displayed in figure 6 c.

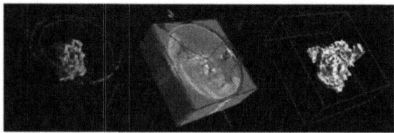

Figure 6. Shows (a.) mostly-transparent image of high density projectile fragments contained in circular rings on TiB2 S1wo impact surface as well as in the sample penetration cavity (left), (b.) fully opaque image of TiB2 S2w (center), and (c.) the corresponding fully transparent image showing only the consolidated projectile fragments (right).

In addition to the isolation of the high density tungsten alloy fragments, the demonstrated capability to segregate and isolate the zero density impact cracking damage from the opacity of the host target ceramic is equally important. For the case of interface defeat, discussed in the next section, there is no penetration and hence no internal residual penetrator fragment(s) leaving mainly impact cracking damage to be examined. This cracking damage isolation has been accomplished by the use of virtual point clouds in which a reconstruction image of the target is created using only the threshold 3D voxel data for the cracking damage alone. Such capability to nondestructively isolate the internal cracking information and then visualize the 3D morphology and details of that cracking damage independent of the host ceramic opacity is quite unique to XCT and very revealing for impact damage analysis and assessment. Two examples of the earlier point cloud visualizations for the cracking damage in target TiB2 S1wo are shown in figure 7. The image on the left shows the overall cracking morphology whereas the image on the right displays a more detailed profile of the hour-glass shape of the outer damage. Further improvements for these respective images can be seen in figure 8. Of note is the observation of a "screw thread" or spiral cracking morphology on the virtually surfaced image on the right of figure 8. Additional examples of indications of a spiral impact cracking mode have been observed in TiC ceramic and in monolithic Ti-6Al-4V target samples as well.

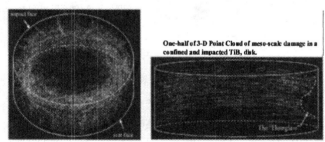

Figure 7. Initial virtual point cloud images of the impact cracking damage in target TiB_2 S1wo.

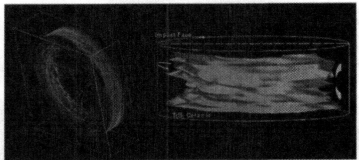

Figure 8. Improved point cloud images of meso-scale cracking damage in TiB₂ sample S1wo showing cracking isolation. Note the hour-glass OD contour (arrow) and the spiral cracking indications on the surfaced point cloud half section (right).

Finally, an example of early results in the use of XCT data for the volumetric quantification and mapping of the impact damage in target TiB₂ S1wo are shown in figure 9. Here a measured localized volumetric damage fraction is plotted in 3D as a function of the target radius and depth. It should be pointed out that the data plotted in these graphs is shown as axi-symmetric cumulative damage and has not been separated as a function of radial angular orientation on a given fixed depth level. Also, the presence of the high density projectile fragments in the center of TiB2 targets created some difficulty in the discrimination of the adjacent cracking damage details due to the localized x-ray absorption by these fragments. A filtering technique was developed and applied by Wheeler et al.[6] to reduce the severity of this effect on the measurement of the adjacent cracking damage. Thus, while the overall damage fraction is similar, some discrete differences can be seen between the two 3D plots, especially near the penetration cavity where the fragments cluster and the effects of the filtering operation have been applied.

Figure 9. Quantitative 3D plot of damage fraction in TiB₂ S1wo sample without (left) & with (right) the penetrator fragments included in quantitative analysis. Filtering of the penetrator fragments reveals more cracking damage.

245

TiC Armor Ceramic

A TiC ceramic target material that was impacted in an interface defeat experiment by Hauver et al.[1] was also examined with XCT. The sample examined was a half disc measuring 72 mm in diameter by 25 mm in thickness. A macro-photograph of the impact surface of this TiC sample is shown in figure 10 along with a reconstructed XCT image of this same surface. The increased surface topology detail in the XCT image is readily apparent. Additional 3D solid

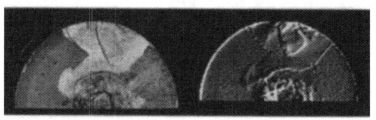

Figure 10. Macro-photograph of the impact surface of the TiC interface defeat target (left) is shown beside a reconstructed XCT image (right) of the identical surface.

object reconstructed images of this sample are shown in figure 11 with various impact cracking damage indications visible on several virtually sectioned surfaces. The images on the top contain orthogonal sectioned surfaces while those on the bottom contain parallel sectioned surfaces. While such planar observations of the cracking damage are possible on any arbitrary 2D virtual section, it is more revealing to view the entire volumetric cracking damage morphology in 3D. This is made feasible by utilizing the above mentioned point cloud technique. Examples of such point cloud images obtained for this TiC sample are shown in figure 12. While no indications of penetration are observed, figures 12 and 13 reveal extensive yet distinctly asymmetric meso-cracking damage details.

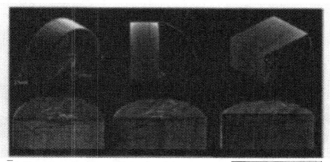

Figure 11. Reconstructed 3D solid images of the impact cracking on selected virtual planar sections.

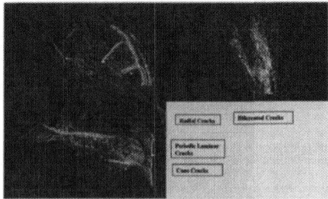

Figure 12. Point cloud images of meso-scale cracking damage in the TiC ceramic sample.

An additional transparent point cloud image, subsequently created and shown in figure 13, reveals an apparent spiral cracking morphology not previously identified in the earlier images for the TiC sample.

Figure 13. Transparent point cloud image showing apparent spiral cracking morphology in TiC interface defeat sample.

Unfortunately, the micro-scale cracking damage in the comminuted zone directly below the impact point is of a size below the resolution of the meso-scale XCT technique utilized here and hence is not observable in these images. Nonetheless, the constrained interface defeat TiC sample obviously sustained significant meso-scale impact cracking damage and yet retained sufficient bulk structural integrity to prevent the penetration of the high speed impacting projectile. It is this author's contention that such meso-scale cracking plays a pivotal role in the interface defeat phenomenon and warrants further detailed investigation.

247

SUMMARY COMMENTS

The nondestructive interrogation, characterization and visualization of ballistic impact damage, other than penetration data, in opaque ceramic targets remains a difficult and challenging task. Conventional NDE modalities have not, to date, proven effective in the detailed 3D impact cracking damage analysis and assessment. Consequently, there is relatively little such meso-cracking damage information and understanding in the ballistic literature, as well as a corresponding lack of predictive modeling capability available for such volumetric impact damage in ceramics. Significant progress has been made, however, in the 3D characterization and visualization of ballistic impact meso-scaled cracking damage with the use of both qualitative and quantitative nondestructive XCT technology.

Several examples of the meso-scale cracking damage results obtained to date with XCT for encapsulated TiB$_2$ and TiC armor ceramic target samples are presented and discussed above. This relatively recent and powerful XCT nondestructive examination capability has demonstrated new details and insights into the 3D character and morphology of meso-scaled cracking. It is highly likely that further technique development and the judicious application of the virtual XCT modality on additional ballistic ceramic targets will provide increased insight into our collective understanding of such ballistic impact damage.

ACKNOWLEDGEMENTS

The author gratefully acknowledges the significant contributions of several collaborators during the course of the earlier joint XCT investigations from which this work has liberally drawn upon. These collaborators include N.L. Rupert, W.H. Green, J.R. Wheeler, and H.T. Miller of ARL, Dr. A.V. Zibarov of GDT Software Group and Dr. C. Reinhart of Volume Graphics Software.

REFERENCES

[1] G.E. Hauver, P.H. Netherwood, R.F. Benck, and L.J. Kecskes, "Ballistic Performance of Ceramic Targets", *Army Symposium on Solid Mechanics, USA*, (1993)

[2] W.H. Green, and J.M. Wells, "Characterization of Impact Damage in Metallic / Non-Metallic Composites Using X-ray Computed Tomography", AIP Conf. Proc. 497, (1999) 622-629.

[3] J.M. Wells, W.H. Green, and N.L. Rupert, "Nondestructive 3-D Visualization of Ballistic Impact Damage in a TiC Ceramic Target Material", Proceedings MSMS2001, Wollongong, NSW, Australia, (2001) 159-165.

[4] W. H Green, K.J. Doherty, N. L. Rupert, and J.M. Wells, "Damage Assessment in TiB2 Ceramic Armor Targets; Part I - X-ray CT and SEM Analyses", Proc. MSMS2001, Wollongong, NSW, Australia, (2001) 130-136.

[5] N.L. Rupert, W.H. Green, K.J. Doherty, and J.M. Wells, "Damage Assessment in TiB2 Ceramic Armor Targets; Part II - Radial Cracking", Proc. MSMS2001, Wollongong, NSW, Australia, (2001) 137-143.

[6] J.R. Wheeler, H.T. Miller, W.H. Green, N. L. Rupert, and J.M. Wells, "Quantitative Evaluation of Damage and Residual Penetrator Material in Impacted TiB2 Targets using X-Ray Computed Tomography", 21st International Symposium on Ballistics, Adelaide, Au , ADPA, Vol. 1, (2004) 729-738.

[7] J. M. Wells, N. L. Rupert, and W. H. Green, "Visualization of Ballistic Damage In Encapsulated Ceramic Armor Targets", 20th Intn'l Symposium on Ballistics, Orlando, FL, ADPA, Vol. 1, (2002) 729-738.

[8] J. M. Wells, N. L. Rupert, and W. H. Green, "Progress in the 3-D Visualization of Interior Ballistic Damage In Armor Ceramics", Ceramic Materials By Design, Ed. J.W. McCauley et. al., Ceramic Transactions, v. 134, ACERS, (2002) 441-448.

Novel Material Concepts

STATIC AND DYNAMIC FRACTURE BEHAVIOR OF LAYERED ALUMINA CERAMICS

Zeming He, J. Ma, Hongzhi Wang
School of Materials Engineering, Nanyang Technological University, Nanyang Avenue, Singapore 639798, Singapore
G.E.B. Tan
DSO National Laboratories, 20 Science Park Drive, Singapore 118230, Singapore
Dongwei Shu, Jian Zheng
School of Mechanical and Production Engineering, Nanyang Technological University, Nanyang Avenue, Singapore 639798, Singapore

Abstract

The preparation of ceramics in the configuration of layered system with weak interfaces is one of the simple ways to achieve tough ceramic components. In the present work, alumina ceramics with alternately dense and porous layers were produced via a tape casting and sintering process. The static and dynamic mechanical properties of the layered systems were characterized, analyzed, and compared with those of monolithic alumina. The effects of the porosity in the porous layer and the strain rate during impact on materials fracture behavior were also investigated. From static test, it could be known that the fracture energy increased with the porosity, and an enhancement on fracture toughness was shown in the prepared layered system. From dynamic test, it was noted that the yield stress of the system increased with increasing the strain rate at a fixed porosity, and for the layered system at a given strain rate, as the porosity increased, the plateau region prolonged and the yield stress decreased.

Introduction

Ceramics have been used for various applications due to their high stiffness and hardness, chemical and biological stability, and thermal and electrical insulation. However, the actual use of ceramic materials has been limited on some occasion because of their brittleness and catastrophic failure mode. It is well accepted that the form of layered system with introduced porous interface is an effective approach to toughen the ceramics [1-3]. The remarkably enhanced toughness, measured under static loading conditions, makes such ceramic system very promising for practical application. It should be noted that the dynamic fracture characterization, especially made close to the real operative conditions, is needed before the actual use of the porous layered ceramic systems. So far, however, the work on the characterization of dynamic mechanical behavior of porous layered ceramic system has not been reported.

In the present work, alumina ceramic systems with alternately dense and porous layers were produced using tape casting and sintering, and their static and dynamic mechanical properties were tested using 4-point bending and split Hopkinson bar. The effects of the porosity in the porous interlayer and the strain rate during impact on materials mechanical properties were investigated.

Experimental procedure

AKP 30 alumina powders with an average particle size of 0.3 μm (Sumitomo, Japan) were used as the raw materials. Polymethyl methacrylate (PMMA) powders with an average particle size of 70 μm (Buehler, USA) were added to generate pores in the porous layers. A Procast tape caster (Unique/Pereny, USA) was used to produce the tapes from the prepared slurries for forming the respective layers in the layered systems. The contents of the prepared slurry were described in our previous work [4], and the added volume percentages of PMMA were 50, 60, and 70, respectively. The layer thickness ratio of the dense to porous was controlled to 1:1 during tape casting. For forming layered systems, the dense and porous layers were stacked alternately and pressed together at room temperature. The stacked green compacts were then subjected to sintering in a high-temperature furnace (Carbolite, UK) at 1550 °C for 3 hours. For property comparison, monolithic dense alumina samples were also produced using identical processing conditions.

Microstructural studies were performed on the cross-sectioned, polished, and then gold-coated samples using a scanning electron microscope (SEM) (JEOL, Japan). A 4-point bending tester (Instron, USA) and a self set-up split Hopkinson bar tester were used to test the static and dynamic mechanical behavior of the prepared samples. Three impact velocities (8.2, 16.5, and 19.1 m/s) were used during dynamic testing.

Results and discussion

Figure 1, as an illustration, shows the SEM image of the prepared layered alumina system with 50 vol% initially added PMMA in the porous layer. It can be seen that the dense and porous layers arrange alternately along the thickness direction, and they are well integrated without any delamination at the interfaces. The relative density of the dense layer is 0.98 [5]. It is also observed that spherical pores uniformly distribute throughout the porous layers, which is resulted from the burnoff of PMMA during sintering.

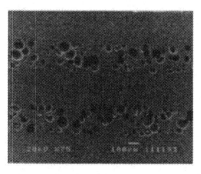

Figure 1 SEM image of layered alumina system with 50 vol% PMMA added

Derived from the static 4-point bending test, the variation of the fracture energy with the added content of PMMA is shown in Figure 2. It can be seen that, compared with monolithic alumina (0 vol% PMMA added), the enhancement of fracture energy has

252

been achieved for the prepared layered systems. The toughening effect of this configuration is mainly attributed to the crack deflection occurring at the weak porous interface, which therefore requires high fracture energy to break the system [4]. With respect to Figure 2, it is also noted that the fracture energy of the layered system increases with the added PMMA content, that is, with the porosity. This could be explained by the fact that, as porosity increases, the distance for crack deflection becomes longer in the porous layer due to lower resistance to against crack propagation.

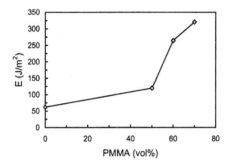

Figure 2 Variation of fracture energy with added volume percentage of PMMA

Figure 3 shows the stress-strain curve obtained at the impact velocity of 16.5 m/s for the layered alumina system with 60 vol% PMMA added. The stress-strain relationship obtained using the Hopkinson bar exhibits four characteristic stages: an initial linear elastic deformation region, a yield point, a plateau region with a slow increase of flow stress to plastically deform and crush the system, and a final densification region compacting the collapsed segments together. The dynamic fracture behavior of the layered alumina system investigated in the present work shows similarity with that of alumium alloy foam [6, 7].

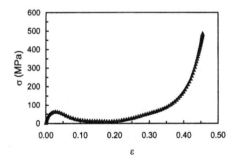

Figure 3 Stress-strain curve of layered alumina system with 60 vol% PMMA added tested at impact velocity of 16.5 m/s

253

To compare the yield stresses of the system tested at different impact velocities, the stress-strain curves of the system with 60 vol% PMMA added are shown in Figure 4. It is noted that the yield stress of the layered system increases as the impact velocity increases. Yu *et al.* [8] explained the effect of the impact velocity, or the commonly adopted strain rate, on the stress variation. It was attributed to the inertia attendant to the fracture process and the time effect provided by the cohesive law.

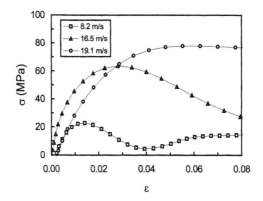

Figure 4 Stress-strain curves of layered alumina system with 60 vol% PMMA added tested at different impact velocities

The stress-strain curves of the layered systems with different volume percentages of PMMA tested at a fix strain rate of 3500 s^{-1} are shown in Figure 5. For the layered systems at a given strain rate, it is noted that the higher the porosity, the longer the plateau region is, but higher porosity implies lower yield stress. These phenomena indicate that the system could deform and yield more easily as the porosity increases. It also can be seen from its stress-strain curve that there exists only one crushing stress at low strain for monolithic alumina, which is different from that of layered system. However, the strength is high for monolithic alumina due to its high density.

Conclusions

In the present work, the static and dynamic fracture properties of the layered alumina systems were investigated. For static behavior, the fracture energy increased with increasing the porosity, and the fracture property of alumina was improved with the configuration of layered system compared with the monolithic alumina. For dynamic behavior, the yield stress increased with increasing the impact velocity for the material at a fixed porosity, and at a fixed strain rate, the plateau region prolonged and the yield stress decreased with increasing the porosity in the layered system.

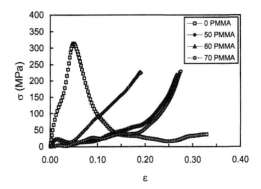

Figure 5 Stress-strain curves of layered alumina systems with different added volume percentage of PMMA tested at strain rate of 3500 s^{-1}

References

1. Clegg WJ, Kendall K, Alford NM, Birchall JD, Button TW. Nature 1990; 347: 455.
2. Blanks KS, Kristoffersson A, Carlstrom E, Clegg WJ. J Eur Ceram Soc 1998; 18: 1945.
3. Davis JB, Kristofferson A, Carlstrom E, Clegg WJ. J Am Ceram Soc 2000; 83: 2369.
4. Ma J, Wang H, Tan GEB. Ceram Eng Sci Proc 2002; 23: 109.
5. Ma J, Wang H, He Z, Tan GEB, Shu D, Zheng J. Ceram Eng Sci Proc 2003; 24: 187.
6. Deshpande VS, Fleck NA. Int J Impact Eng 2000; 24: 277.
7. Yi F, Zhu Z, Hu S, Yi P, He L, Ning T. J Mater Sci Lett 2001; 20: 1667.
8. Yu RC, Ruiz G, Pandolfi A. Eng Fract Mech 2004; 71: 897.

ROCESSING AND BALLISTIC PERFORMANCE OF Al_2O_3/TiB_2 COMPOSITES

A. Gilde and J.W. Adams
S. Army Research Laboratory
erdeen Proving Ground, MD 21005

BSTRACT

Early research on Al_2O_3/TiB_2 composites focused on exploiting their potential as a low cost armor ceramic. Limited llistic data indicated that the microstructure had a dramatic effect on ballistic performance. In some cases, the penetration istance of Al_2O_3/TiB_2 approached that of monolithic TiB_2 ceramics. However, challenges were encountered both in intifying the microstructural details and fabricating the desired microstructure. The large spread in depth of penetration results for these ceramics, coupled with an insufficient number of samples tested to some confusion in accessing the effect microstructure had on the ballistic performance. Our research focused on crostructure control during fabrication and a more thorough ballistic evaluation to correlate microstructure with penetration istance. Composites were made from mixed Al_2O_3 and TiB_2 powders. The composites, prepared with dramatically ferent microstructures had similar ballistic performance.

Results show that the penetration resistance of Al_2O_3/TiB_2 composites is not as good as a hot-pressed silicon carbide.

CKGROUND

Composites of Al_2O_3/TiB_2 were produced by K.V. Logan at Georgia Tech.[1] and the University of Dayton Research titute performed the initial Depth of Penetration (DOP) ballistic evaluations on them.[2] Logan concluded that there could a possible correlation between microstructure and ballistic performance. It was theorized that the composite structure that TiB_2 distributed along the grain boundaries of the Al_2O_3, which formed a continuous distribution of TiB_2, exhibited er ballistic performance than when the TiB_2 was dispersed in the Al_2O_3 matrix. Those results led the Army to continue king at the composite material and the role of microstructure. In some tests the ballistic performance was greater than dicted from the rule of mixtures, as shown in Figure 1, and was high enough to generate interest in these materials as ential armor ceramic.[3] However, there was ambiguity as to whether the observed differences were due to random ballistic

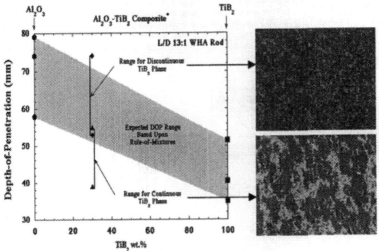

re 1. Early DOP ballistic test data for L/D=13 rod at 1550 m $^{s-1}$ against alumina titanium diboride ceramics

tion or to an actual difference in microstructure.[4] The purpose of this work was to fabricate Al_2O_3/TiB_2 composites with distinctly different microstructures, to correlate microstructural differences with ballistic performance, and to determine well the composite material worked relative to other armor grade ceramics.

EXPERIMENTAL

Processing

Al$_2$O$_3$/TiB$_2$ composites were fabricated with two different microstructures and then ballistically tested to determine t effect of the two different microstructures. The results from these ballistic tests were compared to earlier tests on simi materials made at GTRI. The materials from GTRI were the same thickness and tested under the same conditions by t University of Dayton Research Institute.[2]

Two different powder-processing routes were used to produce composites with the same composition but very differe microstructures. The composition was 77.4 w/o Al$_2$O$_3$ (A16SG, Alcoa, Pittsburgh, PA) and 22.6 w/o TiB$_2$ (Grade D, H. Starck, Newtown, MA). Traditional ball-milling was used to make a composite structure in which the TiB$_2$ was ever dispersed within the Al$_2$O$_3$ matrix. This microstructure will be referred to as manually mixed (MM). The Al$_2$O$_3$ and Ti powders were ball-milled in ethanol for 16 hours using alumina milling media in a polyethylene jar. The slurry was th dried and the powder sieved through a polyester USA Series 60 mesh sieve.

Electrostatic dispersion was used to make a composite structure in which the TiB$_2$ particles surrounded Al$_2$ agglomerates. TiB$_2$ was added to the Al$_2$O$_3$ powder in a polyethylene jar and dry mixed for 30 minutes using a Turbula mi (Turbula Mixer, Glen Mills, Inc., Clifton, NJ). In electrostatic dispersion the TiB$_2$ coats agglomerates of Al$_2$O$_3$ because the positive charge (on TiB$_2$) and negative charge (on Al$_2$O$_3$) that build up on the particles during the dry mixing. The pow was then sieved through a USA Series 60 polyester mesh sieve. The composite microstructure that results whereby the T surrounded areas of AL$_2$O$_3$ was designated as Electrostatically Dispersed (ESD).

Powders from the two different mixing methods were then hot pressed under the same conditions. The composites w hot-pressed in graphite dies in an argon atmosphere. The temperature was raised from 20 to 850 °C at 10 °C /min. and th from 850 to 1650 °C at 4 °C /min. The temperature was held at 1650 °C for four hours and then cooled down at 10 °C/min 20 °C. A pressure of thirty-five MPa was applied to the powder compact at the beginning of the heating cycle $\imath$ maintained through the final hold at 1650 °C, then released prior to cool down.

Characterization

The elastic modulus was measured using the procedures described in ASTM C1259[*] and densities of the compacts w measured using the Archimedes water immersion technique. Samples were sectioned and polished for microstruct analysis. Microscopy was performed on a scanning electron microscope (SEM) using the backscatter mode.

Depth of penetration (DOP) ballistic testing was conducted using a 65g, L/D=10 tungsten rods striking the sampl 1500 m/s. The rod was 7.8 mm in diameter, 78.7 mm long, and made from a tungsten heavy alloy (WHA) which was 9 W/ 4.9% Ni/ 2.1% Fe. The samples tested were 100 mm in diameter and 25 mm thick. These targets were mounted on ro homogeneous armor (RHA) steel, semi-infinite witness blocks, with a thin layer of two-part epoxy. The target configura is shown in Figure 2. The velocity of the projectile and the pitch and yaw were determined using flash x-rays in all C tests. The maximum acceptable pitch-yaw angle was 1.5 degrees.

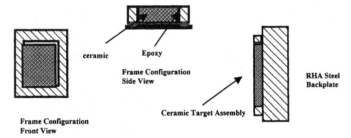

ceramic   Epoxy

Frame Configuration
Side View

RHA Steel
Backplate

Ceramic Target Assembly

Frame Configuration
Front View

Figure 2 Depth of Penitration (DOP) Test configuration

---

[*] ASTM C1259 "Standard Test Method for Dynamic Young's Modulus, Shear Modulus, and Poisson's Ratio for Advanced Ceramics by In Excitation" 1998 Annual Book of ASTM Standards, Vol. 15.01.

From the DOP test data, mass efficiency, (e_m), space efficiency (e_s), and quality factor (q^2), can be defined by the ~owing equations:[5]

$$e_s = \frac{P_{WITN} - P_R}{T_{CER}} \qquad (1)$$

$$e_m = \frac{(P_{WITN} - P_R) \times \rho_{WITN}}{T_{CER} \times \rho_{CER}} = e_s \times \frac{\rho_{WITN}}{\rho_{CER}} \qquad (2)$$

$$q^2 = e_m \times e_s \qquad (3)$$

~re P_{WITN} is the depth of penetration of the projectile into the semi-infinite witness plate without the ceramic facing; P_R is penetration of the projectile into the semi-infinite witness plate with the ceramic mounted to the front face; T_{CER} is the ~kness of the ceramic applied to the face of the witness plate; and ρ refers to the density of the respective materials. It can ~seen that e_m and e_s are dimensionless factors that compare the ballistic performance to RHA steel. The e_m and e_s of the ~rence witness plate are 1; thus a higher number denotes better ballistic performance as compared to the reference backing ~erial. Because both the weight of the armor and the space it takes up are critical factors in designing armors, the armor ~lity factor, q^2, is important to armor designers because it relates both the mass and space efficiencies.

Pieces of the ceramic were recovered after the ballistic test. The pieces were sectioned, mounted and polished. ~rostructural characterization was performed using a scanning electron microscope (SEM) in the backscatter mode.

~SULTS AND DISSCUSSION

This study used the same Al₂O₃/TiB₂ composition to produce two different microstructures and to investigate the effect ~nicrostructure on the ballistic performance. Results show that the only significant difference in the two composites is the ~rostructure. The density and the elastic modulus of the different composites were measured. The density of the ~rostructures were equivalent: both the MM composites and the ESD composites had a density of 4.0 g/cm³. The ~retical density for the composite, as determined by the rule of mixtures, was 4.1 g/cm³. In addition to having a similar ~sity, the elastic moduli were similar.

Figure 3 shows that the microstructure for the ESD composite is very different from the MM composite. In the MM ~posite the TiB₂ is dispersed uniformly within the Al₂O₃ matrix. In the case of the ESD composite the TiB₂ surrounds

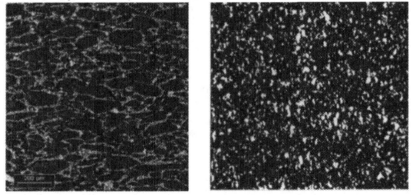

~e 3. Microstructure of ESD composite on left and MM composite on right.

agglomerates of Al₂O₃, forming a continuous TiB₂ phase around islands of Al₂O₃. This structure is developed in the ~rostatic dispersion due to the static electricity that builds up on the particles during the dry Turbula mixing; charge ~rences fix the TiB₂ to the Al₂O₃ agglomerates. When this powder is hot pressed the TiB₂ coating on the Al₂O₃ ~merates are maintained.

Ballistic properties of Al_2O_3/TiB_2 composites impacted with the L/D=10 tungsten alloy rod at 1500 m/s are presented Figure 4. It can be seen that there is no significant difference in the ballistic performance of the two composites. T

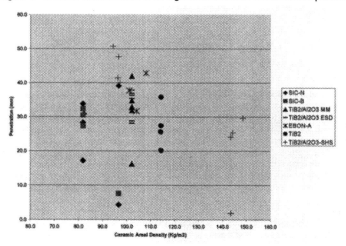

Figure 4. Summary of DOP experiments as described in experimental procedure: Penetration into RHA backing vs. cerany areal density against the L/D=10 rod at 1500 m s[-1]. [4]

average penetration for the MM composite is 32 mm with a standard deviation of 8 mm vs. 34 mm with a standard devia of 4 mm for the ESD composite. The spread in the data is the same for the hot pressed silicon carbide tested, which had average penetration of 29 mm and a standard deviation of 8. This hot pressed silicon carbide is a mature, commerci available, armor ceramic. This variation is typical of ceramic materials tested in a Depth of Penetration test.

Palicka and Rubin reported the average DOP of hot pressed Al_2O_3 as 37.4 mm for an areal density around 108 kg tested under similar conditions.[6] The average depth of penetration of a hot pressed Al_2O_3 is greater as compared to average DOP for Al_2O_3/TiB_2, however, all alumina DOP results fall within the range of DOP results for the composites. not certain that the addition of TiB_2 increases the ballistic performance, although it appears likely that it has some posi effect. The large spread in DOP data for ceramics makes it difficult to draw conclusions based on a small number of Comparisons based on a couple of ballistic tests are very suspect. Unfortunately due to the costs and complexity of ballistic testing, this is often done. . Determining the effect of the TiB_2 on the Al_2O_3 matrix was outside the scope of investigation.

In this study there was no significant difference in ballistic performance of the composites tested despite very diffe microstructures that were produced. When these results are compared to the earlier testing conducted by the Universit Dayton Research Institute on materials furnished by the Georgia Tech. Research Institute the results are similar.[2] Those were done on tiles of the same thickness using the same 65 g WHA test rod striking the target at 1500 m/s. The ta configurations were similar, thus a direct comparison of the ballistic results is possible. The composites furnished by Geo Tech. Research Institute were made using a composite powder formed using a self-propagating high temperature synth (SHS) reaction.[1] The average e_m for the SHS Al_2O_3/TiB_2 composites tested by UDRI was 3.0 and the highest e_m was Logan[1] concluded that the microstructure of the tile that gave the e_m of 4.1 was different from the others tested because it a greater amount of TiB_2 surrounding the Al_2O_3. She postulated that the superior ballistic performance was due to the gr amount of TiB_2 distributed around the Al_2O_3. The average e_m of the Al_2O_3/TiB_2 composites tested in the current study 3.0 and the highest e_m was 4.3. However, the e_m of 4.3 was given by a composite where the TiB_2 is dispersed in the A As can be seen in Figure 5, the microstructure of the composite formed using the SHS-derived powder is different from two microstructures tested in our study. The premise that the difference in ballistic performance is due to the differen microstructure has not been supported by this study. In fact, the observed difference falls within the expected spread of results for ceramics.

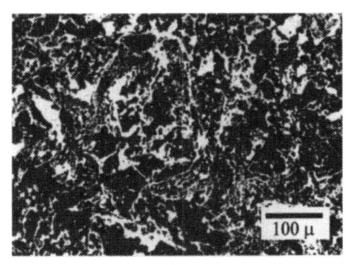

ure 5. Microstructure of Al_2O_3/TiB_2 composite made from SHS derived powder.

In previous work performed in our laboratory composites prepared by Georgia Tech. Research Institute were tested using rger WHA L/D=13 projectile that is 30% longer and twice the mass of the L/D=10 projectile. The ceramic targets were 40 thick vs. 25 mm used in this study. The results are shown in Figure 1. Composite targets were made using both mixed O_3/TiB_2 powders and composite powders formed via a SHS reaction. The best e_m obtained with the larger projectile, e_m = is very close to the best e_m (4.3) in the current study, despite the differences in the testing. The average e_m in that study 3.3 compared to an average e_m of 3.0 in this study. Although it is not possible to make direct comparisons when there differences in the projectile and target thickness, in this case it is safe to conclude that the differences observed in the istic performance were due to the usual random variation in ceramic DOP results and not due to differences in rostructure. In this study our highest e_m was given by the microstructure that had TiB_2 dispersed within the Al_2O_3 (MM) as osed to the previous study where the best e_m was given by the composites made with mixed Al_2O_3/TiB_2 powders where TiB_2 was surrounding the Al_2O_3. This indicates that the observed differences in ballistic performance are due to the rent variability in ceramic DOP tests. Because of the large spread in the depth of penetration results for ceramics, eme care must be taken when analyzing the data, and the sample size must be large enough to get a good indication of the dard deviation.

When the ballistic performance of the Al_2O_3/TiB_2 composite is compared to a hot-pressed silicon carbide, the silicon ide is clearly superior. From Figure 4 it can be determined that silicon carbide has an average e_m of 4.5, e_s of 1.8, and a 8.3 while the Al_2O_3/TiB_2 composite has an average e_m of 3.0, e_s of 1.5, and a q^2 of 4.6.

Although none of the Al_2O_3/TiB_2 composites microstructures tested in this study had an effect on the ballistic rmance, more work is needed to understand the effect of microstructure on ballistic performance. It would be a mistake nclude that because the different microstructures tested in this study had no effect, microstructure will not have an effect allistic performance.

MARY

Our investigation to assess the effect of microstucture on Al_2O_3/TiB_2 composite ballistic performance demonstrated that ictive microstructures could be developed and controlled by a variety of processing methods. A systematic ballistic ation was completed using 65 g L/D=10 projectiles at a velocity of 1500 m/s. None of the microstructures tested had an t on the ballistic properties of the composites. The process of mixing dry powders to electrostatically disperse the TiB_2 nd the Al_2O_3 grains resulted in composite structures that were as effective as those that had the TiB_2 dispersed in the 3, and both proved similar to composites made for powders derived for SHS reactions. The addition of TiB_2 to the Al_2O_3 ix had no deleterious effect on the ballistic performance and may have enhanced the ballistic performance of the Al_2O_3 x. Al_2O_3/TiB_2 composites do not result in e_m, e_s, and q^2 values as high as a state of the art hot pressed silicon carbide t may be concluded that Al_2O_3/TiB_2 composite structures will not be as effective as hot-pressed silicon carbide as a r ceramic.

REFERENCES

[1] K.V. Logan, "Elastic-plastic Behavior of Hot-pressed Composite TiB_2/Al_2O_3 Powders Produced Using Self-propagating High-temperature Synthesis," PhD. Thesis, Georgia Institute of Technology, 1992.

[2] G. Abfalter, N.S Brar, and D. Jurick, "Determination of the Dynamic Unload/Reload Characteristics of Ceramics," University of Dayton Research Institute, Dayton OH, June 1992, Contract No. DAAL03-88-K-0203.

[3] G.A. Gilde, J.W. Adams, M. Burkins, M. Motyka, P.J. Patel, E. Chin, L. Prokurat Franks, M.P. Sutaria and M. Rigali, "Processing of Al2O3/TiB2 Composites for Penetration Resistance," Cer. Eng. Sci. Proc., 22 (2001) 331-342.

[4] J.W. Adams, G.A. Gilde and M. Burkins, "Microstructure Development of Al_2O_3/TiB_2 Composites for Penetration Resistance," in Ceramic Armor Materials by Design, J.W. McCauley et al., Eds., (2002)

[5] P. Woolsey, D. Kokidko and S. Mariano, "An Alternative Test Methodology for Ballistic Performance Ranking of Armor Ceramics," MTL TR 89-43, U.S. Army Materials Technology Laboratory, Watertown, MA, 1989.

[6] R.J. Palicka, J.A. Rubin,"Development, Fabrication and Ballistic Testing of New, Novel Low-Cost High Performance Ballistic Materials," Final Report Contract No. DAAL03-88-C-0032, Jully 1991.

TACTICAL VEHICLE ARMOR SYSTEMS THAT UTILIZE LARGE, COMPLEX-SHAPED REACTION BONDED CERAMIC TILES

M. K. Aghajanian, B. E. Schultz, K. Kremer, T. R. Holmes
M Cubed Technologies, Inc.
1 Tralee Industrial Park
Newark, DE 19711

F. S. Lyons, J. Mears
Simula, Inc.
7822 South 46th Street
Phoenix, AZ 85044

ABSTRACT
The vast majority of armor systems for commercial and tactical vehicles are metal-based. Few ceramic systems are utilized due to cost and structural constraints. Typical ceramic-based armor systems for vehicles consist of a mosaic of small, precision ground ceramic tiles bonded to a metallic and/or polymeric backing. The present work examines a ceramic-based vehicle armor system that consists of large, 1-piece, complex shaped reaction bonded ceramic tiles with a backing of ballistic metal and/or polymer. The use of a 1-piece ceramic tile yields many advantages. First, the large ceramic tile provides structure to the door, thus allowing other weight to be removed (tile is not simply parasitic). Second, performance is enhanced since there are no tile to tile seams and there are no weight adding raised tile edges. Third, the system provides lower cost than competitive ceramic-based systems (no assembly of ceramic tile mosaic and no edge grinding of tiles required). The concept is applied to the design and fabrication of armored doors for the F350 Truck and the HMMWV. Herein, the processing, microstructures and properties of the large ceramic panels are presented; ballistic results are summarized; and example armor designs are provided. The work was supported by TACOM under contract numbers DAAE07-02-R-0002 and W56HZV-04-C-0023.

INTRODUCTION
The US Military and its representatives use very large numbers of light vehicles (commercial vehicles, trucks, light tactical vehicles, etc.). Recent events in Iraq have demonstrated that these vehicles are at risk in many assignments. Numerous examples of this exist in the press. For instance, before the Iraq conflict started, two US civilians, working as contractors for the US military, were driving an SUV outside of Camp Doha in Kuwait. At a stoplight, a terrorist fired multiple shots at the vehicle with a Kalashnikov assault rifle, killing one man and critically wounding the other [1]. Moreover, since the conflict began, there have been many instances where occupants of light vehicles were killed due to small arms fire, RPG attack, road side improvised explosive device (IED) blasts, and land mine explosions during patrols and in convoys [2]. In addition, fatalities have occurred as a result of roll over accidents in HMMWVs and other light vehicles [3]. The propensity for roll-overs is increased by placing heavy appliqué armor above the center of gravity of the vehicle. The US Military clearly needs better armor solutions for its light vehicles (better protection and lighter weight).

During the conflict in Iraq, the Army addressed the need for HMMWV armor by fielding over 10,000 armor kits by the end of 2004, with plans to field an additional 4,000 kits by March of 2005 [4, 5]. These kits, which are steel based, are manufactured by both the private sector (primarily Armor Holdings) and the Army Depots [5]. The steel kits have many advantages, including readily available raw material, low cost relative to most other solutions, good multi-hit performance and good durability. However, steel-based armor systems are heavy. In a recent Army publication [6], the high weight of the steel systems was sited as a serious problem, with an example provided where the vehicle's drive shaft broke due to the added weight to the vehicle. Moreover, the steel-based solutions have been described as non-ideal, temporary fixes, as they add thousands of pounds to the vehicles, thereby reducing payload capability and causing engine failures [7]. Thus, the need exists for higher performance ceramic-based armor systems on the Military's light vehicles. Such systems can provide the same protection as steel-based systems at significantly lower weight, which leads to vehicles with better performance, improved fuel economy, lower maintenance cost and longer life. Issues with ceramic-based armor systems are cost, manufacturing lead time and multi-hit performance.

The present work examines the fabrication of ceramic-based armor systems that contain large, monolithic reaction bonded ceramic tiles. Such systems provide weight advantages as compared to traditional metal-based armor systems. Moreover, the use of reaction bonded ceramics provides cost advantages relative to high performance sintered and hot pressed ceramics. Furthermore, the use of large tiles reduces assembly cost for the armor system (as compared to more traditional ceramic-based systems that utilize an array of small tiles). Major challenges for such a system are (1) the ability to produce large, complex shaped ceramic tiles, and (2) achieving the desired durability and multi-hit performance.

REACTION BONDED CERAMICS FOR LIGHT VEHICLE ARMOR APPLICATIONS

As described in previous work [8, 9], reaction bonded ceramics for armor applications are made with two primary steps. First, preforms of ceramic particles (e.g., SiC or B_4C) plus carbon are fabricated. Second, the preforms are reactively infiltrated with molten Si. The infiltrating Si reacts with the carbon, forming SiC that bonds the ceramic particles together into an interconnected network. Residual Si fills the remaining intersticies, yielding a fully dense ceramic composite material. Typical microstructures of reaction bonded SiC and reaction bonded B_4C are provided in Figure 1. In both materials, a particle size distribution of ceramic particles is utilized, providing good packing.

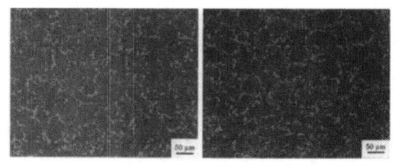

Figure 1. Optical Photomicrographs of Reaction Bonded SiC (left) and B_4C (right)

So long as the scale of the microstructural features is relatively fine (e.g., well less than 100 microns – Figure 1), reaction bonded ceramics perform very well versus small arms fire and soft, fragment simulating rounds. In particular, reaction bonded B$_4$C provides a high level of performance due to two primary reasons. First, B$_4$C particles have superior properties to SiC particles, as is summarized in Table 1. Second, the residual Si phase in reaction bonded B$_4$C fails in a ductile manner, whereas the Si in the reaction bonded SiC ceramic fails in a brittle manner (Figure 2). This difference in behavior of the Si phase provides high toughness to the reaction bonded B$_4$C relative to both reaction bonded SiC and most other armor ceramics, as is provided in Table 2.

Table 1. Property Comparison Between SiC and B$_4$C Reinforcement Particles [10]

Property	SiC Particles	B$_4$C Particles
Density (g/cc)	3.21	2.54
Knoop Hardness (kg/mm^2)	2480	2750
Young's Modulus (GPa)	450	480
CTE from RT to 100°C (ppm/K)	2.9	4.5

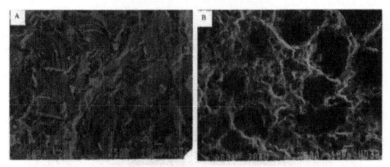

Figure 2. SEM Fractographs of Reaction Bonded SiC (A) and Reaction Bonded B$_4$C (B) – Note Ductile-Like, Knife Edge Failure of Si Phase in Reaction Bonded B$_4$C

Table 2. Property Comparison of Typical Armor Ceramics [11]

	Hardness (GPa)	Density (g/cc)	Young's Modulus (GPa)	Flexural Strength (MPa)	Fracture Toughness (MPa-m$^{1/2}$)
Sintered Al$_2$O$_3$	14	3.81	275	310	3.4
Hot Pressed SiC	23	3.20	450	634	4.3
Hot Pressed B$_4$C	32	2.50	460	410	2.5
Reaction Bonded SiC	22	3.06	384	284	3.9
Reaction Bonded B$_4$C	28	2.57	382	278	5.0

The ductile-like behavior of the Si phase in the reaction bonded B$_4$C ceramic, which provides high toughness, is attributed to the residual micro-stress generated during processing. During processing of reaction bonded SiC (Si + SiC composite), little stress is generated upon cooling after processing because Si and SiC have nominally the same CTE (about 2.9 ppm/K).

However, the case is very different for reaction bonded B₄C (Si + SiC + B₄C composite). The CTE of B₄C is far higher than those of Si and SiC (Table 2), which leads to a high level of CTE mismatch stress. The presence of this micro-stress, keeps the Si in its ductile polytype, which is usually only seen at higher temperatures.

Based on these positive attributes (lower density, higher hardness, greater toughness), reaction bonded B₄C is the desired ceramic for use in armor systems aimed at stopping light threats. Examples are personnel and helicopter seat armor. For tactical vehicles, which are exposed to similar threats, reaction bonded B₄C is also the desirable choice based on performance goals. However, reaction bonded B₄C is more costly than reaction bonded SiC due to higher raw material cost (B₄C powders are about 5 times the cost of SiC powders). For vehicle applications, where cost goals are more aggressive than those for personnel and helicopter armor applications, the use of reaction bonded B₄C is difficult to justify. To this end, a series of "hybrid" reaction bonded ceramics made with preforms containing both SiC and B₄C particles were developed. The goal was to maintain the desirable features of the reaction bonded B₄C ceramic (e.g., high toughness), while significantly reducing raw material cost. The results of the effort are shown in Figure 3. Preforms of ceramic particles plus carbon were produced where the ceramic particles ranged from 100% B₄C to 100% SiC were produced, then reactively infiltrated with molten Si to yield reaction bonded ceramics. The results demonstrated that high toughness of reaction bonded B₄C could be maintained, even with reduction of the B₄C content. Fracture toughness measurements were made using the four-point-bend chevron notch technique [12].

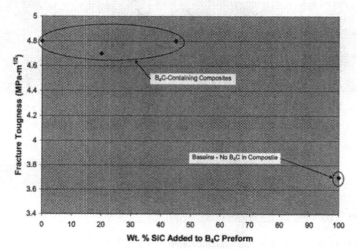

Figure 3. Fracture Toughness of Reaction Bonded Ceramics Containing Differing Percentages of B₄C and SiC Particles

BALLISTIC PERFORMANCE
Previous work [8] demonstrated excellent performance of reaction bonded B₄C versus small arms threats, such as the 7.62 mm M80 Ball and 7.62 mm M2 AP rounds. The goals of the present study were twofold, namely (i) to determine the single shot ballistic behavior of the

lower cost "hybrid" ceramics (SiC/B_4C mixtures) being proposed for tactical vehicle armor and (ii) to demonstrate attractive multi-hit behavior as a result of the relatively high fracture toughness.

For single-shot testing, multiple targets were produced with each ceramic material variant. The targets consisted of a 100 mm x 100 mm tile backed with a 200 mm x 200 mm ballistic polymer. All targets, which are not fully described herein, had constant areal density. Following MIL-STD-662, the targets were shot with a 7.62 mm AP round, and V_{50} velocity was determined. All of the SiC/B_4C mixture materials that were tested (0 to 45 wt. % SiC added to preform) performed very well – all better than reaction bonded SiC. Thus, the reduced density of SiC/B_4C mixtures relative to 100% SiC and the higher fracture toughness attained with B_4C-containing composites (Figure 3) has a positive affect on ballistic performance versus small arms threats. The V_{50} results for the mixture materials are shown in Figure 4. Additions of SiC, which lead to lower cost, up to about 30 wt. % appear to help performance (this needs further study). Above this level, the effect of increasing density begins to have an affect on mass efficiency, thus decreasing performance. Nonetheless, even at 45 wt. % SiC added, the result is good, far better than a 100% SiC material.

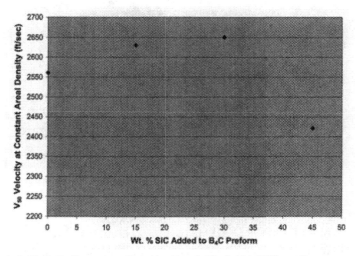

Figure 4. Ballistic Performance of Reaction Bonded B_4C with Different Percentages of SiC

Specifications for light tactical vehicle armor have existed for some time, such as the 1993 Operational Requirements Document (ORD) for the Tactical Wheeled Vehicle Crew Protection Kit (CPK) [13]. However, recent data from the conflicts in Afghanistan and Iraq demonstrate that these specifications are inadequate. For instance, the aggressive IED threats, discussed above, require greater protection levels for tactical vehicles than originally thought. To defeat these more aggressive threats with a traditional metallic armor system requires a very high areal density, which is driving the need for improved, more mass efficient tactical vehicle armor systems.

Three sets of ballistic tests have been completed to date to assess the performance of reaction bonded ceramic tiles versus these new, more aggressive threats, and more testing is currently in-process. Specifically, V_{50} testing was done versus the 0.50 cal. and 20 mm FSP threats (to simulate the IED threat), and multi-hit testing was conducted versus the 7.62 mm LPS. The targets are not fully described herein, but in each case they contained a monolithic reaction bonded ceramic tile (SiC/B_4C mixture with 45 wt. % SiC added) on the strike face and a structural ballistic backing. The results of the testing are provided in Table 3, and the multi-hit target after testing is shown in Figure 5. In each case, the performance goals were met with low areal density targets.

Table 3. Ballistic Test Results

Test	Target Areal Density (psf)	Result	Goal
0.50 cal FSP, Single-Shot	8.82	4,201 ft/sec V_{50}	3,500 ft/sec V_{50}
20 mm FSP, Single Shot	14.94	4,075 ft/sec V_{50}	3,500 ft/sec V_{50}
7.62 mm LPS Multi-Hit	9.35	Stopped 4 shots at 2850 ft/sec	4 shots at 2840 ft/sec

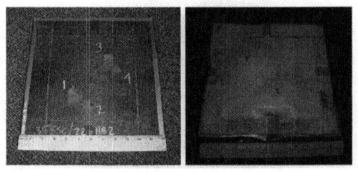

Figure 5. Multi-Hit Target After Testing; Front of Target Left and Rear of Target Right (shot spacing per NATO STANAG 4569 [14])

TACTICAL VEHICLE ARMOR SYSTEMS UTILIZING REACTION BONDED CERAMICS

As discussed above, the use of monolithic ceramic tiles in tactical vehicle armor systems, as opposed to an array of small tiles, yields many advantages. For instance, the large ceramic tiles provide structure to the armor system, minimizing the need for heavy structural members. In addition, no tile to tile seams exist that create weak points in the armor system. Moreover, assembly of the armor system is eased as many small tiles do not need to be handled, edge ground and built into a mosaic. Furthermore, weight of the system is reduced as raised edges at tile interfaces are not needed. Finally, curved shapes can be easily fabricated (i.e., as opposed to the assembly of many flat tiles where faceted geometries are created).

In the present work, two types of vehicle armor systems were fabricated, namely (i) a concealed system for a commercial vehicle, and (ii) a door kit for a HMMWV. The Ford F350 was selected as the platform for the commercial vehicle system. As shown in Figure 6, a system consisting of a curved, monolithic reaction bonded ceramic tile with a ballistic backing was designed such that it fit within the window cavity of the door. Features of the armor system are that it provides a high level of coverage and no weak points at tile to tile seams, its complex geometry allows insertion without vehicle modification, and it provides cost effective manufacture (no assembly or grinding of small ceramic tiles is needed; and monolithic tile acts as fixturing for applying backing, thus reducing tooling cost and lead time).

Figure 6. Concealed Armor System for Ford F350 Truck
Left: Design Concept with Armor System being Inserted into Window Cavity
Right: Monolithic Reaction Bonded Ceramic Tiles

The armor concept for the HMMWV door kit is shown in Figure 7. Again, it uses large reaction bonded ceramic tiles (2 pieces), which yielding a system with enhanced performance and reduced cost. Moreover, the ceramic provides structure to the door, thus eliminating the need for heavy struts or frames.

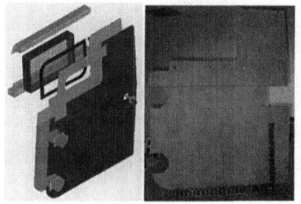

Figure 7. Door Kit Concept for HMMWV using Large Monolithic Ceramic Tiles
Left: Design Concept
Right: Reaction Bonded Ceramic Tiles used in System

SUMMARY

Previous work has demonstrated the utility of reaction bonded ceramics, particularly reaction bonded B_4C, in high performance armor systems for personnel and helicopters. The present work expanded the use of these ceramic materials to tactical vehicle armor systems. In particular, a new SiC/B_4C mixed material was developed that reduces cost relative to 100% B_4C, which is required for cost critical vehicle applications. Static and dynamic testing of these materials showed positive results, with ballistic requirements for tactical vehicles being met. Finally, armor systems were designed and built that utilized large, monolithic reaction bonded ceramic tiles. The use of the large tiles, as opposed to an array of smaller tiles, led to cost and performance advantages.

REFERENCES

1. "Two Americans Shot in Kuwait", Associated Press, 21 January 2003.
2. United States Central Command, MacDill AFB, FL (www.centcom.mil), News Releases 03-05-04, 03-05-32, 03-05-95, 03-06-91, 03-07-05, 03-07-14, 03-07-26, 03-07-32, 03-07-61, 03-07-65, 03-07-74, and 03-07-81; 2003.
3. United States Central Command, MacDill AFB, FL (www.centcom.mil), News Releases 03-05-77, 03-06-84, and 03-07-62; 2003.
4. Major D. S. Rusin, "HMMWV Armor Survivability Kit", Presented at Light Armored Vehicle Conference, Washington, DC, December 1-2, 2004.
5. T. Shanker and E. Schmitt, "Armor Scarce for Heavy Trucks Transporting US Cargo in Iraq", *New York Times*, December 10, 2004.
6. M. Cast, "DTC Oversees Testing of Up-Armored HMMWV Armor Kits", *Aberdeen Proving Ground News*, August 12, 2004.
7. S. I. Erwin, "Survival in Combat Zones Requires Layers of Protections", *National Defense*, December 1, 2004.
8. M. K. Aghajanian, B. N. Morgan, J. R. Singh, J. Mears, R. A. Wolffe, "A New Family of Reaction Bonded Ceramics for Armor Applications", in Ceramic Armor Materials by Design, *Ceramic Transactions*, **134**, J. W. McCauley et al. editors, 527-40 (2002).
9. P. G. Karandikar, M. K. Aghajanian and B. N. Morgan, "Complex, Net-Shape Ceramic Composite Components for Structural, Lithography, Mirror and Armor Applications, *Ceram. Eng. Sci. Proc.*, **24** [4] 561-6 (2003).
10. *Engineered Materials Handbook, Vol. 4, Ceramics and Glasses*, (ASM International, Metals Park, OH, 1991).
11. J. D. Norwood, "Interceptor Body Armor", Presentation at the IDGA 2nd Annual Conference on Lightweight Materials for Defense, Washington, DC, July 28, 2004.
12. D. G. Munz, J. L. Shannon and R. T. Bubsey, "Fracture Toughness Calculations from Maximum Load in Four Point Bend Tests of Chevron Notch Specimens", *Int. J. Fracture*, **16**, R137-41 (1980).
13. "Operational Requirements Document (ORD) for the Tactical Wheeled Vehicle Crew Protection Kit (CPK)", Department of Army, 6 October 1993.
14. "Standardization Agreement (STANAG): Procedures for Evaluating the Protection Levels of Logistic and Light Armored Vehicles for KE and Artillery Threats", NATO Military Organization for Standardization, No 4569, Annex C, Edition 1, January 2004.

MEANS OF USING ADVANCE PROCESSING TO ELIMINATE ANOMALOUS DEFECTS ON SIC ARMOR

Chris Ziccardi, Volkan Demirbas, Richard Haber, and Dale Niesz
Rutgers University
607 Taylor Road
Piscataway, NJ, 08854-8065

J. Mccauley
Army Research Lab
Aberdeen, MD

ABSTRACT
Microstructural detects are known to be present in commercial silicon carbide armor materials. Isolated pores, low density regions, large grains and inclusions are common defects. Microstructural quality can be improved by improving both the chemical and physical uniformity of the body. The increased mixedness of sintering additives greatly improves chemical inhomogenieties which can give rise to localized density variations. Improved powder processing and green forming can significantly reduce the presence of isolated pores, large grains and inclusions.
This study will show how commercial SiC powders can be improved by removing coarse inclusions. Colloidal processing will be shown as a means of improving the green microstructure of SiC compacts. The chemical uniformity of the final microstructure will be shown to be improved by the use of an aqueous based surfactant additive system.

INTRODUCTION
Currently available commercial silicon carbide armor materials exhibit a high variability in their ballistic performance. Depth of penetration (DOP) variations as high as a factor of eight on nominally identical tiles have been observed for SiC, with variations of a factor of 2.5 being typical. Although this high variability may be partially attributable to the ballistic test, most of it has been attributed to the silicon carbide armor material. While the fundamental cause of this variability has not been identified, this variability may be caused by microstructural defects, plastic deformation mechanisms or a combination of both.
Microstructural defects are known to be present in commercial silicon carbide armor materials. These include isolated pores, porous areas, large grains, variation in grain boundary composition due to non-uniform distribution of impurities or hot pressing aids, and inclusions of foreign material. These defects lead to a variation in quasistatic strength and may be related to variation in ballistic performance. Microstructural defects can be separated into two classes. The first class includes the normal distribution of high frequency defects such as small, isolated pores and large grains that cause the strength variation that determines the Weibull modulus obtained from bending strength data from four-point bend tests on standard-size specimens.[1]
The second class can be considered anomalous defects and include pores that are much greater than the grain size, foreign inclusions or a very low percentage of porous areas within a dense matrix. These anomalous defects are often present in such a low

volume percentage that they are often not detected in general microstructural examination on a random polished surface or the fracture surface of a standard four-point bend specimen. These low probability critical defects are usually assumed to be insignificant with regard to ballistic penetration since silicon carbide exhibits extensive cracking prior to ballistic penetration. However, this assumption has not been tested either experimentally or analytically. Thus, these anomalous defects may affect ballistic performance if, for example, they initiate macroscopic cracking early in the ballistic event.

Perhaps the best evidence of the importance of microstructural quality on ballistic performance is the results of Krell on alumina.[2] He reported a 50% improvement in ballistic performance over Coors AD 995. The microstructural quality of his material is demonstrated by its transparency. Transparency requires a density of at least 99.99% of theoretical and a grain size significantly less than 0.5 micrometers. The hardness of his material correlated with both porosity and grain size, as would be expected, and also with ballistic performance. Similar improvements in the ballistic performance of silicon carbide would be expected with similar improvements in microstructural quality over current commercial materials.

The objective of this study is to develop a better understanding of how enhanced powder processing can improve the microstructural quality of silicon carbide used as an armor material. The results are expected to identify the powder characteristics and processing parameters required to significantly improve microstructural quality.

EXPERIMENTAL AND DISCUSSION

The composition used as the basis for this study is the one reported by Ness and Rafaniello, that used 0.9 wt% boron carbide as a boron source and 5.5 wt. % phenolic resin as a carbon source.[3] In this study the carbon was added to remove the silica passivation layer from the silicon carbide particles, and the boron and carbon were both added as sintering aids.[4,5]

Commercial SiC powders are available with varying average particle sizes. In this study a series of powders were obtained and examined. Figure 1 shows the frequency distribution for four powders. While most powders are narrow is particle size distribution, all demonstrated a small percentage of coarse particles on the tail of the particle size distribution curve. Powders were classified using Stoke's settling to determine whether coarse particles could be separated from the powder. The settling was performed using a 13 cm diameter acrylic tube, 0.6 m in length. Starting powders were dispersed in water using Darvan C (R.T. Vanderbilt, Norwalk, CT) in a 20 volume percent suspension. Powders were ball milled for one hour in a polyethylene bottle. The <2um fraction was decanted off after a preset time based upon the settling height of the cylinder. Figure 2 shows that for a coarser starting powder, Superior Graphite 490, no more than two settling cycles were required to remove 99.8% of the >2um particles.

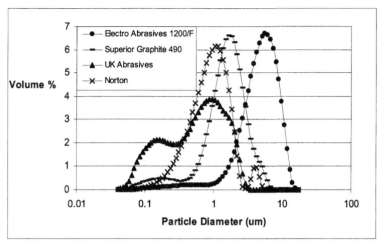

Figure 1: Commercially available SiC powders

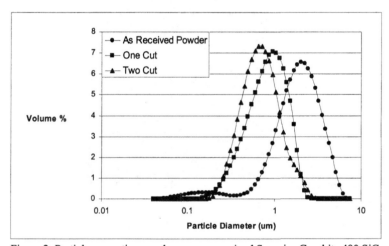

Figure 2: Particle separations made on an as-received Superior Graphite 490 SiC.

For the purpose of this paper, Norton SiC (Niagara Falls, NY) was used for subsequent processing. The Stoke's settling process showed that no coarse, > 2um particles were present in this powder. The suspension was centrifuged to concentrate the suspension to 40 volume percent. Rather than drying the powder and dry milling B_4C and phenolic resin into the batch, an alternative processing route was developed. A proprietary aqueous amine based surfactant (HX1) developed by Huntsman Chemical (Austin, Texas) was added to the SiC slurry. This additive contains both highly cross linked boron and has a substantial percentage of carbon. It was determined that 3 wt. %

of additive based on the dry weight of SiC replicated the sintering behavior of the B₄C/phenolic resin additive system referred to earlier.

The SiC slurry containing the B/C additive was filter pressed in a Baroid filter press at 2 bar air pressure (Fann Instrument Company, Houston, Texas). Fisher #Q5 filter paper, having a pore size of 0.7um was used as the filtration membrane. The resultant filter cake was 6 cm in diameter by 2 cm in thickness. It was found that the quality of the filter cake was highly dependent upon the slurry viscosity. As a result slurry viscosity was maintained between 200-500 centipoise. Higher viscosity slurries lead to a number of casting defects including isolated pores, preferred pore channels, and cracks. Figures 3-4 show micrographs of some typical defects found in the green compacts.

A repressing procedure was employed to homogenize the wet compact. In this step, the wet cake was removed from the filter press and placed into a uniaxial die of similar diameter. Five sheets of filter paper were placed on the top and bottom of the wet cake. The filter cake was then mechanically pressed to 250 psi (1.7 Mpa). Figure 5 shows the improved homogeneity provided by the repressing. Figures 6a and 6b show the schematic of the filter press assembly along with the repressing assembly.

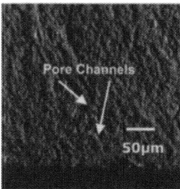

Figure 3: Shows an optical microscope image of the cross section of a filter cake. Pore channels are noted. (100X Mag)

274

Figure 4: Shows an optical microscope image of the cross section of a filter cake. Spherical pores are noted. (100X Mag)

Figure 5: Shows a filter pressed sample of SiC which was repressed to improve the green body microstructure.

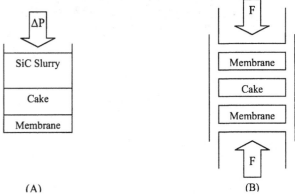

(A) (B)

Figure 6: Shows schematics for the filter pressing and the repressing operation.

275

The formed green disks were densified by hot pressing. Dried disks were loaded in a graphite die. Using a Vacuum Industries Model 4-2078 hot press, samples were heated at a rate of 25 C per hour to 1500 C under vacuum. The sample was held for one hour at 1500 C. This dwell was added to allow the carbon present from the surfactant to reduce the native SiO_2 layer present on the surface of the SiC powder. After one hour the sample was heated to 1800 C where the hot press was then backfilled with argon and the sample pressed to 5.6 Ksi (38 MPa). The sample was then heated to 2300C for 15 minutes. The power was immediately turned off and the sample allowed to cool.

The resulting sample was sectioned and the density determined by Archimedes method. A density of 3.20 g/cm³ was achieved. Figures 7 and 8 contrast the resulting microstructures of the processed SiC with Cercom SiC-N(Vista, CA). The later being a higher grade commercially available hot pressed SiC. In both cases a pore free microstructure is observed on the fracture surface.

Experimentally, the two factors that were found to most profoundly affect density and microstructure were time at maximum temperature and percent of surfactant added. Soak times greater than 15 minutes resulted in a reduction in density ranging from 3.08 to 3.15 g/cm³. Additive dose levels less than 3% resulted in significantly lower densities when fired to 2300 C for 15 minutes. Density variations ranging from 2.90 to 3.08 were observed for 1 and 2% addition levels.

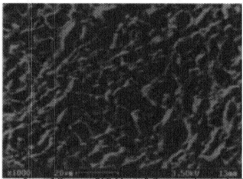

Figure 7: Fracture surface of Cercom SiC-N. A 5-7 μm grain size is observed. No porosity is noted.

Figure 8: Fracture surface of the processed SiC containing Norton SiC and Huntsman HX1 sintering aid. A 2-4 μm grain size is shown. No porosity is observed.

SUMMARY
It was found that the microstructure of hot pressed SiC used as an armor material can be enhanced by use of improved powder processing. A combination of narrower particle size distribution and better green forming resulted in near theoretically dense microstructures. It was found that the use of a surfactant based sintering aid provided a means of increasing the mixedness of the boron and carbon, resulting in an improved microstructure

REFERENCES
1 M.P Bakas, V.A. Greenhut, D.E Niesz, G.D. Quinn, J.W. McCauley, A.A Wereszczak, and J.J. Swab, "Anomalous Defects and Dynamic Material Properties in the Performance," International Journal of Applied Ceramic Technology, Vol.1[3] 211(2004).
2 A. Krell, "Processing of High-Density Submicrometer Al_2O_3 for New Applications," J. Am. Ceram. Soc., 86 [4] 546-553(2003).
3 E. Ness and W. Rafaniello, "Origin of Density Gradients in Sintered β–Silicon Carbide Parts," J. Am. Ceram. Soc., 77 [11] 2879-84(1993).
4 W. Van Rijswiik and D. J. Shanefield, "Effects of Carbon as a Sintering Aid in Silicon Carbide," J. Am. Ceram. Soc., 73 [1] 148-49(1990).
5 De Angelis, C. Rizzo, E. Ferretti, and S.P. Howlett, "Sintering of –SiC: Dispersion of the Carbon Sintering Aid," Ceramics Today – Tomorrow's Ceramics, 1415-1423(1991).
6 F. Lange and K. Miller, "Pressure Filtration: Consolidations Kinetics and Mechanics," J. Am. Ceram. Soc. Bull., 66 [10] 1498-1504(1987).
7 H. Bleier, "Fundamentals of Preparing Suspensions of Silicon and Related Ceramic Powders," J. Am. Ceram. Soc., 66 [1] 79-81(1983).

BALLISTIC PROPERTIES OF PRESSURELESS SINTERED SiC/TiB$_2$ COMPOSITES

Henry Chu
Idaho National Engineering and Environmental Laboratory
Mail stop 0325, PO Box 1625
Idaho Falls, ID 83415 USA

T. M. Lillo
Idaho National Engineering and Environmental Laboratory
Mail stop 2218, P.O. Box 1625
Idaho Falls, ID 83415 USA

B. Merkle, D. W. Bailey, M. Harrison
Superior Graphite Company
4059 Calvin Drive
Hopkinsville, KY 42240 USA

ABSTRACT
    Pressureless sintering of ceramics for armor applications offers the potential of greatly reduced cost and increased production volume. Previously it was shown that pure SiC could be made by pressureless sintering while achieving a ballistic performance slightly less than commercial SiC made by pressure-assisted densification (PAD). Additions of titanium diboride were made to pin the SiC grain size during pressureless sintering to achieve a final grain size closer to that found in PAD SiC and achieve improved ballistic performance. Silicon carbide/titanium diboride composites of various compositions were blended by various means, consolidated and pressureless sintered to near theoretical density. Additions of TiB$_2$ were $\leq$10% by volume and increased the density of the material by less than 3% over that of pure SiC. Variations in the mixing techniques yielded composites with a range of TiB$_2$ particle sizes. TiB$_2$ additions hindered SiC grain growth and the formation of elongated grains during high temperature pressureless sintering. The microstructure of the composites is documented and compared to commercially available SiC material. The SiC/TiB$_2$ composites demonstrated improved ballistic properties in Depth-of-Penetration (DOP) tests over pure, pressureless-sintered SiC material and approach that of SiC made by hot pressing.

INTRODUCTION
    Advanced ceramics for armor applications are highly attractive due to their relatively low density, high hardness and increased ballistic efficiency, compared to traditional steel armor materials. Boron carbide tends to be the most desirable ceramic for armor applications due its combination of high hardness (superior to all other ceramics except diamond) and low density (2.52 g/cm^3). Unfortunately, affordable processing of this material has not progressed sufficiently to be used in most armor applications. Silicon carbide is generally accepted as an economical alternative to boron carbide. Major drawbacks with using silicon carbide, as well as other ceramic materials, for armor application, in addition to the lack of multi-hit capability, include difficult fabrication methods, limited armor tile geometries and a high cost for the finished tile. Most of these drawbacks arise from the use of relatively expensive pressure-assisted densification methods (PAD), e.g. hot pressing and hot isostatic pressing. Recently it

has been demonstrated that a grade of silicon carbide powder produced by Superior Graphite Company, can be pressureless sintered, a relatively common and inexpensive industrial fabrication method for producing a wide variety of commercial ceramics, to a high relative density (>98% of theoretical density)[1]. Ballistic testing by the Depth of Penetration (DOP) test showed this pressureless sintered silicon carbide ceramic exhibited ballistic properties only slightly less than hot pressed silicon carbide[1], Table 1. Microstructural characterization showed a larger grain size compared to the hot pressed material. (The pressureless sintered material also exhibited a slightly lower relative density compared to hot pressed material which was virtually 100% dense.) Additions of inert particles, TiB_2, were made to the pressureless sintered SiC in an effort to pin the grain size at a value similar to that found in hot pressed material. Other researchers had used TiB_2 as inert reinforcement particles in SiC previously[2]. Subsequent preliminary ballistic testing showed the pressureless sintered composite had improved ballistic behavior over pressureless sintered SiC.

Table I. DOP with Mass Efficiency Results

Target ID	Process	Density, g/cc	Thickness, cm	Velocity, m/sec	DOP, cm	Mass Efficiency	Ballistic Limit Velocity*, m/sec
502-SP-4	HSC490, pressureless sintered	3.17	0.45	861	0.335	5.7	728
SiC-B	Pressure assisted densification	3.22	0.44	847	0.147	6.5	750

*Test condition of the ballistic limit (V50) experiment: Test rounds were 14.5 mm BS41 WC core. Target configuration was (100 x 100 x 12.7) mm SiC tile glued to (120 x 120 x 25.4) mm Al-5083 plate.

Pressureless sintering offers the potential of relatively low cost SiC/TiB_2 composites with a ballistic efficiency as good or better than hot pressed SiC without a significant penalty in density (~3% greater than pure SiC). Furthermore, pressureless sintering is amenable to fabrication of more complex geometries than is capable with PAD methods. This work reports on efforts to improve the TiB_2 particulate distribution in SiC/TiB_2 composites and fully characterize the microstructure and ballistic behavior of these composites.

EXPERIMENTAL METHODS

Sample Fabrication
    Two fabrication methods were explored to fabricate SiC/TiB_2 composites with a high relative density. These methods are described below:

Method 1 – Directly sinterable SiC powder (490DP) was obtained from Superior Graphite Company and dry mixed with TiB_2 powder (Cerac, -325 mesh, 10 micron average particle size). The powders were mixed to in the proper ratios to obtain a SiC/TiB_2 composite of either 7% or 10% TiB_2 by volume. The powders were placed in a plastic bottle with SiC mixing balls (~3 mm in diameter) and mixed for 12 hours using Turbula, Shaker-Mixer (model #T2C). The mixed powders were then loaded into a 76.2 mm diameter die and uni-axially pressed at 35 MPa and then cold isostatically pressed at 400 MPa. The green density after pressing was greater than 60% of theoretical.

280

Method 2 – At this time details of this method are proprietary. Directly sinterable SiC powder was also used in this method, however, the additives and mixing methods used substantially differed from those used in Method 1. Samples were made to have 3.5% TiB_2 by volume. The goal of this method was to produce a finer and more homogenous distribution of TiB_2 particles in the sintered compact.

Sintering of samples made by both methods was carried in a resistance-heated furnace with graphite insulation and furnace elements. Flowing high-purity argon was used as a cover gas. (It is critical that the oxygen concentration in the sintering atmosphere be controlled and kept to a low value.) The samples were sintered by one of two different sintering profiles:

Sintering Profile 1: This is a proprietary sintering profile developed by Superior Graphite Company and was carried out in their facilities.

Sintering Profile 2: The temperature was raised to 1750°C at 15°C/minute and held for 3 hours. The temperature was then raised to 2200°C at 15°C/minute and held for 1 hour and then furnace cooled.

Sintering Profile 2 was developed to allow the SiO_2 on the surface of the SiC powder to convert to SiC and CO gas which was then purged from the system by the flowing UHP argon.

Characterization

The sintered density was determined by water immersion techniques (Archimedes principle). Samples were also prepared for metallographic analysis and etched using a molten mixture of potassium hydroxide – 10 wt% potassium nitrate at 400°C for 5-10 minutes. The microstructure was further characterized by x-ray diffraction to establish the presence of TiB_2 and potential other secondary phases, e.g. TiO_2, SiO_2, B_4C, excess carbon, etc.. The ballistic efficiency of the sintered samples was evaluated through depth of penetration (DOP) tests. A projectile is shot into a witness block (AL6061 in the T6 condition) protected by a ceramic tile placed in front of the block. A projectile at the same velocity was shot into an unprotected witness block of the same material. A comparison of the depth of penetration for the two configurations provides a relative measure of the protection afforded by the ceramic tile. The protective capability of the ceramic material, referred to as the mass efficiency (M.E.), is then calculated from:

$$M.E. = \frac{\rho_{witnessblock} \bullet P_{BWB}}{(\rho_{ceramic} \bullet T_{ceramic}) + (\rho_{RWB} \bullet P_{RWB})} \qquad \text{Eqn. 1}$$

where the variables correspond to density of the witness block, $\rho_{witness\ block}$, density of the ceramic, $\rho_{ceramic}$, ceramic tile thickness, $T_{ceramic}$, penetration into bare witness block, P_{BWB}, and residual penetration in the ceramic-protected witness block, P_{RWB}. The projectile used was .30 caliber armor piercing (AP) M2 round. The projectile velocity was set at 848 ± 7.6m/s. The standoff distance from the muzzle of the gun to the target surface was set at 6.24 meters in accordance to Mil-STD-662. All projectiles were carefully measured to 164 ± 0.5 grain and each powder load was accurately weighed to within 0.1 grain to ensure the kinetic energy of each impact was identical. The targets were bonded between the machined ceramic target plate and

281

the witness block with approximately 0.5mm of epoxy. All ceramic targets were thinned via surface grinding to the thickness ranging from 4.0 to 6.5 mm to obtain a significantly measurable residual penetration in the witness block. Each ceramic target has sufficient impact surface area to maintain a minimum of 38 mm between point of impact and the free edge of the material.

RESULTS AND DISCUSSION
Microstructural Characterization
    Figure 1 shows the unetched microstructure of pressureless sintered SiC – 7 vol% TiB_2, SiC – 10 vol% TiB_2 made by Method 1, Sintering Profile 1 and SiC-3.5 vol% TiB_2 made by Method 2, Sintering Profile 1. The density, as determined by the water immersion technique, for these were 3.21, 3.24 and 3.24 grams/cm^3, respectively, which corresponds to a relative density of 97%, 97% and 99%, respectively. Metallographic preparation of the SiC-3.5 vol% TiB_2, Fig. 1c, appears to have resulted in significant pullout of the TiB_2 (bright particles in this micrograph). In general, both fabrication methods result in agglomerates of TiB_2 particles, although fabrication Method 2 appears to produce finer TiB_2 particles. The results from x-ray diffraction analysis show only the presence of α-SiC, TiB_2 and possibly a minor amount of residual carbon (graphite). Potential impurity phases of SiO_2, B_4C and TiO_2 are either not present or below the limits of detection.

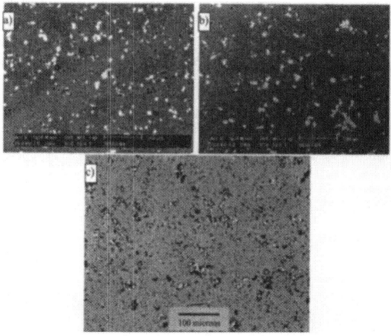

Figure 1. Unetched microstructures of a) SiC – 7 vol% TiB_2 and b) SiC – 10 vol% TiB_2 produced by Method 1, Sintering Profile 1 and c) SiC – 3.5 vol% TiB_2 produced by Method 2, Sintering Profile 2. Bright particles in the micrographs are TiB_2.

282

Figure 2 compares the etched microstructures of pressureless sintered SiC – 7 vol% TiB$_2$ made by Method 1, Sintering Profile 1 and SiC-3.5 vol% TiB$_2$ made by Method 2, Sintering Profile 1 with pressureless sintered SiC (the control sample - no TiB$_2$) Sintering profile 1 and hot

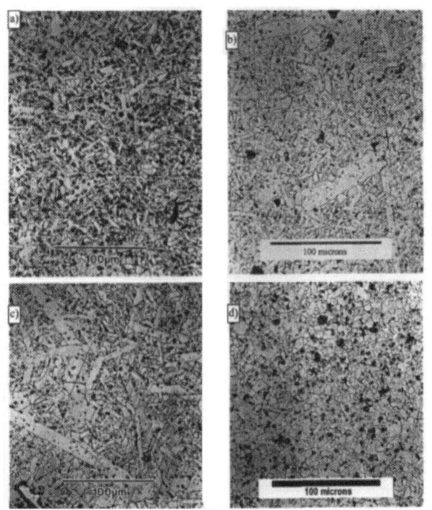

Figure 2. Microstructure of SiC composites, a) 7 vol% TiB$_2$ – Method 1, Sintering Profile 1, b) 3.5 vol% – Method 2, Sintering Profile 1, c) pressureless sintered SiC and d) hot pressed SiC. The molten salt etchant removed the TiB$_2$ particles.

pressed SiC (Ceracom SiC-B, no TiB$_2$ particles). (Etching removed the TiB$_2$ particles from the surface of the samples leaving behind black holes.) Addition of TiB$_2$ particles has reduced the

283

formation of elongated grains, compare Figs. 2a & 2b with Fig. 2c. A TiB$_2$ content of 7% by volume seems to be more effective at reducing the prevalence of elongated grains compared to the sample with 3.5% TiB$_2$ particles by volume, compare Figs. 2a and 2b. However, the grain shape is still more elongated than that of hot pressed SiC, Fig. 2d. The grain size of the hot pressed material is also smaller than the SiC/TiB$_2$ composites.

The microstructural characterization appears to show that TiB$_2$ particles are relatively stable in the SiC matrix during high temperature sintering. Furthermore the TiB$_2$ particles appear to be responsible for controlling grain shape, especially preventing the formation of highly elongated SiC grains, and at the same time reducing the average grain size. Fabrication Method 2 appears to produce a finer TiB$_2$ particle size, however, both fabrication methods result in agglomerates of TiB$_2$ particles which probably limit the effectiveness of the grain boundary pinning by the TiB$_2$ particles. Improved grain boundary pinning and the associated reduction in grain growth could be accomplished with a more uniform dispersion of the TiB$_2$ particles.

Ballistic Characterization

At this time, only material made by fabrication Method 1 has been subjected to ballistic testing, specifically the DOP test. Table 2 reports the results of the DOP test for pressureless sintered SiC (control) as well as the SiC/TiB$_2$ composites sintered by the two sintering profiles outlined above. In general, Sintering Profile 2 consistently yields a slightly higher relative density, regardless of the TiB$_2$ content. Also the relative density of the composites is on the same order as that of the control samples.

Table 2. Ballistic Behavior of SiC/TiB$_2$ Composites

Sample Description	Fabrication Method	Sintering Profile	Density*, g/cm^3	Relative Density, %	Mass Efficiency*
SiC (control)	1	1	3.07 ± 0.02	95	5.32
SiC (control)	1	2	3.09 ± 0.02	96	4.74
SiC - 7 vol% TiB$_2$	1	1	3.16 ± 0.01	95	6.27
SiC - 7 vol% TiB$_2$	1	2	3.19 ± 0.02	96	5.45
SiC - 10 vol% TiB$_2$	1	1	3.19 ± 0.00	95	5.13
SiC - 10 vol% TiB$_2$	1	2	3.23 ± 0.01	96	5.6

* The Density and Mass Efficiency values were calculated as averages of three DOP test samples.

The mass efficiency of the composites, generally, exceeds that of the control sample for a given sintering profile. (The value of pressureless sintered SiC given in Table 1 is for a single DOP test preformed in a preliminary investigation while the value for the same material given in Table

2 is the average of three DOP tests.) The 7 vol% TiB_2 sample exhibited the highest mass efficiency, 6.27, of all the samples and approached the mass efficiency of hot pressed SiC (SiC-B) given in Table 1 as 6.5. A V_{50} ballistic test (the velocity of the projectile at which there is a 50% it will completely penetrate the material) will be required to determine whether the SiC/TiB_2 composite provides ballistic protection equivalent to hot pressed SiC.

No clear trend of ballistic efficiency with density (and/or sintering profile) is evident in Table 2. One would expect that a higher relative density would exhibit a higher mass efficiency since porosity would be expected to decrease the ballistic properties. However, it must be kept in mind that these composite samples were made by manual mixing methods and the various samples most likely contained local variations in the TiB_2 content. These local variations may cause significant variation in the ballistic behavior and significant scatter in the mass efficicency. Future work will utilize more automated mixing methods and advanced mixing techniques to improve the TiB_2 distribution.

Finally, the reason for improved ballistic performance of the pressureless sintered SiC/TiB_2 composites over pressureless sintered SiC is also not clear at this time. The only differences evident are the modification of the grain shape and a slight reduction in the average grain size in the composites compared to pressureless sintered SiC. It is possible, however, that there may be a slight solubility of titanium in SiC that altered the ballistic response (or is even responsible for the observed modification of the grain shape). Further research is necessary to elucidate the mechanism responsible for the improved ballistic response of the composite material.

CONCLUSIONS

Additions of TiB_2 to pressureless sintered SiC appears to improve the ballistic performance without significantly altering the density. The reason for the improvement in ballistic properties is not clear at this time but modification of the grain shape compared to pure pressureless sintered SiC was noted. Specifically the SiC/TiB_2 composites had a reduce prevalence of elongated SiC grains and a somewhat smaller grain size as compared to the control sample that did not contain TiB_2 particles.

Of the two fabrication methods explored, a proprietary method was found to produce finer TiB_2 particles, however, agglomerations of TiB_2 particles were prevalent in samples made by both fabrication methods. Improved mixing methods will be utilized in the future to reduce the number of agglomerations and produce a more uniform distribution of TiB_2 particles. This should lead to more effective grain boundary pinning and a microstructure more closely resembling that of hot pressed SiC. It is assumed the ballistic properties of the pressureless sintered composite will then be at least equivalent to relatively expensive, low production volume hot pressed SiC.

REFERENCES:
[1]T.Lillo, H. Chu, D. Bailey, W. Harrison, D. Laughton, "Development of a Pressureless Sintered Silicon Carbide Monolith and Special-shaped Silicon Carbide Whisker Reinforced Silicon Carbide Matrix Composite for Lightweight Armor Application", *Ceramic Engg. And Sci. Proceedings*, **24**, 359-364 (2003*).*
[2]Murata, et al, "Sintered Silicon Carbide - Titanium Diboride Mixtures and Articles Thereof", US Patent 4,327,186, April 27, 1982.

ACKNOWLEDGEMENTS

The authors would like to thank Dr. Steve Frank at ANL-West for the x-ray diffraction analysis. Prepared for the U.S. Department of Energy through the INEEL LDRD Program under DOE Idaho Operations Office Contract DE-AC07-99ID13727.

IMPROVED BALLISTIC PERFORMANCE BY USING A POLYMER MATRIX COMPOSITE FACING ON BORON CARBIDE ARMOR TILES

S. D. Nunn, J. G. R. Hansen, B. J. Frame, and R. A. Lowden
Oak Ridge National Laboratory
P. O. Box 2008
Oak Ridge, TN 37831

ABSTRACT
Ballistic impact testing was used to compare the performance of boron carbide armor tiles with and without a polymer matrix composite (PMC) facing. This facing layer was in addition to a Spectra Shield Plus® spall cover and backing plate. The armor tiles were tested against a 7.62 mm armor piercing round fired from a universal receiver at varying velocities. Variations in the PMC facing were examined to evaluate the effect on performance. Facing variables included: the type of fiber, the number of plies in the PMC layer, and the orientation of the fibers in the individual plies. The use of a PMC facing improved the ballistic performance by as much as 40% compared to tiles without the facing layers. Although preliminary, these results suggest that the use of a PMC facing is an effective way to improve the ballistic penetration resistance of ceramic armor.

INTRODUCTION
Ceramic materials have been studied in ballistic armor applications for many years.[1,2] The high hardness of ceramics makes them particularly beneficial for defeating armor piercing projectiles. Boron carbide (B_4C) ceramic has the highest hardness and lowest weight of the commonly used armor ceramic materials. Its light weight makes it especially attractive for aircraft and personnel armor, where the overall armor weight is a critically important factor. Ceramic armor assemblies usually incorporate other types of materials to form composite structures. These can range from a simple ballistic fabric wrap[3] to a complex layering system incorporating ceramic, metal, and organic composite materials.[4] In general, the ceramic is usually located at or near the impact surface to blunt, fracture, deflect, and erode the incoming projectile. A more elastic material, which can deform to absorb residual energy, is located behind the ceramic to stop fragments of the projectile and fractured ceramic. In the present study, a fiber reinforced polymer matrix composite facing was bonded to the faces of B_4C ceramic armor tiles to evaluate the effect on penetration resistance in ballistic impact tests.

EXPERIMENTAL
Armor tiles of PAD B_4C were obtained from Cercom Inc., Vista, CA. The tiles had machined surfaces and measured 102 x 102 x 6.2 mm (4 x 4 x 0.245 in.). In the baseline configuration shown in Fig. 1a), a layer of Honeywell Spectra Shield Plus® was bonded directly to the faces of the B_4C tile. The spall cover on the impact side was 1 mm (0.04 in.) thick, while the backing plate was 7.25 mm (0.285 in.) thick. For the armor tiles with a PMC facing, Fig. 1b), the PMC layer was cured on the faces of the B_4C tile prior to applying the Spectra Shield Plus® spall cover and backing plate. The fibers that were used to form the PMC layer are listed in Table I. These were primarily carbon fibers, but also included the polymer fiber Zylon®. The fiber types varied in elastic modulus and tensile strength as indicated in the table. Armor tiles

were prepared with from 2 to 8 PMC layers on each face of the ceramic tile. The orientation of the carbon fibers in each layer was at 0°, 45°, or 90° to the vertical sides of the tile. The Zylon® material consisted of a 0/90° woven fabric and was oriented with the fibers parallel to the edges of the tile. The polymer matrix used to form the PMC layers was epoxy resin for the carbon fibers and vinylester for the Zylon® fiber.

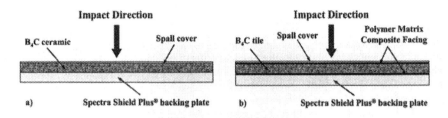

Fig. 1. Schematic diagram of the armor tile cross-section. a) Baseline armor tile without PMC facing layers; b) armor tile with PMC facing.

Table I. Identification and characteristics of fibers used to form the PMC facing layers.

Sample Number	Fiber	Material	Elastic Modulus	Tensile Strength
2-6	Toray T700[a]	Carbon	Intermediate	High
7	Granoc XN-05[b]	Carbon	Low	Low
8	Toray M46J[a]	Carbon	High	High
9	Granoc CN-80[b]	Carbon	Ultra-high	Low
10	Zylon® (PBO)[c]	Polymer	Intermediate	High

[a]Toray Carbon Fibers America, Inc.; [b]Nippon Graphite Fiber Corp.; [c]Toyobo Company, Ltd.

The ballistic impact testing was conducted versus the armor piercing 7.62 mm AP M61 (NATO .308) round. The powder charge in the cartridge was adjusted to produce varying impact velocities at the target location. The ceramic tile thickness that was selected, 6.2 mm, was chosen to assure that complete penetration of the armor tiles could be achieved within the range of velocities available. The armor targets were mounted on the back surface of a steel plate (relative to the impact direction) using a bolted-on window frame holder that applied a uniform clamping force around the perimeter of the armor tile. This arrangement is shown in Fig. 2. The central 76 x 76 mm (3 x 3 in.) area of the back face of the armor tile was unsupported during the test. The steel plate with the mounted armor tile was held in a rigid frame at a muzzle-to-target distance of 10 m (30 ft.). A universal receiver on a fixed pedestal was used to fire the rounds at the target. After the bullet was fired, the armor tiles were examined to determine whether the impact resulted in a complete penetration or a partial penetration, in which the armor is partially penetrated, but the projectile is stopped within the armor system. Every effort was made to be consistent in tile preparation, mounting, and testing to assure valid side-by-side comparison of the ballistic impact performance.

Fig. 2. Clamping arrangement for holding the armor tile during ballistic impact testing. a) Armor tile and picture frame clamp; b) armor tile clamped to steel plate; c) impact side of assembly; d) exit side of assembly.

RESULTS AND DISCUSSION

The results of the ballistic impact tests are summarized in Table II. In all cases where a partial penetration was recorded, the armor tiles having a PMC facing showed improved ballistic impact performance compared to the baseline armor tile without the PMC facing. Although the areal density of the tiles was generally increased by the addition of the PMC facing, this was more than offset by the improvement in penetration resistance. For example, the areal density of sample number 4 with 8 PMC layers was 9% higher than the baseline armor tile, but the apparent ballistic V_{50} was increased by more than 40%.

Table II. Armor tile variations and ballistic impact results.

Sample Number	PMC Fiber	PMC Plies	Fiber Orientation	Areal Density, psf	"V_{50}" *, fps	"V_{50}" Increase
1	No PMC	-	-	5.26	2050	-
2	T700	2	0/90	5.20	>2175	>6%
3	T700	4	0/90/0/90	5.53	2550	24%
4	T700	8	0/90/0/90	5.73	>2880	>40%
5	T700	4	+45/-45/+45/-45	5.41	>2625	>28%
6	T700	4	0/-45/+45/90	5.44	no partial	-
7	XN-05	4	0/90/0/90	5.35	2500	22%
8	M46J	4	0/90/0/90	5.42	no partial	-
9	CN-80	4	0/90/0/90	5.45	>2610	>27%
10	Zylon® (PBO)	4	0/90/0/90	5.43	>2730	>33%

*For most variations, the number of samples tested was insufficient to determine a true ballistic V_{50} value. V_{50} is the velocity at which 50% of impacts are complete penetrations and 50% are partial penetrations.

The test results are plotted in Figs. 3 – 5, where each data point represents an individual ballistic impact of one bullet on each armor tile. These plots compare the penetration resistance of tiles with variations in the PMC facing configuration to the baseline armor tile assembly without a PMC facing.

With the limited number of test samples of each PMC variation that were evaluated in the tests, it is difficult to draw many absolute conclusions regarding the different configurations. It is clear that increasing the number of plies in the PMC facing increased the penetration resistance of the armor tile for the range of values tested. As shown in Fig. 3, ballistic performance improved monotonically as the number of plies was increased from 0 to 8. It also seems apparent that the orientation of the fibers in the PMC plies had an effect on the test results (Fig. 4). Samples 3, 5, and 6 all had 4 plies of PMC with T700 carbon fibers, but sample 5 with the fibers in a +45/-45/+45/-45 orientation showed better penetration resistance than sample 3 with a 0/90/0/90 orientation, which in turn was better than sample 6 with the 0/-45/+45/90 orientation.

From the test results shown in Fig. 5 comparing the types of fibers used in the PMC facing layers, it is difficult to identify a clearly superior fiber among those evaluated. All of the fibers showed a benefit in improving the penetration resistance of the armor tiles (except the M46J fiber, for which no partial penetrations were obtained). The CN-80 carbon fiber and the Zylon® (PBO) fiber appear to show the best performance for the 4-ply, 0/90/0/90 PMC configuration, but more tests will be needed to confirm these results.

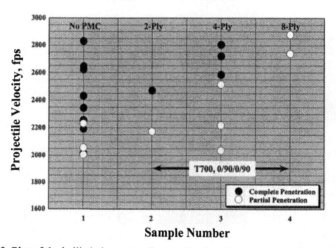

Fig. 3. Plot of the ballistic impact testing results showing complete and partial penetration velocities for armor tiles having a different number of plies in the PMC facing layer. From left to right, the number of plies was 0, 2, 4, and 8 on each face of the tile. All facing layers were made up of T700 fibers in a 0/90/0/90 configuration.

290

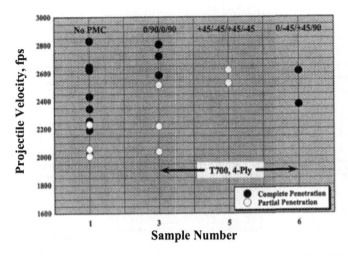

Fig. 4. Plot of the ballistic impact testing results showing complete and partial penetration velocities for armor tiles having different orientations of the fibers in the individual plies of the PMC facing layer. The sample numbers correspond to the descriptions given in Table II.

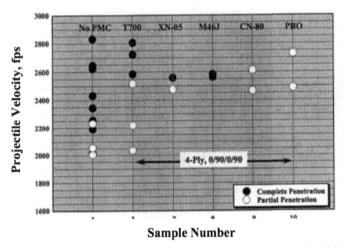

Fig. 5. Plot of the ballistic impact testing results showing complete and partial penetration velocities for armor tiles having different fibers in the PMC facing layer. For a more complete description of the samples, see the corresponding sample number in Table II.

291

The reason for the improvement in ballistic impact performance when the PMC facing layers were present is not yet fully understood. It may be speculated that the composite layers act to delay the onset of fracture and fragmentation of the ceramic material, but the actual mechanisms are not clear. The composite layers may provide a lateral constraint on the ceramic tile, which could slow the spread of cracks and the separation of tile fragments. Based on the observed effect of the fiber orientation, it is also possible that the PMC layers may provide a form of acoustical damping that affects the propagation of stress waves in the ceramic tile resulting in delayed fracture. Other mechanisms are possible, but further tests combined with computer modeling of the tile configuration will be needed to clarify the test results.

It is not known whether similar effects would be observed with B_4C from other suppliers or with alternative ceramic armor materials such as Al_2O_3, SiC, and Si_3N_4. Additionally, there are many other types of fibers and possible PMC configurations available which may further improve the enhancement in ballistic impact performance that was observed in this study.

CONCLUSIONS

Ballistic impact testing was used to compare the performance of boron carbide armor tiles with and without a polymer matrix composite facing. The armor tiles were tested against the 7.62 mm AP M61 armor piercing round. The use of a PMC facing improved the ballistic performance by as much as 40% compared to tiles without the facing layers. The penetration resistance of the armor tiles increased steadily with an increasing number of plies in the PMC layer up to the maximum of 8 that was tested. It also appears that the orientation of the reinforcing fibers in the individual plies of the PMC can have an effect on ballistic performance. Of the different fibers used in the PMC facing layers, the CN-80 carbon fiber and the Zylon® (PBO) fiber showed the best ballistic impact performance based on a limited number of test samples. Although preliminary, these results suggest that the use of a PMC facing is an effective way to improve the ballistic performance of B_4C ceramic armor. Further investigation is needed to understand and to optimize the effect.

REFERENCES

[1]D. J. Veichnicki, M. J. Slavin, and M. I. Kliman, "Development and Current Status of Armor Ceramics," *Am. Ceram. Soc. Bull.*, **70** [6] 1035-39 (1991).

[2]S. R. Skaggs, "A Brief History of Ceramic Armor," *Ceram. Eng. Sci. Proc.*, **24** [3] 337-349 (2003).

[3]A. F. Pivitt, D. K. Rock, and N. S. Sridharan, "Composite Ceramic Armor and Method for Making Same," U. S. Patent No. 4,911,061, Mar. 27, 1990.

[4]T. J. Madison, "Composite Tactical Hard Body Armor," U. S. Patent No. 5,306,557, Apr. 26, 1994.

ANALYSIS OF SCATTERING SITES IN TRANSPARENT MAGNESIUM ALUMINATE SPINEL

Guillermo Villalobos, Jasbinder S. Sanghera, and Ishwar D. Aggarwal
U.S. Naval Research Laboratory
4555 Overlook Ave
Washington, DC 20375

Robert Miklos
SFA Inc.
9315 Largo Dr. West
Largo, MD 20774

ABSTRACT
Scattering sites greatly degrade the transmission of magnesium aluminate spinel. It appears that some of the scattering sites are caused by reactions with the LiF traditionally used as a sintering aid. LiF is used to etch spinel particles during the hot pressing process. The LiF was found to react with the aluminum in the spinel structure thereby leaving magnesium rich regions behind that do not sinter well and result in opaque regions in the otherwise transparent matrix. Avoiding the conditions where the reaction is most aggressive allows the densification of spinel that transmits throughout the entire transmission range at 100% of the theoretical bulk value.

INTRODUCTION
Magnesium aluminate spinel has good mechanical and optical properties that will allow its use as a transparent armor and as a window and dome material for missiles and pods [1]. Its mechanical properties are comparable to polycrystalline aluminum oxide and, since it has a cubic structure (i.e. no birefringence), polycrystalline samples can transmit from 200 nm to 5.5 μm with no optical distortion. Spinel has been studied for over 40 years [2,3]. Unfortunately, the literature has very little information on its sintering behavior, and the material still cannot be reproducibly sintered to transparency [4,5]. It is difficult to make transparent spinel without the formation of a host of scattering sites that degrade its transmission.

Spinel is generally densified with the use of sintering aids, the most common being LiF. Without LiF sintering aid the material tends to be translucent and gray. Previous researchers have proposed that the LiF etches and removes impurities from surface of the spinel particles thereby enhancing diffusion. It is also believed that the molten LiF can aid initial compaction by lubricating the particles and allowing better packing. The LiF must be removed from the material before complete consolidation or it will manifest itself as white precipitates [6].

The spinel hot press schedule is designed around two successive thermal treatments. The first at 950°C is used to allow the LiF to react with the particle surfaces, and the second at 1200°C to allow the LiF to volatilize and be removed from the material before the pore structure collapses and traps the LiF sintering aid.

In this work we will attempt to understand the origin of the scattering sites and issues for developing magnesium aluminate spinel with high optical transparency. Specifically we will identify reactions between the sintering aid and the spinel and determine methods to significantly reduce the number of scattering sites to make the material useful for DoD and commercial applications.

EXPERIMENTAL PROCEDURE

Spinel powder was purchased from Ceralox (Tucson, AZ). The powder was mixed using a mortar and pestle with 0.5, 2 and 10 weight percent LiF (Sylvania, Towanda, PA) sintering aid. Ten grams of powder/sintering aid mixture were loaded into a 25 mm diameter I.D. graphite die (Poco Graphite, Decatur, TX) lined with grafoil (Polycarbon Inc., Valencia, CA) to minimize carbon diffusion and extend die life. The samples were hot pressed in a vacuum hot press (Electrofuel, Toronto, Canada) at a heating rate of 10°C/minute to 1600°C and held for 2 hours. There were also two 30-minute holds, one at 950°C and the other at 1200°C. Pressure schedule consisted of maintaining 200 psi until the end of the 1200°C hold. The pressure was then slowly increased to 4000 psi and held until the end of the 1600°C treatment. The resulting samples were 25 mm diameter x 2 mm thick.

Powder samples were mixed in a mortar at a 50/50 weight ratio of spinel/LiF, Al_2O_3/LiF, and MgO/LiF and heat treated at 950°C and 1200°C in a vacuum furnace to identify reactions during the densification procedure. The 50/50 ratio was used to increase the amount of reacted material and thereby enable detection of the reaction products by XRD. The 50/50 mixtures were also run in an SDT (TA Instruments SDT 2960, New Castle, DE) to determine the reaction temperatures and weight loss. The SDT was run in flowing argon at a 10°C/min rate to 1500°C. The powder and densified samples were also analyzed using optical microscopy, SEM/EDS (LEO 1550 LEO Electron Microscopy Inc., Throrwood, NY), and XRD (Sintag XDS 2000, Sunnyvale, CA).

RESULTS AND DISCUSSION

Figure 1a is an XRD pattern of a 50:50 mix of spinel and LiF heat treated at 1200°C for ½ hour. It shows the formation of $LiAlO_2$, and MgO. Figure 1b is the XRD pattern of a 50:50 mix of MgO and LiF also heat-treated at 1200°C for ½ hour. There was no evidence of any reaction in the MgO/LiF sample. XRD of a 50:50 mix of Al_2O_3 and LiF (not shown) also shows the formation of $LiAlO_2$. LiF was not observed in any of these experiments since it has a high

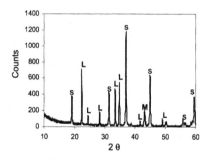 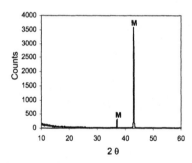

Figure 1a. XRD of 50:50 mix of Spinel (S) and LiF showing the formation of LiAlO₂ (L), b. XRD of 50:50 mix of MgO (M) and LiF both samples were heat treated at 1200 °C for ½ hour.

294

vapor pressure and evaporates at the elevated temperatures.

These results match well with a thermodynamic study of the $LiF-Al_2O_3-MgO$ system which shows that while MgO does not appreciably react with LiF, Al_2O_3 can react to form various lithium aluminum containing compounds. Of all the Li-Al containing compounds listed in the JANAF tables [7], only $LiAlO_2$ appears to be a solid throughout the temperature range of the hot press schedule, the other reaction products are either vapors or solids with high vapor pressures at the hot pressing temperature and help to explain the results of the 50/50 mixture experiments.

Unfortunately, it is difficult to identify $LiAlO_2$ regions in the hot pressed samples by EDS because EDS cannot detect lithium, and it is difficult to distinguish Al associated with $LiAlO_2$ inclusions and Al associated with spinel due to the small size of the precipitates and the penetration depth of the electron beam. However, it is possible to detect MgO or Mg-rich phases both by SEM and EDS. Figure 2 is a photograph of a spinel disk showing clear and opaque

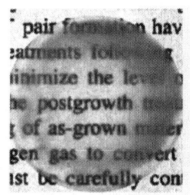 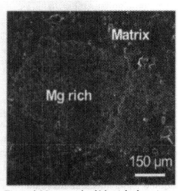

Figure 2 Spinel disk (1" dia.) showing clear and opaque regions

Figure 3 Micrograph of Mg-rich phase in

regions. Figure 3 is a high resolution SEM micrograph of an opaque area showing that it is composed of small (300-500nm) crystals that have not sintered as well as the surrounding material. EDS analysis shows that while the transparent area has the expected 2:1 atomic ratio of Al and Mg, the small grains have a 1:1 ratio of Al and Mg. This suggests that the small grains are made up of an MgO rich phase. MgO is considerably more refractory than spinel and would not be expected to sinter as readily at the lower sintering temperatures used for spinel.

Reactions between LiF and spinel can also result in the formation of discrete scattering sites at the spinel grain boundaries. These discrete sites can take the form of spherical or planar

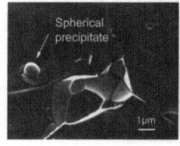 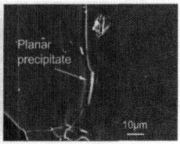

Figure 4a. Spherical Mg-rich precipitate. b. Planar Mg-rich precipitate

precipitates (Figures 4a and b). The spherical precipitates are optical scattering sites leading to the opaque regions shown in Figure 2. The planar sites have higher aspect ratios and so will scatter light more strongly. This contributes to the dark regions at the grain boundaries shown in Figure 5. We initially thought that the dark regions were caused by carbon at the grain boundaries, but EDS and SEM analysis could find no evidence of carbon.

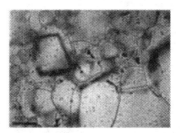

Figure 5. Spinel sample showing light scattering at grain boundaries.

It is apparent that LiF does react with the particle surfaces as is generally thought, but it preferentially reacts with the Al thereby leaving Mg rich areas behind. From a combination of hot press runs interrupted at intermediate temperatures and SDT experiments, it appears that the reaction between aluminum and LiF begins soon after LiF melts and becomes more aggressive in the early stages of vaporization. Figure 6 is an SDT trace showing the melting and vaporization behavior of LiF under flowing argon and heating rate of 10°C/min.

Changing the processing conditions during densification can greatly decrease the amount

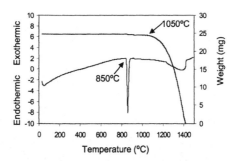

Figure 6. SDT of LiF under flowing argon showing melting at 850 ℃ and vaporization beginning at 1050 ℃.

of scattering sites that are formed by reactions between the sintering aid and the spinel. Figure 7 is a photograph of a 50mm diameter x 6mm thick spinel sample that shows excellent clarity and no distortion. Figure 8 is the transmission plot of a spinel sample that achieves 100% of theoretical bulk transmission. Due to reflection losses based on index of refraction a material with an index of 1.7 (such as spinel) would have a maximum transmission of 87% of the incident beam as is shown in Figure 8. Anti-reflection coatings would increase the total transmission close to 100%.

Figure7. Photgraph of 50mm diameter x 6mm thick spinel.

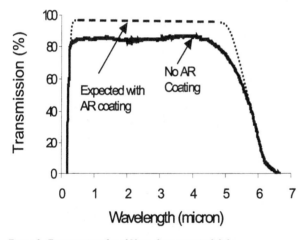

Figure 8. Transmission plot of 50mm diameter spinel disk.

CONCLUSION

Although LiF is necessary to densify spinel to transparency, it also reacts with the aluminum in the spinel matrix to create magnesium rich areas and precipitates that cause degradation in the transmission. Optimization of the densification parameters to avoid rapid degradation results in a reproducible process which produces uniform material that has high optical transparency.

ACKNOWLEDGEMENTS
Richard Brown, MVA Scientific Consultants (Norcross, GA)

REFERENCES
[1] M.C.L. Patterson, J.E. Caiazza, D.W. Roy and G. Gilde, "Transparent Spinel Development", *SPIE 45th International Symposium on Optical Science and Technology* 30July-4August 2000, San Diego, CA.
[2] D.W. Roy, M.C.L. Patterson, J.E. Caiazza, and G. Gilde, "Progress in the Development of Large Transparent Spinel Plates", *8th DoD Electromagnetic Windows Symposium*, April 2000, Colorado Springs, CO.
[3] D.W. Roy, D.R. Johnson and D.L. Mann, "Fabrication and Properties of Transparent $MgAl_2O_4$", *American Ceramic Society Bulletin*, **52**, [4], 372-373 (1973).
[4] M.C.L. Patterson, D.W. Roy and G. Gilde, "Improvements in Optical Grade Spinel", *9th DoD EM Windows Symposium*, May 2002, Huntsville, AL.
[5] M.Shimada, T. Endo, T. Saito, and T. Sato, "Fabrication of Transparent Spinel Polycrystalline Materials", *Matrials Letters*, **28**, [8], 413-415, (1996).
[6] D.W. Roy private communications 2002
[7] M.W. Chase ed, *NIST-JANAF Thermochemical tables 4th edition*, NIST, Gaitherburg, 1998.

Author Index

Rafaniello, W., 131
Ray, D., 131
Rendtel, A., 161
Rickter, B., 11, 203
Roberson, C.J., 143, 151

Sadangi, R., 67
Sanghera, J.S., 293
Schoenfeld, S.E., 59
Schultz, B.E., 263
Schwetz, K.A., 161
Shu, D., 251
Steckenrider, J.S., 215

Tan, G.E.B., 251

Villalobos, G., 293

Walker, J.D., 27, 35, 109
Wang, H., 251
Wells, J.M., 51, 239
Wheeler, J., 215
Wright, T.W., 59

Zheng, J., 251
Zhou, M., 123
Ziccardi, C., 271